Lecture Notes in Physics

For information about Vols. 1–44, please contact your bookseller or Springer-Verlag.

Vol. 45: Dynamical Concepts on Scaling Violation and the New Resonances in e^+e^- Annihilation. Edited by B. Humpert. VII, 248 pages. 1976.

Vol. 46: E. J. Flaherty, Hermitian and Kählerian Geometry in Relativity. VIII, 365 pages. 1976.

Vol. 47: Padé Approximants Method and Its Applications to Mechanics. Edited by H. Cabannes. XV, 267 pages. 1976.

Vol. 48: Interplanetary Dust and Zodiacal Light. Proceedings 1975. Edited by H. Elsässer and H. Fechtig. XII, 496 pages. 1976.

Vol. 49: W. G. Harter and C. W. Patterson, A Unitary Calculus for Electronic Orbitals. XII, 144 pages. 1976.

Vol. 50: Group Theoretical Methods in Physics. 4th International Colloquium. Nijmegen 1975. Edited by A. Janner, T. Janssen, and M. Boon. XIII, 629 pages. 1976.

Vol. 51: W. Nörenberg and H. A. Weidenmüller. Introduction of the Theory of Heavy-Ion Collisions. 2nd enlarged edition. IX, 334 pages. 1980.

Vol. 52: M. Mladjenović, Development of Magnetic β-Ray Spectroscopy. X, 282 pages. 1976.

Vol. 53: D. J. Simms and N. M. J. Woodhouse, Lectures on Geometric Quantization. V, 166 pages. 1976.

Vol. 54: Critical Phenomena. Sitges International School on Statistical Mechanics, June 1976. Edited by J. Brey and R. B. Jones. XI, 383 pages. 1976.

Vol. 55: Nuclear Optical Model Potential. Proceedings 1976. Edited by S. Boffi and G. Passatore. VI, 221 pages. 1976.

Vol. 56: Current Induced Reactions. International Summer Institute, Hamburg 1975. Edited by J. G. Körner, G. Kramer, and D. Schildknecht. V, 553 pages. 1976.

Vol. 57: Physics of Highly Excited States in Solids. Proceedings 1975. Edited by M. Ueta and Y. Nishina. IX, 391 pages. 1976.

Vol. 58: Computing Methods in Applied Sciences. Proceedings 1975. Edited by R. Glowinski and J. L. Lions. VIII, 593 pages. 1976.

Vol. 59: Proceedings of the Fifth International Conference on Numerical Methods in Fluid Dynamics. 1976. Edited by A. I. van de Vooren and P. J. Zandbergen. VII, 459 pages. 1976.

Vol. 60: C. Gruber, A. Hintermann, and D. Merlini, Group Analysis of Classical Lattice Systems. XIV, 326 pages. 1977.

Vol. 61: International School on Electro and Photonuclear Reactions I. Edited by C. Schaerf. VIII, 650 pages. 1977.

Vol. 62: International School on Electro and Photonuclear Reactions II. Edited by C. Schaerf. VIII, 301 pages. 1977.

Vol. 63: V. K. Dobrev et al., Harmonic Analysis on the n-Dimensional Lorentz Group and Its Application to Conformal Quantum Field Theory. X, 280 pages. 1977.

Vol. 64: Waves on Water of Variable Depth. Edited by D. G. Provis and R. Radok. 231 pages. 1977.

Vol. 65: Organic Conductors and Semiconductors. Proceedings 1976. Edited by L. Pál, G. Grüner, A. Jánossy and J. Sólyom. 654 pages. 1977.

Vol. 66: A. H. Völkel, Fields, Particles and Currents. VI, 354 pages. 1977.

Vol. 67: W. Drechsler and M. E. Mayer, Fiber Bundle Techniques in Gauge Theories. X, 248 pages. 1977.

Vol. 68: Y. V. Venkatesh, Energy Methods in Time-Varying System Stability and Instability Analyses. XII, 256 pages. 1977.

Vol. 69: K. Rohlfs, Lectures on Density Wave Theory. VI, 184 pages. 1977.

Vol. 70: Wave Propagation and Underwater Acoustics. Edited by J. Keller and J. Papadakis. VIII. 287 pages. 1977.

Vol. 71: Problems of Stellar Convection. Proceedings 1976. Edited by E. A. Spiegel and J. P. Zahn. VIII, 363 pages. 1977.

Vol. 72: Les instabilités hydrodynamiques en convection libre forcée et mixte. Edité par J. C. Legros et J. K. Platten. X, 202 pages. 1978.

Vol. 73: Invariant Wave Equations. Proceedings 1977. Edited by G. Velo and A. S. Wightman. VI, 416 pages. 1978.

Vol. 74: P. Collet and J.-P. Eckmann, A Renormalization Group Analysis of the Hierarchical Model in Statistical Mechanics. IV, 199 pages. 1978.

Vol. 75: Structure and Mechanisms of Turbulence I. Proceedings 1977. Edited by H. Fiedler. XX, 295 pages. 1978.

Vol. 76: Structure and Mechanisms of Turbulence II. Proceedings 1977. Edited by H. Fiedler. XX, 406 pages. 1978.

Vol. 77: Topics in Quantum Field Theory and Gauge Theories. Proceedings, Salamanca 1977. Edited by J. A. de Azcárraga. X, 378 pages 1978.

Vol. 78: Böhm, The Rigged Hilbert Space and Quantum Mechanics. IX, 70 pages. 1978.

Vol. 79: Group Theoretical Methods in Physics. Proceedings, 1977. Edited by P. Kramer and A. Rieckers. XVIII, 546 pages. 1978.

Vol. 80: Mathematical Problems in Theoretical Physics. Proceedings, 1977. Edited by G. Dell'Antonio, S. Doplicher and G. Jona-Lasinio. VI, 438 pages. 1978.

Vol. 81: MacGregor, The Nature of the Elementary Particle. XXII, 482 pages. 1978.

Vol. 82: Few Body Systems and Nuclear Forces I. Proceedings, 1978. Edited by H. Zingl, M. Haftel and H. Zankel. XIX, 442 pages. 1978.

Vol. 83: Experimental Methods in Heavy Ion Physics. Edited by K. Bethge. V, 251 pages. 1978.

Vol. 84: Stochastic Processes in Nonequilibrium Systems, Proceedings, 1978. Edited by L. Garrido, P. Seglar and P. J. Shepherd. XI, 355 pages. 1978

Vol. 85: Applied Inverse Problems. Edited by P. C. Sabatier. V, 425 pages. 1978.

Vol. 86: Few Body Systems and Electromagnetic Interaction. Proceedings 1978. Edited by C. Ciofi degli Atti and E. De Sanctis. VI, 352 pages. 1978.

Vol. 87: Few Body Systems and Nuclear Forces II, Proceedings, 1978. Edited by H. Zingl, M. Haftel, and H. Zankel. X, 545 pages. 1978.

Vol. 88: K. Hutter and A. A. F. van de Ven, Field Matter Interactions in Thermoelastic Solids. VIII, 231 pages. 1978.

Vol. 89: Microscopic Optical Potentials, Proceedings, 1978. Edited by H. V. von Geramb. XI, 481 pages. 1979.

Vol. 90: Sixth International Conference on Numerical Methods in Fluid Dynamics. Proceedings, 1978. Edited by H. Cabannes, M. Holt and V. Rusanov. VIII, 620 pages. 1979.

Lecture Notes in Physics

Edited by J. Ehlers, München, K. Hepp, Zürich
R. Kippenhahn, München, H. A. Weidenmüller, Heidelberg
and J. Zittartz, Köln
Managing Editor: W. Beiglböck, Heidelberg

120

Nonlinear Evolution Equations and Dynamical Systems

Proceedings of the Meeting
Held at the University of Lecce
June 20–23, 1979

Edited by
M. Boiti, F. Pempinelli, and G. Soliani

Springer-Verlag Berlin Heidelberg GmbH 1980

Editors

M. Boiti
F. Pempinelli
G. Soliani
Università degli Studi di Lecce
Istituto di Fisica
73100 Lecce Via Arnesano
Italy

ISBN 978-3-540-09971-0 ISBN 978-3-540-39197-5 (eBook)
DOI 10.1007/978-3-540-39197-5

Originally published by Springer-Verlag Berlin Heidelberg New York in 1980

2153/3140-543210

CONTENTS

FOREWORD

This volume contains the Proceedings of the Meeting on "Nonlinear Evolution Equations and Dynamical Systems" held at the University of Lecce (Italy) from June 20th-23rd, 1979.

The main emphasis of the meeting was on recent advances and new trends in nonlinear evolution equations, their application to physics and the mathematical development which has followed from their study. These research areas have developed rapidly over the last decade from the pioneering period, when solitons were discovered, to a much more professional and specilized stage. The collection of papers in this volume is a representative selection of work carried out recently. These papers are an expanded version of must of the talks given at the meeting and include many additional details and results not presented then.

We are particularly pleased to thank the authors of these papers for their hard work and cooperation in preparing the manuscripts and, in general, to all partecipants whose confidence in the meeting made its success possible.

The International Scientific Committee of the meeting was composed of Proff. R.K.Boullough, F.Calogero, G.F.Dell'Antonio, F.Pirani and T. Regge. We would like to thank each of them for their collaboration.

We are especially grateful to Prof.Calogero for his helpful advice in the early stages of the organization of the meeting.

Financial support from the University of Lecce, from the Italian National Research Council (CNR), from the Banca Piccolo Credito Salentino and from the Banca Fratelli Vallone made the meeting possible. Contributions from the Banca Agricola Popolare di Matino e Lecce, Banca Popolare di Parabita, Banca Venturi, Banca Commerciale Italiana (Lecce) and Banca dei Paschi di Siena (Lecce) were equally important in meeting our most immediate financial obbligations.

The social events were highlighted by a visit to Otranto and the coast of Salento organised by the Ente Provinciale del Turismo of Lecce (the ufficial tourist organization).

We would like to thank all the Organizations and Institutions mentioned above for having contributed to the Meeting's success.

Finally we want to express here our thank and appreciation to Proff. M.Leo, R.A.Leo, A.Luches and P.Rotelli of the Local Organizing Committee for their aid in overcoming all kinds of serious difficulties, to the secretaries Mrs. G.Mazzotta and Mrs. A.Rossi for their invaluable help in the preparation and organization of the meeting and to Mr. C. Laddomada for his excellent typing of the photo-ready copy of the manuscripts.

Lecce, January 1980

M. Boiti

F. Pempinelli

G. Soliani

NOTE. In order to speed up the pubblication of the Proceedings most of the proofs have not been revised by the authors.

NONLINEAR EVOLUTION EQUATIONS SOLVABLE BY THE SPECTRAL TRANSFORM: SOME RECENT RESULTS

F. Calogero

*Department of Applied Mathematics,
Queen Mary College, London

**Istituto di Fisica, Università di Roma

ABSTRACT

This is a terse survey of some recent results. Except for the last item, it is meant to serve rather as a guide to the literature (where more detailed treatments can be found) than as a complete self-contained presentation.

0. INTRODUCTION

The material in this section is by now standard, and the presentation here is therefore confined to the mere essentials. The basic idea is of course due to Gardner, Greene, Kruskal and Miura [1]. For more

---

* To June, 1980.

** Permanent address.

extensive presentations see for instance [2, 3].

0.1 - Solution of linear evolution equations via the Fourier transform

Consider the prototypical linear evolution equation

$$u_t(x,t) = -i\omega(-i\frac{\partial}{\partial x})\, u(x,t) \quad , \tag{1}$$

where $\omega(z)$ is a polynomial in z (say, an odd polynomial, $\omega(z)=-\omega(-z)$, so that (1) is real).

Focus attention on the Cauchy problem with

$$u(x,0) = u_0(x) \tag{2}$$

given, and $u(x,t)$ to be determined for $t>0$. Assume moreover $u_0(x)$ (and therefore $u(x,t)$ as well; see below) to be regular for all (real) x and to vanish (fast enough!) as $x\to\pm\infty$,

$$u(\pm\infty,t) = 0 \quad . \tag{3}$$

The solution is then given by the formulae

$$u(x,t) = (2\pi)^{-1}\int_{-\infty}^{+\infty} dk\, \exp(ikx)\, \hat{u}(k,t) \quad , \tag{4}$$

$$\hat{u}(k,t) = \int_{-\infty}^{+\infty} dx\, \exp(-ikx)\, u(x,t) \quad , \tag{5}$$

$$\hat{u}_t(k,t) = -i\omega(k)\hat{u}(k,t) \quad , \tag{6a}$$

$$\hat{u}(k,t) = \hat{u}(k,0)\, \exp[-i\omega(k)t] \quad , \tag{6b}$$

as indicated by the following schematic diagram:

$$u(x,0) \overset{(5)}{\Rightarrow} \hat{u}(k,0) \overset{(6)}{\Rightarrow} \hat{u}(k,t) \overset{(4)}{\Rightarrow} u(x,t) \quad . \tag{7}$$

The technique of solution whose essentials have been outlined here is of fundamental importance in physics, and indeed also in many other exact sciences; not only does it provide the theoretical framework and language, but it determines the experimental methods as well.

0.2 - Solution of nonlinear evolution equations via the spectral transform

The (simplest) class of nonlinear evolution equations solvable via the spectral transform read

$$u_t(x,t) = \alpha(\underline{L})\, u_x(x,t) \quad , \tag{1}$$

again with $\alpha(z)$ a polynomial in z and with the operator $\underline{L}$ defined by

$$\underline{L}\, f(x) = f_{xx}(x) - 4u(x,t)f(x) + 2u_x(x,t)\int_x^{\infty} dx' f(x') , \qquad (2)$$

where f(x) is a generic function (vanishing as $x\to+\infty$). The simplest non-trivial equation of this class is of course the KdV equation

$$u_t(x,t) + u_{xxx}(x,t) - 6u_x(x,t)u(x,t) = 0 , \qquad (3)$$

corresponding to $\alpha(z)=-z$.

We focus again attention on the Cauchy problem with

$$u(x,0) = u_0(x) \qquad (4)$$

given, and u(x,t) to be determined for t>0; again assuming $u_0(x)$ (and therefore u(x,t) as well) to be regular for all (real)x and to vanish (fast enough!) as $x\to\pm\infty$:

$$u(\pm\infty,t) = 0 . \qquad (5)$$

The solution is then given by the bijective correspondence between u(x,t) and its *spectral transform*

$$S(t) : \{R(k,t),\ -\infty<k<\infty;\ p_n,\ \rho_n(t),\ n=1,2,\dots,N\} , \qquad (6)$$

and by the fact that, as u evolves according to (1), its spectral transform evolves in a completely known fashion:

$$R(k,t) = R(k,0)\exp\left[2ik\,\alpha(-4k^2)t\right] , \qquad (7a)$$

$$p_n(t) = p_n(0) = p_n , \qquad (7b)$$

$$\rho_n(t) = \rho_n(0)\exp\left[-2p_n\alpha(4p_n^2)t\right] . \qquad (7c)$$

Moreover the relationship between u(x) and its spectral transform S, although itself nonlinear, is defined by a linear (spectral) problem. Indeed, given u(x), the spectral transform S is defined by the formulae

$$-f_{xx}^{(\pm)}(x,k) + u(x)f^{(\pm)}(x,k) = k^2 f^{(\pm)}(x,k) , \qquad (8)$$

$$\lim_{x\to\pm\infty}\left[f^{(\pm)}(x,k)\exp(\mp ikx)\right] = 1 , \qquad (9)$$

$$R(k) = -\frac{\left[f_x^{(-)}(x,k)f^{(+)}(x,-k) - f^{(-)}(x,k)f_x^{(+)}(x,-k)\right]}{\left[f_x^{(-)}(x,k)f^{(+)}(x,k) - f^{(-)}(x,k)f_x^{(+)}(x,k)\right]} , \quad k^2>0 , \qquad (10)$$

$$f_x^{(-)}(x,ip_n)f^{(+)}(x,ip_n) = f^{(-)}(x,ip_n)f_x^{(+)}(x,ip_n) , \quad p_n>0 , \qquad (11)$$

$$\rho_n = \left(\int_{-\infty}^{+\infty} dx\left[f^{(+)}(x,ip_n)\right]^2\right)^{-1} ; \qquad (12)$$

and, given the spectral transform S, u(x) is defined by

$$M(x) = \sum_{n=1}^{N} \rho_n \exp(-p_n x) + \frac{1}{2\pi}\int_{-\infty}^{+\infty} dk R(k) \exp(ikx) \quad , \tag{13}$$

$$K(x,y) + M(x+y) + \int_{y}^{\infty} dz K(x,z) M(z+y) = 0 \quad , \quad y>x \quad , \tag{14}$$

$$u(x) = -2 \frac{d}{dx} K(x,x+0) \quad . \tag{15}$$

In these last equations, (8-15), we have omitted to indicate any time dependence, since it would be purely parametric.

As in the case of the preceding subsection, the technique of solution we have just outlined can be synthetized by the scheme

$$u(x,0) \overset{(8-12)}{\Rightarrow} S(0) \overset{(7)}{\Rightarrow} S(t) \overset{(13-15)}{\Rightarrow} u(x,t) \quad . \tag{16}$$

As in the case of the linear evolution equations solvable via the Fourier transform discussed in the preceding subsection, the importance of the technique of solution we have just outlined is mainly in the insight it yields on the qualitative behaviour of the solutions of the nonlinear evolution equations of the class (1) - and therefore also on the natural phenomena that can be described by these equations. Of primary relevance in that respect is of course the phenomenology associated with *solitons* [1-3].

1. NONLINEAR EVOLUTION EQUATIONS WITH LINEARLY x-DEPENDENT COEFFICIENTS

The results reported in this section have all been obtained in collaboration with A.Degasperis; a more detailed treatment than that given below can be found in the original literature [4-9]. The idea of looking at nonlinear evolution equations of this kind was first introduced by A.Newell [10].

The class of nonlinear evolution equations that are tersely treated here are those [4,5,9] that can be solved by the spectral transform outlined in the preceding section 0.2, namely that associated with the Schroedinger spectral problem. The analogous classes of nonlinear evolution equations [6-10] that are instead solvable via the spectral transform associated with the generalized Zakharov-Shabat spectral problem will not be discussed here.

We write the class of nonlinear evolution equations in the form

$$u_t = \alpha(\underline{L})\, u_x + \beta(\underline{L})\left[x u_x + 2u\right] , \quad u \equiv u(x,t) \ . \tag{1}$$

The operator $\underline{L}$ is that defined above, see (0.2-(2)); $\alpha(z)$ and $\beta(z)$ are polynomials. For $\beta=0$, the class (0.2-(1)) is recovered.

The time evolution of the spectral transform associated with $u(x,t)$, corresponding to (1), reads

$$R_t(k,t) = 2ik\alpha(-4k^2)R(k,t) - k\beta(-4k^2)R_k(k,t) \ , \tag{2}$$

which can be formally integrated to yield

$$R(k,t) = R_0\left[k_0(t,k)\right] \exp\left[2i\int_0^t dt' \, \chi\alpha(-4\chi^2)\right] , \tag{3}$$

where

$$R_0(k) = R(k,0) \ , \tag{4}$$

$$\chi \equiv \chi\left[t', k_0(t,k)\right] \ , \tag{5}$$

and the function $\chi(t,k)$ is defined by the (ordinary nonlinear) differential equation

$$\chi_t(t,k) = \chi(t,k)\beta\left[-4\chi^2(t,k)\right] \tag{6}$$

and by the boundary condition

$$\chi(0,k) = k \ , \tag{7}$$

while the function $k_0(t,k)$ is (implicitly) defined by the formula

$$\chi(t,k_0) = k \ . \tag{8}$$

These equations characterize the time-evolution of the part of the spectral transform corresponding to the continuous spectrum. The discrete spectrum part evolves according to the equations

$$\dot{p}_n(t) = p_n(t)\beta\left[4p^2(t)\right] , \tag{9}$$

$$\rho_n(t) = \rho_n(0)\chi_k(t;ip_n(0)) \exp\left\{-2\int_0^t dt' p_n(t')\alpha\left[4p_n^2(t')\right]\right\} . \tag{10}$$

These equations, together with the appropriate equations relating $u(x,t)$ to its spectral transform (see section 0.2), provide *in principle* the solution of the Cauchy problem for the class of evolution equations (1). In particular the corresponding behaviour of the solitons can be investigated, and turns out to be rather rich [4-9]. On the other hand, for solutions that are not purely solitonic, a pathology generally appears as soon as the time evolution begins; this is manifested by the fact that, as soon as $t>0$, the property of $k_0(t,k)$ to diverge as k diverges, is lost; and as a consequence also lost is the property of $R(k,t)$ to vanish as k diverges. This indicates the emergence of a singularity of $u(x,t)$ as soon as the time-evolution is turned on, for *all* solutions that are not purely solitonic. The exact nature of this phenomenon remains to be understood. To clarify this point it shall probably be necessary to make sense of questions such as: "What is the (direct and inverse) spectral transform of a constant?". The task to legitimize such a question is clearly analogous to that of legitimizing the analogous question in the context of the Fourier transform -which has led to the introduction of *distributions*. Whether this framework will be sufficient to provide an appropriate setting for the question mentioned above in the spectral transform context as well, or whether some extension will be required, remains at present an open problem -clearly a very interesting one, in view of the nonlinear character of the spectral transform and of its usefulness to solve certain classes of nonlinear evolution equations.

2. THE "CYLINDRICAL" KORTEWEG-DE VRIES EQUATION

The results reported in this section have all been obtained in collaboration with A.Degasperis, and a somewhat more detailed description of them than the terse outline reported below can be found in the original papers [11-13].

The "cylindrical" KdV equation reads

$$u_t + u_{xxx} - 6u_x u + (2t)^{-1}u = 0 , \quad u \equiv u(x,t) . \tag{1}$$

For a justification of the name, and some indications of the applicative relevance of this equation, see [14]. The relation of this equation to the Schroedinger spectral problem on the line with an additional linear potential present was first brought to our attention by Dryuma [15]. This result implies the possibility to transform the cylindrical KdV equation (1) into the ordinary KdV equation (0.2-(3)), a result that had in fact been discovered several years ago by Lugovtzov and Lugovtzov [16]. However, under this transformation, the property of a function to vanish asymptotically ($x\to\pm\infty$) is not preserved. Therefore, to solve the Cauchy problem for the cylindrical KdV equation (1) within the more interesting class of functions (those that vanish asymptotically), the procedure described in section 0.2 to solve the Cauchy problem for the ordinary KdV equation is of no use, since such a technique is only applicable to solutions that are themselves asymptotically vanishing (see 0.2-(5)), and that correspond therefore to asymptotically (linearly) divergent solutions of (1).

To solve the Cauchy problem for (1) in the class of functions $u(x,t)$ that vanish asymptotically ($x\to\pm\infty$) it is thus necessary to introduce a novel spectral transform. This is defined by the formula

$$f(z) = \pi\left[F_x(x,z)\Phi(x,z) - F(x,z)\Phi_x(x,z)\right] , \qquad (2)$$

where $F(x,z)$ and $\Phi(x,z)$ are the two solutions of the "Schroedinger equation"

$$-\Psi_{xx} + \left[x + u(x)\right]\Psi = z\Psi \quad , \qquad \Psi \equiv \Psi(x,z) \quad , \qquad (3)$$

characterized by the asymptotic boundary conditions

$$\lim_{x\to+\infty} \left[\Phi(x,z)/Ai(x-z)\right] = 1 \quad , \qquad (4a)$$

$$\lim_{x\to-\infty} \left\{F(x,z)/\left[Bi(x-z) - iAi(x-z)\right]\right\} = 1 \; , \qquad (4b)$$

where Ai and Bi are the usual Airy functions [17]; note that (4) are consistent with the requirement that $u(x)$ in (3) vanish asymptotically, since the Airy functions in (4) are solutions of (3) for $u=0$.

It has been shown in [11] that a bijective correspondence between $u(x)$ and $|f(z)|$ holds, and a procedure to solve the "inverse problem" -to retrieve $u(x)$ given $|f(z)|$, $-\infty<z<\infty$- has moreover been provided [11], which is closely analogous to that outlined in section 0.2. (A more rigorous although less complete treatment of this inverse problem has been given, almost simultaneously and certainly independently, by Li Yi Shen [18]).

It has moreover been shown [12] that, as $u(x,t)$ evolves according to

$$u_t + (12t)^{-1}\left[u_{xxx} - 6u_x u - 4xu_x - 2u\right] = 0 \; , \qquad (5)$$

f(z,t) evolves according to the completely explicit formula

$$f(z,t) = f(z(t/t_0)^{1/3},t_0) \quad ; \tag{6}$$

while by a simple change of variables (5) can be transformed into (1). The Cauchy problem for (1) -given $u(x,t_0)$, find u(x,t) for $t>t_0$- is thereby solved in principle. This solution can moreover be used to analyze the asymptotic behaviour (as $t\to\infty$ and as $t\to 0$) of the solutions of (1) [19-20], and also to exhibit an infinite sequence of conserved quantities associated to (1) [13]. An amusing implication of the latter findings is the discovery that the "center-of-mass"

$$\bar{x}(t) = \int_{-\infty}^{+\infty} dx\, x\, u(x,t) \Big/ \int_{-\infty}^{+\infty} dx\, u(x,t) \tag{7}$$

of the generic solution of the cylindrical KdV equation (1) evolves in time according to the simple law

$$\bar{x}(t) = a + b|t|^{\frac{1}{2}} \quad , \tag{8}$$

with a and b time-independent. This may be contrasted with the analogous result for the ordinary KdV equation (0.2-(3)), that reads

$$\bar{x}(t) = A + Bt \quad , \tag{9}$$

where A and B are also time-independent.

Note that (8) and (9) are very different for large t, which contrasts with the naive expectation that might have been promoted by the observation that the difference between the cylindrical KdV equation (1) and the ordinary KdV equation (0.2-(3)) disappears as t diverges.

3. KdV EQUATION WITH DAMPING

The KdV equation with damping reads

$$u_t + u_{xxx} - 6u_x u - cHu_x = 0 \quad , \quad u \equiv u(x,t) \ , \tag{1}$$

with H the Hilbert operator,

$$Hf(x) \equiv \frac{P}{\pi}\int_{-\infty}^{+\infty} dy\, f(y)\Big/(x-y) \quad . \tag{2}$$

The last term in (1), for c>0, accounts for (Landau) damping. The interest to study (1) has been emphasized by Maxon [14]. Little is known about this equation, other than the obvious existence (for the class of solutions vanishing asymptotically, $u(\pm\infty,t)=0$, to which our consideration is always restricted) of the conserved quantity

$$c_0 = \int_{-\infty}^{+\infty} dx\, u(x,t) \quad . \tag{3}$$

The marginal progress to be tersely mentioned here has been achieved in collaboration with M.A.Olshanetsky and A.M.Perelomov [21]. It consists in the observation that there exist a class of rational solutions of (1), namely

$$u(x,t) = 2 \sum_{j=1}^{n} \left[x - x_j(t)\right]^{-2} \quad . \tag{4}$$

The restrictions on the number n of poles, on their locations and on their time evolution, that are required in order that (4) satisfy (1), are closely related [21] to the analogous conditions that must be satisfied in order that (4) satisfy the ordinary KdV equation (0.2-(3)); these latter requirements have been spelled out in considerable detail by Airault, McKean and Moser [22]. Unfortunately none of the solutions of (1) that can be obtained in this manner is real, and this constitutes clearly a major drawback with respect to their applicative relevance.

4. ON SOME NOVEL NONLINEAR EVOLUTION EQUATIONS

The results reported in this section have been obtained in collaboration with A.Degasperis; they have not been published so far. We first discuss a nonlinear evolution equation and we indicate how it is possible to solve it using the spectral transform of section 0.2. We next show that a more straightforward way to solve the Cauchy problem for this equation is also available (this was discovered in a rather roundabout way using the technique mentioned above, but it can be also verified directly, in a very straightforward manner). It is moreover shown that the nonlinear evolution equation under consideration possesses a nondenumerable infinity of conserved quantities. Finally a more general class of evolution equations, also possessing a nondenumerable infinity of conserved quantities, is exhibited.

The equation reads

$$v_t = \alpha(t)v_x + \beta(t)v - \gamma(t)\left[2v^2 - v_x\int_x^\infty dx' v(x',t)\right], \quad v \equiv v(x,t). \quad (1)$$

The attention will be focused on the Cauchy problem, with $v(x,0)$ given and $v(x,t)$ to be determined, say for $t>0$, in the class of functions vanishing (sufficiently fast) at spatial infinity,

$$v(\pm\infty,t) = 0 \quad . \quad (2)$$

While (1) is integrodifferential, it can be recast in pure differential form introducing, in place of v, the dependent variable

$$w(x,t) = \int_x^\infty dx' v(x',t) , \quad w_x(x,t) = -v(x,t) , \quad (3)$$

so that (1) reads

$$w_{xt} = \alpha(t)w_{xx} + \beta(t)w_x + \gamma(t)\left[2w_x^2 + w_{xx}w\right] . \quad (4)$$

The method of solution via the spectral transform described below provides, in addition to the solution of (1), or (4), the solution of the Cauchy problem for the associated (non-homogeneous linear) equation

$$u_t = \left[\alpha(t)+\gamma(t)w\right]u_x + 2\gamma(t)w_x u + \eta(t)w_x - \tfrac{1}{2}\gamma(t)w_{xxx} , \quad (5)$$

$$u\equiv u(x,t), \quad u(\pm\infty,t)=0, \quad u(x,0) \text{ given.}$$

Another form of (1), that might possibly be more interesting for applications, obtains from the change of variable

$$r(x,t) = \left[v(x,t)\right]^{\frac{1}{2}} \quad (6)$$

and it reads

$$r_t = \tfrac{1}{2}\beta(t)r + r\left[\alpha(t)-\gamma(t)\int_x^\infty dx' \, r^2(x',t)\right]_x \quad . \quad (7)$$

In these equations $\alpha(t)$, $\beta(t)$, $\gamma(t)$ (and $\eta(t)$) indicate arbitrary functions of t; but it is easily seen that (1), (4), and (7) can be reduced to the case $\alpha=\beta=0$, $\gamma=1$ by the following change of variables:

$$x' = x + \int_0^t dt_1\alpha(t_1) , \quad t' = \int_0^t dt_1\gamma(t_1)\exp\left[\int_0^{t_1} dt_2\beta(t_2)\right] ,$$
$$w'(x',t') = w(x,t)\exp\left[-\int_0^t dt_1\beta(t_1)\right] . \quad (8)$$

We therefore restrict hereafter attention to the case with $\alpha=\beta=0$, $\gamma=1$, so that (1) and (5) read

$$v_t = -2v^2 + v_x \int_x^\infty dx' \, v(x',t) \,, \qquad v \equiv v(x,t) \,, \tag{9}$$

$$u_t = wu_x + 2w_x u + \eta(t) w_x - \tfrac{1}{2} w_{xxx} \,. \tag{10}$$

The technique of solution via the spectral transform obtains from the remark that, if the "potential"

$$u(x,y,t) = u(x,t) + yv(x,t) \tag{11}$$

evolves in time according to (9) and (10), the corresponding evolution of its spectral transform (defined as in section 0.2, with y and t playing a purely parametric role) is given by the simple explicit formulae

$$R(k,y,t) = R_0(k,\ y-2k^2 t - \int_0^t dt' \eta(t')) \,, \tag{12}$$

$$p(y,t) = p_0(y+2p^2 t - \int_0^t dt' \eta(t')) \,, \tag{13}$$

$$\rho(y,t) = \rho_0(y+2p^2 t - \int_0^t dt' \eta(t')) \,, \tag{14}$$

where of course

$$R_0(k,y) = R(k,y,0) \,, \tag{15}$$

$$p_0(y) = p(y,0) \,, \tag{16}$$

$$\rho_0(y) = \rho(y,0) \,, \tag{17}$$

and we have omitted for notational simplicity to label with an index n the discrete spectrum parameters.

These equations provide the basis for the solution of the Cauchy problem for (9) and (10), along the lines sketched in section 0.2. It is actually possible, by introducing a vector variable $\vec{y}$ in place of the scalar used above, to solve in the same manner the nonlinear vector evolution equation

$$\vec{w}_{xt} = \alpha(t)\vec{w}_{xx} + \beta(t)\vec{w}_x + 2\left[\vec{\gamma}(t)\cdot\vec{w}_x\right]\vec{w}_x + \left[\vec{\gamma}(t)\cdot\vec{w}\right]\vec{w}_{xx} \,. \tag{18}$$

This equation is a generalization of (4), and equations generalizing in an analogous fashion (1) or (5) can be easily written; note that $\vec{w}$ and $\vec{\gamma}$ in (18) are vectors of arbitrary (but equal) dimension, with the standard convention for the scalar product. Hereafter we limit however consideration to the scalar case, and to the simplified version (9) of the nonlinear evolution equation.

It is then easy to verify that a straightforward solution of the Cauchy problem for this nonlinear evolution equation is provided by the explicit formula

$$v(x,t) = v_0[z(x,t)] / \{1 + 2tv_0[z(x,t)]\} \quad , \tag{19}$$

where of course

$$v_0(x) = v(x,0) \tag{20}$$

and the function $z(x,t)$ is (implicitly) defined by the formula

$$x = z + \int_z^{\infty} dz' \{1 - [1 + 2tv_0(z')]^{\frac{1}{2}}\} \tag{21}$$

implying

$$z_x(x,t) = [1 + 2tv_0(z)]^{-\frac{1}{2}} \quad . \tag{22}$$

It is moreover easily verified that, associated with the nonlinear evolution (9), there exist a nondenumerable infinity of constants of the motion, whose explicit definition reads

$$C(p) = \int_{-\infty}^{+\infty} dx [v(x,t)]^p [1-2tv(x,t)]^{\frac{1}{2}-p} \quad . \tag{23}$$

In this formula p is arbitrary, except for the requirement that the integral in the r.h.s. of (23) converges; since $v(x,t)$ vanishes as $x\to\pm\infty$, a necessary condition for this is that Re p be positive; and if $v(x,t)$ vanishes faster than any power of x, this condition is also sufficient.

There actually exists a more general class of nonlinear evolution equations with a nondenumerable infinity of constants of the motion, that includes (9) as a special case. The equations of this class read

$$v_t = f[v,t]v_x + v^2 + \mu v_x \int_x^{\infty} dx' v(x',t) \quad , \quad v \equiv v(x,t) \quad , \tag{24}$$

or equivalently (see (3))

$$w_{xt} = f[-w_x,t]w_{xx} - w_x^2 + \mu w_{xx} w \quad , \quad w \equiv w(x,t) \quad , \tag{25}$$

or

$$r_t = \{g[r,t] + \mu r \int_x^{\infty} dx' [r(x',t)]^{-1/\mu}\}_x \quad , \quad r \equiv r(x,t) \quad , \tag{26}$$

with

$$r(x,t) = [v(x,t)]^{-\mu} \quad . \tag{27}$$

The function f in (24) and (25) is arbitrary, as well as the function g in (26), being related to f by

$$g_r(r,t) = f(v,t) \quad , \quad r = v^{-\mu} \quad . \tag{28}$$

The constants of the motion read now

$$C(p) = \int_{-\infty}^{+\infty} dx [v(x,t)]^p \, [1 + tv(x,t)]^{-\mu-p} \quad , \tag{29a}$$

$$C(p) = \int_{-\infty}^{+\infty} dx \, r(x,t) \{t + [r(x,t)]^{1/\mu}\}^{-\mu-p}, \tag{29b}$$

again with p arbitrary except for the requirement that the integral in the r.h.s. of (29) converges.

REFERENCES

[1] C.S.Gardner, J.M.Greene, M.D.Kruskal and R.M.Miura: Method for solving the Korteweg-de Vries equation. Phys.Rev.Lett. 19, 1095-1097 (1967).

[2] F.Calogero: Nonlinear evolution equations solvable by the inverse spectral transform. In: *Mathematical problems in theoretical physics* (G.Dell'Antonio, S.Doplicher and G.Jona Lasinio, eds.), Lecture Notes in Physics 80, Springer, Heidelberg, 1978.

[3] F.Calogero and A.Degasperis: *Spectral transform and solitons: tools to solve and investigate nonlinear evolution equations*. North-Holland, Amsterdam, 1980.

[4] F.Calogero and A.Degasperis: Extension of the spectral transform method for solving nonlinear evolution equations. Lett.Nuovo Cimento 22, 131-137 (1978).

[5] F.Calogero and A.Degasperis: Exact solution via the spectral transform of a nonlinear evolution equation with linearly x-dependent coefficients. Lett.Nuovo Cimento 22, 138-141 (1978).

[6] F.Calogero and A.Degasperis: Extension of the spectral transform method for solving nonlinear evolution equations, II. Lett.Nuovo Cimento 22, 263-269 (1978).

[7] F.Calogero and A.Degasperis: Exact solution via the spectral transform of a generalization with linearly x-dependent coefficients of the modified Korteweg-de Vries equation. Lett.Nuovo Cimento 22, 270-273 (1978).

[8] F.Calogero and A.Degasperis: Exact solution via the spectral transform of a generalization with linearly x-dependent coefficients of the nonlinear Schroedinger equation. Lett.Nuovo Cimento 22, 420-424 (1978).

[9] F.Calogero and A.Degasperis: Conservation laws for classes of nonlinear evolution equations solvable by the spectral transform. Comm.Math.Phys. 63, 155-176 (1978).

[10] A.Newell: Near integrable systems, nonlinear tunnelling and solitons in slowly changing media. In: *Nonlinear evolution equations solvable by the spectral transform* (F.Calogero, ed.), Research Notes in Mathematics 26, Pitman, London, 1978, pp.127-179.

[11] F.Calogero and A.Degasperis: Inverse spectral problem for the one-dimensional Schroedinger equation with an additional linear potential. Lett.Nuovo Cimento 23, 143-149 (1978).

[12] F.Calogero and A.Degasperis: Solution by the spectral transform method of a nonlinear evolution equation including as a special case the cylindrical KdV equation. Lett.Nuovo Cimento 23, 150-154 (1978).

[13] F.Calogero and A.Degasperis: Conservation laws for a nonlinear evolution equation that includes as a special case the cylindrical KdV equation. Lett.Nuovo Cimento 23, 155-160 (1978).

[14] S.Maxon: Cylindrical and spherical solitons. Rocky Mountain Journ.Math. (special issue edited by H.Flaschka and D.W.McLaughlin) 8, 269-282 (1978).

[15] V.S.Dryuma: Isv.Akad.Nauk Mold. SSR, 3, 87 (1976) (in Russian).

[16] A.A.Lugovtzov and B.A.Lugovtzov, in *Dynamics of continuous media*, 1, Nauka, Novosibirsk, 1969 (in Russian).

[17] *Handbook of mathematical functions* (M.Abramowitz and I.Stegun, eds.), New York, Dover, 1965.

[18] Li Yi Shen: On a special inverse problem for second order differential equations on the whole axis. University of Science and Technology of China, Hefei, Anhui, 1979 (preprint, to be published).

[19] P.M.Santini: Asymptotic behaviour (in t) of solutions of the cylindrical KdV equation, I. Nuovo Cimento

[20] P.M.Santini: Asymptotic behaviour (in t) of solutions of the cylindrical KdV equation, II. Nuovo Cimento

[21] F.Calogero, M.A.Olshanetsky and A.M.Perelomov: Rational solutions of the KdV equation with damping. Lett.Nuovo Cimento 24, 97-100 (1979).

[22] H.Airault, H.P.McKean and J.Moser: Rational and elliptic solutions of the KdV equation and a related many-body problem. Comm.Pure Appl.Math. 30, 95-148 (1977).

QUANTIZATION OF COMPLETELY INTEGRABLE HAMILTONIAN SYSTEMS

B. Kostant

Department of Mathematics,
Massachusetts Institute of Technology
Cambridge, USA

(Manuscript not received)

REDUCTION TECHNIQUE FOR MATRIX NONLINEAR EVOLUTION EQUATIONS

A. Degasperis

Istituto di Fisica - Università di Roma, 00185 Roma-Italy

Istituto Nazionale di Fisica Nucleare, Sezione di Roma

1. INTRODUCTION

The Spectral Transform method was introduced in mathematical physics, for the first time, as a tool for solving the Korteweg-de Vries equation [1]. However, since that historical step, many other nonlinear partial differential equations have been discovered to be solvable via the Spectral Transform [2], many of these equations being of relevance to physics [3]. Such a proliferation of solvable nonlinear evolution equations is the result of several extensions and generalizations of the original definition of the Spectral Transform. In fact, some fruitful generalizations [4] have led to fairly large classes of matrix nonlinear partial differential equations, the dependent variables being therefore the entries of a $N\times N$ matrix valued function of the space and time coordinates. Looking now at this system of N^2 nonlinear partial differential equations, we may reasonably ask whether any subsystem of $M<N^2$ equations exist; this amounts to ask whether N^2-M relationship between the N^2 dependent variables can be found which are compatible with the time evolution, so that the number of dependent variables can be reduced from N^2 to M. Answering this question is an appealing mathematical problem as well as an interesting task for the purpose of applications.

In the following, after a brief description of the Spectral Transform method, I will present a general technique to reduce the number of dependent variables of a matrix nonlinear evolution equation solvable by the Spectral Transform associated to the multichannel Schroedinger scattering problem (2d). The results given here are mostly based on the content of Ref.5, which one should refer to for further details.

In our notation, the σ_μ's are N^2 linearly independent hermitian matrices; greek indeces take the values $0,1,2,\ldots,N^2-1$, while latin indeces run from 1 to N^2-1; repeated indeces are summed upon. Furthermore $\sigma_0=1$, $[A,B]\equiv AB-BA$ and $\{A,B\}\equiv AB+BA$.

2. SOLVABLE MATRIX NONLINEAR EVOLUTION EQUATIONS[6]

The Spectral Transform of a N×N matrix valued function Q(x), which vanishes at infinity sufficiently fast, is defined through the following Schroedinger linear problem

$$\psi_{xx} = [Q-k^2]\psi \ , \qquad \psi \equiv \psi(x,k), \quad Q \equiv Q(x) \tag{2.1}$$

together with the "scattering" asymptotic conditions for real k

$$T(k)\exp(-ikx) \underset{x\to-\infty}{\longleftarrow} \psi(x,k) \underset{x\to+\infty}{\longrightarrow} \exp(-ikx)+R(k)\exp(ikx), \tag{2.2}$$

which uniquely define the N×N matrix solution ψ. Here T(k) and R(k) are the N×N matrix valued transmission and reflection coefficients, respectively. The discrete part of the spectrum, if it exists, is characterized by n positive numbers p_ℓ's, corresponding to the negative eigenvalues $k^2=-p_\ell$, $\ell=1,2,\ldots,n$, and to the column vector solutions $\psi_\ell(x)$ of (2.1) (with $k=ip_\ell$) satisfying the normalization condition

$$\int_{-\infty}^{+\infty} dx\, \psi_\ell^\dagger(x)\psi_\ell(x) = 1 \ , \qquad \ell=1,2,\ldots,n \ ; \tag{2.3}$$

here, and in the following, the dagger means hermitian conjugation. The asymptotic behaviour of the solution ψ_ℓ defines the column vector c_ℓ, namely

$$\lim_{x\to+\infty} [\exp(p_\ell x)\, \psi_\ell(x)] = c_\ell \ . \tag{2.4}$$

It is convenient to introduce, together with the vector c_ℓ, the positive number ρ_ℓ and the N×N projector P_ℓ via the relations

$$c_\ell c_\ell^\dagger = \rho_\ell P_\ell \ , \quad \rho_\ell = c_\ell^\dagger c_\ell \ , \quad P_\ell^2 = P_\ell \ . \tag{2.5}$$

The Spectral Transform of the matrix $Q(x)$ is then defined as the collection of quantities

$$ST[Q] = \{R(k),\ p_\ell,\ \rho_\ell,\ P_\ell\} \quad . \tag{2.6}$$

The reason for this definition relies on the fact that not only, of course, Q uniquely determines $ST[Q]$ through the solution of the Schroedinger equation (2.1), but also $\{R(k),\ p_\ell,\ \rho_\ell,\ P_\ell\}$ (satisfying appropriate but general conditions) is the Spectral Transform of one and only one matrix $Q(x)$. This one-to-one correspondence is the main result of the theory of the inverse problem for the Schroedinger spectral problem and the actual construction of Q, from the reflection coefficient R and the spectral parameters p_ℓ, ρ_ℓ and P_ℓ, is obtained by solving a linear integral (Fredholm type) equation. This equation, known as Gel'fand-Levitan-Marchenko equation, reads

$$K(x,y)+M(x+y)+\int_x^{+\infty} dx'\ K(x,x')M(x'+y) = 0, \quad x \le y \ , \tag{2.7}$$

where the $N\times N$ matrix $M(z)$ is given in terms of the input quantities by the following definition

$$M(z) = (2\pi)^{-1}\int_{-\infty}^{+\infty} dk \exp(ikz)\, R(k) + \sum_{\ell=1}^{n} \rho_\ell \exp(-p_\ell z)\, P_\ell \ , \tag{2.8}$$

and the solution $K(x,y)$ yields the corresponding matrix Q via the simple expression

$$Q(x) = -2\frac{d}{dx} K(x,x) \quad . \tag{2.9}$$

The remarkable property of the mapping $Q \leftrightarrow ST[Q]$ is that a large class of *nonlinear partial differential* evolution equations exists such that if the time-dependent matrix $Q(x,t)$ satisfies an equation of this class, then its corresponding Spectral Transform $ST[Q]=\{R(k,t),p_\ell(t),\rho_\ell(t),P_\ell(t)\}$ satisfies a *linear ordinary differential* equation. This peculiar class of nonlinear evolution equations is given by the compact formula [7]

$$Q_t = \alpha_n(L)[\sigma_n,Q] + \beta_\nu(L)G\sigma_\nu \ , \quad Q \equiv Q(x,t) \quad , \tag{2.10}$$

where the action of the operators L and G on a generic x-dependent matrix F is defined by the expressions

$$LF(x) = F_{xx}(x) - 2\{Q(x,t),F(x)\} + G\int_x^{+\infty} dx'\, F(x') \quad , \tag{2.11}$$

$$GF(x) = \{Q_x(x,t),F(x)\} + \left[Q(x,t), \int_x^{+\infty} dx' \left[Q(x',t),F(x')\right]\right] . \tag{2.12}$$

In eq.(2.10) the operator L enters as the argument of the N^2-1 functions $\alpha_n(z)$ as well as of the N^2 functions $\beta_\nu(z)$; these $2N^2-1$ functions may be regarded as arbitrary polynomials.

The Spectral Transform of a solution $Q(x,t)$ of the nonlinear equation (2.10) satisfies the following linear equations

$$R_t = \left[A(-4k^2),R\right] + 2ik\{B(-4k^2),R\} \quad , \qquad R \equiv R(k,t) \quad , \tag{2.13}$$

$$\frac{d}{dt} p_\ell = 0 \quad ,$$

$$\frac{d}{dt} c_\ell = \left[A(4p_\ell^2),c_\ell\right] - 2p_\ell\{B(4p_\ell^2),c_\ell\}, \qquad c_\ell \equiv \rho_\ell(t)P_\ell(t), \tag{2.14}$$

where

$$A(z) \equiv \alpha_n(z)\sigma_n \quad , \qquad B(z) \equiv \beta_\nu(z)\sigma_\nu \quad . \tag{2.15}$$

Note that the discrete eigenvalues are t-independent (isospectral flow).

The Cauchy problem is the determination of$Q(x,t)$ for $t>0$ given $Q(x,0)=\bar{Q}(x)$, and its solution is obtained by the following 3-step procedure: i) compute $ST[\bar{Q}]=\{\bar{R}(k),\bar{p}_\ell,\bar{\rho}_\ell,\bar{P}_\ell\}$ by solving the *linear* Schroedinger problem, ii) find $\{R(k,t),\bar{p}_\ell,\rho_\ell(t),P_\ell(t)$ by solving the *linear* equations (2.13) and (2.14) with $R(k,0)=\bar{R}(k)$, $\rho_\ell(0)=\bar{\rho}_\ell$, $P_\ell(0)=\bar{P}_\ell$, iii) recover $Q(x,t)$ from $\{R(k,t),\bar{p}_\ell,\rho_\ell(t),P_\ell(t)\}$ by solving the *linear* Gel'fand-Levitan-Marchenko equation (2.7) (and using (2.9)). By definition, the nonlinear equation (2.10) is solvable since the Cauchy problem can be solved by means of linear operations. This solvable matrix nonlinear equation is the system of N^2 coupled equations which we want to reduce to a smaller subsystem.

Finally it should be emphasize that the generalization to the matrix case of the Spectral Transform method brings certain novelties in the world of solitons. In fact, due to the coupling of $P_\ell(t)$ (known as the soliton polarization matrix) to the soliton velocity through eq.(2.14), the position of the soliton being $\xi(t)=(2p_\ell)^{-1}\ln\left[(2p_\ell)^{-1}\rho_\ell(t)\right]$, the soliton acquires a nonvanishing acceleration (with the exception of special cases). Solitons which behave as a boomerang are named "boomerons" and those which oscillate in a confined region are named "trappons"[8]. The simplest equation which exhibits soliton solutions with both boomeron and trappon behaviours is known as the "Boomeron equation"[9] and obtains from (2.10) by letting the matrices (2.15), A and B, be independent of z.

3. REDUCTION TECHNIQUE

For matrices of any rank, a trivial reduction to an evolution equation involving one field q(x,t) is obtained by asking that $Q(x,t)=q(x,t)\sigma_0$; this ansatz is compatible with the time evolution only if the N^2-1 functions $\beta_n(z)$ in eq.(2.10) vanish (the α_n's, being irrelevant, may vanish as well), thus obtaining the well-known KdV family. For 2×2 matrices, it has already been shown by Jaulent and Leon [10] that the class of two coupled nonlinear evolution equations solvable via the Spectral Transform associated to the generalized Zakharov-Shabat spectral problem [2-a,b,c] is a subcase of the system of 4 coupled equations (2.10). Further reductions to one field equations of the Zakharov-Shabat class yield well-known equations such as the Modified KdV and the Sine-Gordon equations.

For matrices of rank 4, the simpler equation of the class solvable via the Schroedinger Spectral Transform has been analyzed by Bruschi, Levi and Ragnisco [11]. This equation involves 16 fields; reduced versions involving respectively 10, 8, 6, 5 and 4 fields have also been obtained, by identifying the cases in which some of the 16 fields, if vanishing at the initial time, will also vanish at any time.

All these reductions, although producing non trivial evolution equations, are rather simple and easily guessed by inspection; they can be treated by the technique described below, that is however considerably richer and allows for a systematic exploration.

Our technique of reduction is based again on the general rule of the Spectral Transform method, namely that of approaching problems by dealing with the Spectral Transform of Q, ST[Q] given by (2.6), rather than with Q itself. Indeed, transforming nonlinear problems into linear ones is the main virtue of this methods. Important instances of such semplification are the time evolution itself, eqs.(2.10), (2.13) and (2.14), the Bäcklund transformations (2-d) and the conservation laws (12).

The basic connection between a solution Q(x,t) of (2.10) and its Spectral Transform (for sake of simplicity, I will consider only the continuous part R(k,t) of ST[Q]) is given by the wronskian-type formulae (2-d)

$$2ik\left[F(-4k^2),R(k,t)\right] = \int_{-\infty}^{+\infty} dx\,\bar{\psi}(x,k,t)\{f_n(L)\left[\sigma_n,Q(x,t)\right]\}\psi(x,k,t) \qquad (3.1)$$

$$(2ik)^2\{H(-4k^2),R(k,t)\} = \int_{-\infty}^{+\infty} dx\,\bar{\psi}(x,k,t)\{h_\mu(L)G\sigma_\mu\}\psi(x,k,t) \qquad (3.2)$$

where L and G are the operators defined by (2.11) and (2.12), respectively, and the functions $f_n(z)$ and $h_\mu(z)$ are polynomials defining the matrix valued functions

$$F(z) \equiv f_n(z)\sigma_n \quad , \quad H(z) \equiv h_\mu(z)\sigma_\mu \qquad (3.3)$$

entering the l.h.s. of (3.1) and (3.2), respectively. $\bar{\psi}$ and ψ are appropriate matrix solutions of the Schroedinger spectral problems (2-d); also note that the variable t is only a parameter in these relationships.

The transformations (3.1) and (3.2) easily suggests what properties of Q(x,t) can be immediately translated into the corresponding properties of R(k,t), for they imply that, if the matrix Q(x,t) satisfies the *non-linear integro-differential* equation

$$f_n(L)\left[\sigma_n, Q(x,t)\right] + h_\mu(L)G\sigma_\mu = 0 \quad , \qquad (3.4)$$

the corresponding matrix R(k,t) satisfies the *linear algebraic* equation

$$\left[F(-4k^2),R(k,t)\right] + 2ik\{H(-4k^2),R(k,t)\} = 0 \quad , \qquad (3.5)$$

for given $2N^2-1$ arbitrary polynomials $f_n(z)$ and $h_\mu(z)$. For given matrices F and H, the matrix equation (3.5) is a linear homogeneous system of N^2 equations for the N^2 components of R; therefore for generic F and H it is satisfied only by the trivial solution R=0. But for appropriate choices of F and H, the condition (3.5) merely implies a reduction of the number of the independent components of R; this equation (3.5) will be referred to as the reduction-equation in k-space. The corresponding reduction-equation in x-space (3.4) is nonlinear and integro-differential, and finding its general solution may be a very difficult task; however, if the polynomials $f_n(z)$ and $h_\mu(z)$ are of very low-order (zero or possibly one), (3.4) can be solved and the N^2 elements of Q can be explicitly written in terms of $M<N^2$ independent fields. Examples of such solutions will be given below in the case N=2.

One should realize that this technique is merely providing a way of translating certain properties of a matrix Q into the corresponding properties of its Spectral Transform and viceversa, and it has nothing to do with the time evolution. But of course if Q (and so R) evolves in time, the question of compatibility arises: if at the initial time Q satisfies the reduction-equation (3.4), shall it satisfy it at any subsequent time? Again, it is easier to answer this question for the matrix R than for Q, and once this compatibility problem has been solved for R, the correspondence between R and Q will imply the compatibility of the time evolution of Q described by (2.10) with the reduction equation (3.4).

Therefore, let us define

$$Z(k,t) = \left[F(-4k^2),R(k,t)\right] + 2ik\{H(-4k^2),R(k,t)\} \qquad (3.6)$$

in order to ascertain when $Z(k,t)=0$ is compatible with (2.13). Differentiating with respect to t and using (2.13) one obtains

$$Z_t = \left[A(-4k^2),Z\right] + 2ik\{B(-4k^2),Z\} + C(k,t), \quad Z \equiv Z(k,t), \qquad (3.7)$$

with

$$C(k,t) = \left[R,([A,F]-4k^2[B,H])\right] - 2ik\{R,([B,F]+[A,H])\} \quad . \qquad (3.8)$$

Since $Z(k,0)=0$ implies $Z(k,t)=0$ for $t>0$ if $C(k,t)=0$, the compatibility equation for the matrices A and B (2.15), characterizing the nonlinear evolution equation (2.10), reads

$$\left[R,([A,F]-4k^2[B,H])\right] - 2ik\{R,([B,F]+[A,H])\} = 0, \qquad (3.9)$$

where $R\equiv R(k,t)$ is any solution of the reduction equation (3.5) corresponding to the matrices $F\equiv F(-4k^2)$ and $H\equiv H(-4k^2)$ (here $A\equiv A(-4k^2)$ and $B\equiv B(-4k^2)$). Note that at least one evolution equation of the class (2.10) satisfies the compatibility condition (3.9), namely the "scalar" one corresponding to $A(z)=0$ and $B(z)=\beta_0(z)\sigma_0$.

In conclusion, this reduction method can be summarized by this 4-step procedure:

1) for given matrices F(z) and H(z), find the general solution R(k,t)

of the reduction-equation (3.5); after this step, one knows the number M of independent fields be will end up with, which is the number of independent elements of the general matrix solution R;

2) find the general matrices A(z) and B(z) (2.15) which satisfy the compatibility equation (3.9);

3) solve the reduction-equation in x-space; at this point one obtains the expression of M independent fields in terms of the components of Q(x,t);

4) write explicitly the nonlinear evolution equations for these M fields, by inserting in (2.10) the results of steps 2) and 3).

Note that the first step requires solving a purely algebraic problem, which for matrices of rank > 2 may be fairly complicate; on the other hand the most difficult analytical problems are met at step 3).

Of course, this process of reduction can be applied more than once, each further reduction corresponding to a different choice of the matrices F(z) and H(z).

As a final remark, it is clear that the nonlinear evolution equations for the M independent fields obtained by reduction from the N×N matrix equation (2.10) enjoy all the nice properties of the solvable nonlinear evolution equations; therefore for these equations the Cauchy problem can be solved via the Spectral Transform method, the (scalar) Bäcklund Transformation (2-d) exists, the set of conserved quantities can be obtained from that known in the unreduced N×N case, and so on. But, of course, among all the solutions of the reduced evolution equations, only those that satisfy appropriate asymptotic conditions can be investigated by this method; in each particular case these asymptotic conditions are obtained by requiring that the Spectral Transform of the original N×N matrix Q exists.

4. RESULTS IN THE 2×2 CASE

In dealing with matrices of rank 2 it is convenient to identify the matrices σ_μ with the Pauli basis

$$\sigma_0=\begin{pmatrix}1&0\\0&1\end{pmatrix},\quad \sigma_1=\begin{pmatrix}0&1\\1&0\end{pmatrix},\quad \sigma_2=\begin{pmatrix}0&-i\\i&0\end{pmatrix},\quad \sigma_3=\begin{pmatrix}1&0\\0&-1\end{pmatrix},\tag{4.1}$$

and to write Q(x,t) and R(k,t) in this basis

$$Q = Q_\mu \sigma_\mu = Q_0\sigma_0 + Q_n\sigma_n, \quad R = R_\mu\sigma_\mu = R_0\sigma_0 + R_n\sigma_n . \tag{4.2}$$

The results reported below follow from a systematic investigation of the reduction equation in k-space (3.5) by choosing the matrix F(z) to be either constant or linear in z and the matrix H(z) to be constant. This limitation is mainly dictated by the structure of the reduction equation in x-space (3.4), which would become too complicated if F and H were higher order polynomials.

First of all, it has been found that no reduction exists to M=3 independent fields. There exist instead several possibilities to reduce to M=2 independent components. Further reductions to M=1 are also possible. Because of the known symmetry properties of the Pauli matrices, it is easy to show that different choices of F and H can yield the same system of 2 coupled equations. Taking into account this sort of equivalent choices of F and H, it turns out that one can discuss all the interesting reductions to M=2 fields by considering the following case

$$F(z) = -i(\gamma_0+\gamma_1 z)\sigma_1 , \quad H(z) = \sigma_3 . \tag{4.3}$$

The general solution of the reduction equation (3.5) is then found to be

$$R(k,t) = -(2ik)^{-1}(\gamma_0-4k^2\gamma_1)R_2(k,t)\sigma_0+R_1(k,t)\sigma_1+R_2(k,t)\sigma_2, \; R_3=0, \tag{4.4}$$

which explicitly displays the fact that R contains now only 2 independent components. The corresponding reduction equation for Q can also be solved, with some labor, and its general solution reads

$$Q(x,t) = Q_0(x,t)\sigma_0+Q_1(x,t)\sigma_1+Q_2(x,t)\sigma_2, \quad Q_3=0, \tag{4.5}$$

$$Q_0 = (1+2\gamma_1 W_2)^{-2}\left[\gamma_1(1+2\gamma_1W_2)W_{2xx}-\gamma_1^2W_{2x}^2+\gamma_1^2W_2^4+2\gamma_1W_2^3- -4\gamma_1UW_1+(1+\gamma_0\gamma_1)W_2^2+W_1^2+4\gamma_1^2U^2+\gamma_0W_2\right] \tag{4.6}$$

$$W_n \equiv W_n(x,t) = \int_x^{+\infty} dx'\, Q_n(x',t) , \quad U \equiv U(x,t) = -\int_x^{+\infty} dx' Q_1(x',t)W_2(x',t). \tag{4.7}$$

Considering next the compatibility equation (3.9), its solution is easily found to be

$$\alpha_1(z)=\alpha_2(z)=\beta_2(z)=\beta_3(z)=0, \quad iz\beta_1(z)=(\gamma_0+\gamma_1 z)\alpha_3(z) ; \tag{4.8}$$

thus the reduced class of nonlinear evolution equations for a matrix Q satisfying equations (4.5) and (4.6) reads

$$Q_t = 2\beta_0(L)Q_x + \beta_1(L)G\sigma_1 + \alpha_3(L)[\sigma_3, Q] \tag{4.9}$$

where $\alpha_3(z)$ and $\beta_1(z)$ are related to each other according to (4.8).

It is convenient to introduce the two fields

$$q(x,t)=-[W_1(x,t)-iW_2(x,t)], \quad r(x,t)=-[W_1(x,t)+iW_2(x,t)] \tag{4.10}$$

so that the reduced matrix Q takes the following structure

$$Q = \begin{pmatrix} Q_0 & q_x \\ r_x & Q_0 \end{pmatrix} , \tag{4.11}$$

where Q_0 is the expression (4.6), with W_1, W_2 and U expressed in terms of q and r.

As for a further reduction to M=1 field, among other possibilities, I consider now that obtained by choosing

$$F(z) = \sigma_2 \quad , \qquad H(z) = 0 , \tag{4.12}$$

since this one will be shown to yield an interesting novel equation. A different reduction will be discussed below. Both reduction equations for R and Q, with F and H given by (4.12), are merely algebraic and their solutions obtain with $R_1=Q_1=0$; considering then the expressions (4.4) and (4.5), in this case we have

$$R(k,t) = [-(2ik)^{-1}(\gamma_0-4k^2\gamma_1)\sigma_0 + \sigma_2]R_2(k,t) \tag{4.13}$$

$$Q(x,t) = Q_0(x,t)\sigma_0 + Q_2(x,t)\sigma_2 \quad . \tag{4.14}$$

This last expression can be rewritten in the form (4.11) with r=-q

$$Q = \begin{pmatrix} Q_0 & q_x \\ -q_x & Q_0 \end{pmatrix} \tag{4.15}$$

with

$$Q_0 = -(1-2i\gamma_1 q)^{-2}[q^2+i\gamma_0 q+\gamma_0\gamma_1 q^2+i\gamma_1(q_{xx}-2q^3) + \gamma_1^2(2qq_{xx}-q_x^2-q^4)] \quad . \tag{4.16}$$

The evolution equation compatible with the structure (4.15) is easily found by solving the compatibility equation (3.9) corresponding to the choice (4.12), and it is given by (4.9) with the further condition

$$\beta_1(z) = \alpha_3(z) = 0 \quad . \tag{4.17}$$

Proceeding now to discuss particular examples of these reductions, let us start with the simplest case $\gamma_0=\gamma_1=0$. The evolution equations obtained for the fields q and r defined by (4.10) are in this case those which are already known to be solvable via the Spectral Transform associated to the generalized Zakharov-Shabat spectral problem (2.10). The connection formulae between the reduced Schroedinger and the Zakharov-Shabat case are

$$Q=\begin{pmatrix} qr & q_x \\ r_x & qr \end{pmatrix}, \quad R=R_1\sigma_1+R_2\sigma_2=\begin{pmatrix} 0 & \alpha^{(-)}(-k,t) \\ \alpha^{(+)}(k,t) & 0 \end{pmatrix} \tag{4.18}$$

(for this notation see, for instance, Ref.(2-c)); note that now the compatibility condition (4.8) does not imply any restriction on $\alpha_3(z)$, so that the evolution equations read

$$Q_t = 2\beta_0(L)Q_x + \alpha_3(L)[\sigma_3,Q] \tag{4.19}$$

$$R_t = 4ik\beta_0(-4k^2)R + \alpha_3(-4k^2)[\sigma_3,R] \tag{4.20}$$

which are equivalent to

$$\sigma_3 v_t + \gamma(L_{ZS})v = 0, \qquad v \equiv v(x,t), \tag{4.21}$$

$$\alpha_t^{(\pm)} \pm \gamma(k)\alpha^{(\pm)} = 0, \qquad \alpha^{(\pm)} \equiv \alpha^{(\pm)}(k,t) \tag{4.22}$$

with

$$v(x,t) \equiv \begin{pmatrix} r(x,t) \\ q(x,t) \end{pmatrix} \tag{4.23}$$

$$\gamma(k) = -4ik\beta_0(-4k^2) + 2\alpha_3(-4k^2), \tag{4.24}$$

the matrix integro-differential operator L_{ZS} being defined by the formula

$$L_{ZS}\begin{pmatrix} U^{(1)}(x) \\ U^{(2)}(x) \end{pmatrix} = \frac{1}{2i}\begin{pmatrix} U_x^{(1)}(x) \\ -U_x^{(2)}(x) \end{pmatrix} + i\int_x^{+\infty} dx'\left[r(x',t)U^{(2)}(x')-q(x',t)U^{(1)}(x')\right]\begin{pmatrix} r(x,t) \\ q(x,t) \end{pmatrix}. \tag{4.25}$$

To display an explicit example, we set

$$\beta_0(z) = \tfrac{1}{2}(c+dz), \qquad \alpha_3(z) = (2i)^{-1}(a+bz); \tag{4.26}$$

then the nonlinear evolution equations read

$$q_t=-iaq-ib[q_{xx}-2(qr)q]+cq_x+d[q_{xxx}-6(qr)q_x], \tag{4.27a}$$

$$r_t= iar+ib[r_{xx}-2(qr)r]+cr_x+d[r_{xxx}-6(qr)r_x] \tag{4.27b}$$

with the asymptotic conditions (here and in the following $f(x)\to 0$ means the vanishing of the function f and of all its derivatives)

$$q(x,t)\xrightarrow[x\to+\infty]{}0, \quad r(x,t)\xrightarrow[x\to+\infty]{}0, \quad q(x,t)r(x,t)\xrightarrow[x\to-\infty]{}0, \tag{4.28}$$

which are actually more general than those required in the Zakharov-Shabat spectral approach.

Since this class of equations has been fully investigated in the literature (2), I limit myself to mention that the well-known reductions of the system (4.27) are easily recovered by the technique described here. For instance, it is easy to verify that the Modified KdV equation

$$q_t = q_{xxx} - 6\eta q^2 q_x \quad , \qquad \eta = \pm 1 \tag{4.29}$$

obtains by performing the further reduction with $F=\sigma_1$ and $H=0$ for $\eta=+1$ and with $F=\sigma_2$ and $H=0$ for $\eta=-1$, which implies $a=b=0$ and $q=\eta r$, and setting, of course, $c=0$ and $d=1$. These reductions to an evolution equation for one field were already known because they can be easily found by inspection. Now I consider a less trivial reduction of the system (4.27) which was not known before as it requires the full power of the reduction technique to be uncovered.

This reduction results from the choice

$$F(z) = \sigma_1 + i(c_0+c_1 z)\sigma_2 \quad , \qquad H(z) = 0 \tag{4.30}$$

where c_0 and c_1 are constant. While the solutions of the reduction equation in k-space (3.5) and of the compatibility equation (3.9) are easily derived, being

$$R(k,t) = R_1(k,t)\left[\sigma_1+i(c_0-4k^2c_1)\sigma_2\right] \ , \tag{4.31}$$

$$\alpha_3(z) = 0 \ , \tag{4.32}$$

the derivation of the corresponding formula for Q from the reduction equation (3.4) that now reads

$$\left[\sigma_1,Q\right] + i(c_0+c_1L)\left[\sigma_2,Q\right] = 0 \ , \tag{4.33}$$

is less elementary. The final result is most conveniently written in terms of the fields W_1 and W_2 defined by (4.7), and is given by the following relation

$$W_2 = i\frac{\left[c_0W_1-c_1(2W_1^3-W_{1xx})\right]}{\left[1-4c_0c_1W_1^2+4c_1^2(W_1^4-W_{1x}^2)\right]^{\frac{1}{2}}} \quad . \tag{4.34}$$

Therefore a class of nonlinear evolution equations for the single field W_1 has been obtained which is solvable by the Spectral Transform technique. This class is written, for example, in the Schroedinger case formalism as

$$Q_t = 2\beta_0(L)Q_x \qquad (4.35)$$

with

$$Q = (W_1^2 + W_2^2)\sigma_0 - W_{1x}\sigma_1 - W_{2x}\sigma_2 \qquad (4.36)$$

with W_2 expressed in terms of W_1 through the formula (4.34). The asymptotic conditions approriate to this class of evolution equations are

$$W_1(x,t)\xrightarrow[x\to+\infty]{}0 \qquad (4.37)$$

as implied by the definitio n (4.7), and

$$W_1(x,t)\xrightarrow[x\to-\infty]{}0 \quad , \qquad \text{if} \quad c_0^2 \neq 1 \; ; \qquad (4.38)$$

in the special case $c_0^2=1$, as it is implied by (4.34) $W_2 \sim iW_1$ as $x\to-\infty$ and therefore $Q_0=W_1^2+W_2^2\to 0$ for any constant value taken by W_1 as $x\to-\infty$, then

$$W_1(x,t)\xrightarrow[x\to-\infty]{}\text{arbitrary const.},\; W_{1x}(x,t)\xrightarrow[x\to-\infty]{}0, \text{ if } c_0^2=1 \; . \qquad (4.39)$$

A simple example of nonlinear evolution equation of this class obtains inserting (4.36) in (4.35) with $\beta_0(z)=\frac{1}{2}(c+dz)$; the following change of dependent and independent variables

$$u(x',t')=(c_1/c_0)^{\frac{1}{2}}W_1(x,t), \quad x'=(c_0/c_1)^{\frac{1}{2}}(x+ct), \quad t'=d(c_0/c_1)^{\frac{3}{2}}\,t \qquad (4.40)$$

yields for u the neater equation

$$u_t = u_{xxx} - 6u_x\{u^2-(u-2u^3+u_{xx})^2/[a-4(u^2-u^4+u_x^2)]\}\,, \qquad (4.41)$$

where, for notational convenience, we have omitted the primes, and we have set $a=1/c_0$. The boundary conditions for u are $((c_0/c_1)^{\frac{1}{2}}>0)$

$$u(x,t)\xrightarrow[x\to+\infty]{}0 \qquad (4.42a)$$

$$u(x,t)\xrightarrow[x\to-\infty]{}0 \text{ if } a^2\neq 1;\; u(x,t)\xrightarrow[x\to-\infty]{}\text{arbit.const.},\; u_x(x,t)\xrightarrow[x\to-\infty]{}0 \text{ if } a^2=1. \qquad (4.42b)$$

Let me conclude the discussion of this case by reporting the single soliton solution of (4.41)

$$u(x,t) = A/\cosh\{[x-\xi(t)]/\lambda\} \qquad (4.43)$$

with the amplitude and position of the soliton given in terms of its width λ by the expressions

$$A=a\lambda/[(1+\lambda^2)^2-a^2\lambda^4]^{\frac{1}{2}}, \qquad \xi(t)=\xi_0-t/\lambda^2 \; . \qquad (4.44)$$

The Spectral Transform of the 2×2 matrix Q corresponding to the soliton solution (4.43) has, of course, R=0 and a two-fold degenerate discrete eigenvalue $-p^2$, whose corresponding spectral parameters are (see sec.2)

$$\rho_j = 2p\exp(2p\xi_j), \qquad j=1,2 \tag{4.45}$$

together with the polarization matrices

$$P_j = \tfrac{1}{2}(\sigma_0 + \hat{n}_m^{(j)}\sigma_m), \qquad j=1,2, \tag{4.46}$$

where

$$\xi_1 = \xi-\lambda\ln\sinh\mu,\ \xi_2 = \xi_1-i\pi\lambda,\ \lambda = (2p)^{-1}, \tag{4.47}$$

$$\hat{n}_1^{(j)}=\cosh[\nu-(-)^j\mu],\ \hat{n}_2^{(j)}=i\sinh[\nu-(-)^j\mu],\ \hat{n}_3^{(j)}=0,\ j=1,2. \tag{4.48}$$

Here I assume that p is a positive constant such that the quantity ν defined by setting

$$\mathrm{tgh}\,\nu = a/(1+4p^2) = a\lambda^2/(1+\lambda^2) \tag{4.49}$$

is real (so that A is real). Note that the constant μ entering in (4.47) and (4.48) does not appear in (4.43) because it is not present in the matrix

$$\rho_1 P_1+\rho_2 P_2 = 2p\exp(2p\xi)\left[\sigma_1\sinh\nu + i\sigma_2\cosh\nu\right] \tag{4.50}$$

which is the relevant matrix in the kernel (2.8) of the Gel'fand-Levitan-Marchenko equation (2.7).

Let us consider now the more general case of our original reduction corresponding to the choice (4.3) with $\gamma_0\neq 0$ but $\gamma_1=0$. In this case the solution of the reduction equation for R and for Q are easily read out of the general expression (4.4) and (4.5) (together with (4.6) and (4.7)) respectively, while the compatibility condition is (4.8) with $\gamma_1=0$. However it is more convenient to introduce the new fields

$$\tilde{q}(x,t) = q(x,t)+i\gamma_0/2,\quad \tilde{r}(x,t) = r(x,t)-i\gamma_0/2 \tag{4.51}$$

and the constant

$$\rho = |\gamma_0|/2 \tag{4.52}$$

so that this reduction is characterized by the expressions

$$Q = \begin{pmatrix} \tilde{q}\tilde{r}-\rho^2 & \tilde{q}_x \\ \tilde{r}_x & \tilde{q}\tilde{r}-\rho^2 \end{pmatrix},\quad R = \begin{pmatrix} i\gamma_0(2k)^{-1}R_2 & R_1-iR_2 \\ R_1+iR_2 & i\gamma_0(2k)^{-1}R_2 \end{pmatrix}. \tag{4.53}$$

As an example of evolution equation of this class, let us choose

$$\beta_0(z)=\frac{1}{2}(c+dz),\quad \beta_1(z)=-\frac{1}{2}\gamma_0 b,\quad \alpha_3(z)=-\frac{1}{2}ibz; \tag{4.54}$$

the system of the two coupled equations then reads

$$\tilde{q}_t = c\tilde{q}_x + d\left[\tilde{q}_{xxx}-6(\widetilde{\widetilde{qr}}-\rho^2)\tilde{q}_x\right]-ib\left[\tilde{q}_{xx}-2(\widetilde{\widetilde{qr}}-\rho^2)\tilde{q}\right] \tag{4.55a}$$

$$\tilde{r}_t = c\tilde{r}_x + d\left[\tilde{r}_{xxx}-6(\widetilde{\widetilde{qr}}-\rho^2)\tilde{r}_x\right]+ib\left[\tilde{r}_{xx}-2(\widetilde{\widetilde{qr}}-\rho^2)\tilde{r}\right] \tag{4.55b}$$

with the boundary conditions

$$\tilde{q}(x,t)\xrightarrow[x\to+\infty]{} i\gamma_0/2,\quad \tilde{r}(x,t)\xrightarrow[x\to+\infty]{} -i\gamma_0/2,\quad \tilde{q}(x,t)\tilde{r}(x,t)\xrightarrow[x\to-\infty]{}\rho^2 . \tag{4.56}$$

An interesting form of (4.55) obtains if c, d and b are real and

$$\gamma_0^* = \eta\gamma_0 \quad , \qquad \eta = \pm 1 ; \tag{4.57}$$

in fact, in this case, the system (4.55) is equivalent to a single equation for the complex field

$$\psi(x,t) \equiv \exp(-iat+i\mu)\,\tilde{q}(x,t) \quad , \tag{4.58}$$

with a and μ real arbitrary constants. This equation, a "generalized Hirota equation" [13], obtains from (4.55a) and reads

$$\psi_t = -ia\psi-ib\left[\psi_{xx}-2\eta(|\psi|^2-\rho^2)\psi\right]+c\psi_x+d\left[\psi_{xxx}-6\eta(|\psi|^2-\rho^2)\psi_x\right] \tag{4.59}$$

with the asymptotic conditions

$$|\psi(-\infty,t)|^2 = |\psi(+\infty,t)|^2 = \rho^2 \quad . \tag{4.60}$$

Of course, the second equation (4.55b) is automatically satisfied with

$$\tilde{r}(x,t) = \eta\exp(-iat+i\mu)\,\psi^*(x,t) \quad . \tag{4.61}$$

Also here, the further reduction corresponding to (4.12) and discussed in the general case, yields a new nonlinear evolution equation for one field . This equation, which can be properly specialized to a sort of generalized MKdV equation, can be easily investigated in analogy with the previous case and will not be reported here [5].

I finally consider the most general case with $\gamma_0\neq 0$ and $\gamma_1\neq 0$. As the evolution equations for the two fields W_1 and W_2, obtained from (4.9) with (4.5) and (4.6), are rather complicated, I will limit myself to the special choice

$$\beta_0(z) = \frac{1}{2}(c+dz) \quad , \qquad \beta_1(z) = \alpha_3(z) = 0 \tag{4.62}$$

which yields the system

$$W_{jt} = c\,W_{jx} + d\left[W_{jxxx} - 6Q_0 W_{jx}\right], \qquad j=1,2 \tag{4.63}$$

with Q_0 given by (4.6). Furthermore, I will discuss only a simpler equation for one field which can be derived from this system by performing the further reduction corresponding to (4.12). Therefore, since $W_1=0$, the matrix Q has the structure

$$Q = \begin{pmatrix} Q_0 & iW_{2x} \\ -iW_{2x} & Q_0 \end{pmatrix} \tag{4.64}$$

with

$$Q_0=(1+2\gamma_1 W_2)^{-2}\{\gamma_1(1+2\gamma_1 W_2)W_{2xx}-\gamma_1^2 W_{2x}^2+W_2(1+\gamma_1 W_2)[\gamma_0+W_2(1+\gamma_1 W_2)]\} \tag{4.65}$$

The corresponding reflection coefficient is given by (4.13).

The evolution equation so obtained is more neatly written for the field

$$u(x,t) = 2\gamma_1 W_2(x,t/d) \tag{4.66}$$

and it reads

$$u_t = \frac{\partial}{\partial x}\left[u_{xx} - \tfrac{3}{2}u_x^2/(1+u) + \tfrac{1}{3}A(1+u)^3 - B/(1+u) + Cu\right] \tag{4.67}$$

where

$$A=-3/(8\gamma_1^2), \quad B=A+3\gamma_0/(2\gamma_1), \quad C=-A-B+c/d\ . \tag{4.68}$$

The boundary condition to be associated to (4.67) is

$$u(x,t)\xrightarrow[x\to+\infty]{}0\ , \qquad u_x(x,t)\xrightarrow[x\to-\infty]{}0\ . \tag{4.69}$$

As far the value of u as $x\to-\infty$, the following 4 possibilities are all compatible with the condition $Q_0(-\infty,t)=0$:

$$u(-\infty,t) = -1 \pm 1\ , \qquad u(-\infty,t) = -1 \pm (B/A)^{\frac{1}{2}}\ . \tag{4.70}$$

Here I assume u to be real, and (B/A) to be positive.

Another interesting version of the nonlinear equation (4.67) obtains setting

$$u(x,t) = \exp\left[\tfrac{1}{2}v(x,t)\right] - 1\ ; \tag{4.71}$$

the new field v satisfies then the equation

$$v_t = v_{xxx} - \tfrac{1}{8}v_x^3 + v_x\left[A\exp(v) + B\exp(-v) + C\right]\ , \tag{4.72}$$

with the boundary condition

$$v(x,t)\xrightarrow[x\to+\infty]{}0,\ v_x(x,t)\xrightarrow[x\to-\infty]{}0,\ v(-\infty,t)=0 \text{ or } v(-\infty,t)=\ln(B/A)\ . \tag{4.73}$$

Let me emphasize that this equation can be investigated by the

Spectral Transform method associated to the 2×2 Schroedinger problem corresponding to the matrix

$$Q(x,t) = \begin{pmatrix} Q_0 & (-i4\gamma_1)^{-1} v_x \exp(\frac{1}{2}v) \\ (i4\gamma_1)^{-1} v_x \exp(\frac{1}{2}v) & Q_0 \end{pmatrix} , \quad v \equiv v(x,dt), \tag{4.74}$$

with

$$Q_0 = \tfrac{1}{4}\left[v_{xx} + \tfrac{1}{4}v_x^2 - \tfrac{2}{3}(A\exp(v) + B\exp(-v) + C) + 2c/(3d)\right] . \tag{4.75}$$

The corresponding reflection coefficient, given by (4.13), evolves according to the simple equation

$$R_{2\,t}(k,t) = 2ik(c-4k^2 d) R_2(k,t) . \tag{4.76}$$

Therefore, for the equation (4.72) one can derive the set of conserved quantities, the Bäcklund transformation and the multi-soliton solutions by standard Spectral Transform techniques (2-d). A special case of equation (4.72), namely

$$\theta_t = \theta_{xxx} + \tfrac{1}{8}\theta_x^3 + \alpha \sin\theta \tag{4.77}$$

has been recently investigated by means of differential geometric techniques and its prolongation structure has been shown to possess interesting algebraic properties [14] .

I end this analysis by displaying a transformation relating the equation (4.72) to the MKdV equation. This transformation, which generalizes a transformation discovered by F.Magri [15] for the equation

$$\phi_t = \phi_{xxx} + \tfrac{1}{8}\phi_x^3 + \tfrac{3}{8}\alpha^2 \phi_x \sin^2(\phi/2) , \tag{4.78}$$

this being again a special case of (4.72), takes the following expression

$$q(x,t) = -\tfrac{1}{4}v_x(x,t) + a\exp[v(x,t)/2] + b\exp[-v(x,t)/2] , \tag{4.79}$$

and is such that, if $v(x,t)$ satisfies (4.72), then $q(x,t)$, given by (4.79), satisfies the MKdV equation

$$q_t = q_{xxx} - 6q^2 q_x + \nu q_x , \tag{4.80}$$

where the constants a, b and ν are given by the formulae

$$A=-6a^2 , \quad B=-6b^2 , \quad C=\nu-12ab . \tag{4.81}$$

REFERENCES

[1] GARDNER, C.S., GREENE, J.M., KRUSKAL, M.D., MIURA, R.M.: "Method for solving the Korteweg-de Vries equation". Phys.Rev.Lett.19, 1095-1097 (1967).

[2] a) ZAKHAROV, V.E., SHABAT, A.B.: "Exact theory of two-dimensional self-focusing and one-dimensional self-modulation of waves in nonlinear media". Sov.Phys.JETP 34, 62-69 (1972) [Zh.Eksp.Teor.Fiz.61, 118(1971)] ; b) ABLOWITZ, M.J., KAUP, D.J., NEWELL, A.C., SEGUR, H.: "The inverse scattering transform-Fourier analysis for nonlinear problems". Studies Appl.Math.53, 249-315(1974); c) CALOGERO, F., DEGASPERIS, A.: "Nonlinear evolution equations solvable by the inverse spectral transform, I". Nuovo Cimento 32B, 201-242 (1976); d) CALOGERO, F., DEGASPERIS, A.: "Nonlinear evolution equations solvable by the inverse spectral transform, II". Nuovo Cimento 39B, 1-54(1977); NEWELL, A.C.: "The general structure of integrable evolution equations". Proc.R.Soc.Lond.A 365, 283-311(1979); CALOGERO, F., DEGASPERIS, A.: "Extension of the Spectral Transform method for solving nonlinear evolution equations, I & II". Lett.Nuovo Cimento 22, 131-137(1978) & 22, 263-269(1978); CALOGERO, F., DEGASPERIS, A.: "Solution by the Spectral Transform method of a nonlinear evolution equation including as a special case the Cylindrical KdV Equation". Lett. Nuovo Cimento 23, 150-154(1978).

[3] BULLOUGH, R.K., CAUDREY, P.J.: "The soliton and its history" in "Solitons" (Bullough R.K. and Caudrey, P.J., eds), Lecture Notes in Physics, Springer, Heidelberg,1979; and also references quoted there.

[4] WADATI, M., KAMIJO, T.: "On the extension of inverse scattering method". Prog. Theor.Phys.52, 397-414 (1974). Ref.(2-d). ZAKHAROV, V.E.:"The inverse scattering method", in "Solitons" (Bullough, R.K. and Caudrey, P.J., eds.), Lecture Notes in Physics, Springer, Heidelberg, 1979.

[5] CALOGERO, F., DEGASPERIS, A.: "Reduction technique for matrix nonlinear evolution equations solvable by the Spectral Transform". To appear in J. of Math.Phys.

[6] For an introductory review, see DEGASPERIS, A.: "Spectral Transform and solvability on nonlinear evolution equations" in "Nonlinear Problems in Theoretical Physics" (Rañada, A.F. ed.) Lecture Notes in Physics, 98, Springer 1979, pp.35-90.

[7] The class of solvable equations is actually larger, as a t-dependent polynomial of L could act also on Q_t and the functions α_n and β_ν could dependent also on t. See Ref.(2-d).

[8] DEGASPERIS, A.: "Solitons, Boomerons, Trappons", in "Nonlinear Evolution Equations solvable by the Spectral Transform" (Calogero F., ed.) Research Notes in Mathematics 26, Pitman Publishing, London, 1978, pp.97-126.

[9] CALOGERO, F., DEGASPERIS, A.: "Coupled nonlinear evolution equations solvable via the inverse Spectral Transform, and solitons that come back: the Boomeron". Lett. Nuovo Cimento 16, 425-433 (1976).

[10] JAULENT, M., LEON, J.J.P.: "Nonlinear evolution equations associated with a massive Dirac system". Lett.Nuovo Cimento 23, 137(1978).

[11] BRUSCHI, M., LEVI, D., RAGNISCO, O.: "Nonlinear evolution equations solvable by the inverse Spectral Transform associated to the matrix Schrödinger equation of rank 4". Nuovo Cimento 43B, 251-270(1978).

[12] CALOGERO, F., DEGASPERIS, A.: "Conservation laws for classes of nonlinear evolution equations solvable by the Spectral Transform". Comm.Math.Phys.63, 155-176 (1978).

[13] HIROTA, R.: "Exact envelope-soliton solutions of a nonlinear wave equation". J.Math.Phys.14, 805(1973).

[14] PIRANI, F., SOLIANI, G. (private communication).

[15] Private communication.

SIMILARITY SOLUTIONS OF THE KORTEWEG-DE VRIES EQUATION, BÄCKLUND TRANSFORMATIONS AND PAINLEVE TRANSCENDENTS

M.Boiti and F.Pempinelli*

Istituto di Fisica dell'Università

Lecce

1. INTRODUCTION

We consider the evolution equation

$$u_t - 6u\,u_x + u_{xxx} = 0 \tag{1.1}$$

introduced by Korteweg and de Vries[1] to describe the lossless propagation of shallow-water waves in one direction only.

This equation (KdV) applies to a vaste range of nonlinear physical systems, when one whishes to include to first-order non linear and dispersive effects[2]. But the most relevant fact about this equation is that it has been used as a key for opening the way in the discovery and investigation of many different and important fields such as solitons, the inverse spectral transform[3], i.e. a generalization of the Fourier transform to nonlinear phenomena[4], and Abelian varieties.[5]

In this lecture we will show that the KdV equation can give another bonus and precisely that in its reduced form, the similarity equation, together with the Bäcklund transformation[6], can be used to obtain hidden symmetry properties of the Painlevé transcendents[7-8], which are new, as far as we know.[9]

The Painlevé transcendents are also related to the reduced similarity-like form of many other completely integrable evolution equations.

---

* Supported in part by I.N.F.N.

This fact has raised new interest in ordinary differential equations free from movable critical points[10], a field of research which seems to have been forgotten by mathematicians and physicists since the beginning of this century.

We think that it is worthwhile to fish out old results on Painlevé transcendents and recast them in a more modern language.

Therefore in this lecture we review also the Boutroux's work[11] on the asymptotic properties of Painlevé transcendents.

2. SIMILARITY EQUATIONS AND FIRST PAINLEVÉ TRANSCENDENTS.

It is convenient to deal with a function $w(x,t)$, which satisfies the equation

$$w_t - 3w_x^2 + w_{xxx} = 0 , \tag{2.1}$$

which we shall refer to as the potential KdV equation (PKdV).

If we set $u = w_x$ the potential function w furnishes the general solution of the KdV equation.

All the properties of the KdV equation can be of course translated in the language of the PKdV equation. For instance the PKdV is invariant under the transformation

$$w(x,t) \longrightarrow w(x-6k^2 t,\ t) - k^2 x + 3k^4 t \quad , \tag{2.2}$$

which becomes the Galilean transformation for $u = w_x$.

The similarity solution of the PKdV equation is more usefully written separating explicitly the possible quadratically diverging part in x

$$w(x,t) = -\frac{\eta}{x}\left[\frac{\eta^2}{2} + 2\phi(\eta)\right] , \tag{2.3}$$

where

$$\eta = \frac{x}{(6t)^{1/3}} . \tag{2.4}$$

The function $\phi(\eta)$ will in the following be called the potential similarity solution.

The u(x,t) hence becomes

$$u(x,t) = -\frac{\eta^2}{x^2}\,(\eta + 2\phi_\eta(\eta)) \;. \tag{2.5}$$

The similarity equation of the PKdV equation is

$$\phi_{\eta\eta\eta} + 6\phi_\eta^2 + 4\eta\phi_\eta - 2\phi = 0 \tag{2.6}$$

and it can be once integrated to

$$\phi_{\eta\eta}^2 + 4\eta\phi_\eta^2 + 4\phi_\eta^3 - 4\phi\phi_\eta = \mu^2 \tag{2.7}$$

with μ an arbitrary constant.

By the combined use of eqs. (2.6) and (2.7) one obtains

$$2\phi_\eta\phi_{\eta\eta\eta} - \phi_{\eta\eta}^2 + 4\eta\phi_\eta^2 + 8\phi_\eta^3 + \mu^2 = 0 \;. \tag{2.8}$$

We have therefore succeded in once integrating also the similarity equation of KdV.

Eq. (2.8), apart from some obvious changes in the definition of ϕ_η and η, is one of the 50 canonical types of second-order differential equations with fixed critical points, listed by Painlevé and Gambier.

Its solutions are related to the Painlevé transcendents by the following equality

$$\phi_\eta(\eta;\mu^2) = -\,2^{-1/3}\,(V_z(z;\mu) + V^2(z;\mu) + \frac{1}{2}z) \tag{2.9}$$

where

$$z = 2^{1/3}\,\eta \tag{2.10}$$

and $V(z;\mu)$ satisfies the second Painlevé equation

$$V_{zz}(z;\mu) = 2V^3(z;\mu) + zV(z;\mu) + \mu - \frac{1}{2} \;. \tag{2.11}$$

Therefore the similarity solution of the KdV equation takes the form

$$u = \frac{1}{(3t)^{2/3}}\left[V_z(z;\mu) + V^2(z;\mu)\right] \;. \tag{2.12}$$

It must however still be proved that eq.(2.12) actually gives the *general* similarity solution, because it may happen that a 1-parameter family of V's furnishes the same u.

Thanks to our particular approach, we can definitively clarify this point drawing an explicit one-to-one correspondence between the potential similarity solutions a of PKdV equation, and the V solutions of the second Painlevé equation.

Precisely from eqs.(2.9), (2.11) and (2.6) one gets

$$V(z;\mu) = 2^{-1/3} \frac{\phi_{\eta\eta}(\eta;\mu^2) + \mu}{2\phi_\eta(\eta;\mu^2)} \tag{2.13}$$

and

$$\phi(\eta;\mu^2) = 2^{-2/3}\left[V_z^2(z;\mu) - \left[V^2(z;\mu) + \tfrac{1}{2}z\right]^2 - 2\mu V(z;\mu)\right] . \tag{2.14}$$

This bijective correspondence fails only for $\phi(\eta;0) \equiv 0$, which is the image of the 1-parameter family of $V(z;0)$ solutions of the Riccati equation

$$V_z^2(z;0) + V(z;0) + \frac{1}{2} z = 0 \quad . \tag{2.15}$$

3. BÄCKLUND TRANSFORMATION OF THE SIMILARITY SOLUTIONS AND OF THE RELATED PAINLEVÉ TRANSCENDENTS

Let w be any solution of the PKdV equation. A new solution $\tilde{w}$ is then obtained by the Bäcklund transformation (BT)

$$\begin{cases} \tilde{w}_x = -w_x + \frac{1}{2}(\tilde{w} - w)^2 \\ \tilde{w}_t = -w_t + 2w_x^2 + w_x(\tilde{w} - w)^2 + 2w_{xx}(\tilde{w} - w) \quad . \end{cases} \tag{3.1}$$

We call this a "pure" BT, because the general BT is obtained by applying the Galilean transformation (2.2) to $\tilde{w}$.

The Galilean transformation for the particular class of the potential similarity solutions

$$\phi(\eta) \longrightarrow \phi(\eta - k^2\frac{x^2}{\eta^2}) \tag{3.2}$$

destroys the similarity character of ϕ. Therefore, if we want to get a new similarity solution from $\phi(\eta)$, we must restrict ourselves to "pure"

BT's.

Actually, if we impose to w and $\tilde{w}$ in eqs.(3.1) to be both similarity solutions of the PKdV equation, we get

$$\begin{cases} \tilde{\phi}_\eta = -\eta - \phi_\eta - (\tilde{\phi} - \phi)^2 \\ \phi = \phi_\eta^2 + \eta\phi_\eta - \phi_\eta(\tilde{\phi} - \phi)^2 + \phi_{\eta\eta}(\tilde{\phi} - \phi) \quad . \end{cases} \tag{3.3}$$

If ϕ is the solution $\phi \equiv 0$, eqs.(3.3) reduce to the first one. If $\phi \not\equiv 0$, eqs.(3.3) (remembering that ϕ satisfies eq.(2.6)) reduce to

$$\tilde{\phi}(\eta;\tilde{\mu}^2) = \phi(\eta;\mu^2) + \frac{\phi_{\eta\eta}(\eta;\mu^2) + \mu}{2\ \phi_\eta(\eta;\mu^2)} \tag{3.4}$$

where

$$\tilde{\mu}^2 = (\mu - 1)^2 \ . \tag{3.5}$$

Eq.(3.4) is the B.T. for the similarity solutions of the PKdV equation and it shows that for the particular class of the similarity solutions even the first B.T. reduces to an algebraic transformation on the derivatives of the starting solution.

The B.T. (3.4) can be translated in the language of the second Painlevé transcendents.

Noting that in eq.(2.13) $V(z;\mu)$ depends on μ and $\phi(\eta;\mu^2)$ on μ^2, one first gets

$$V(z;-\mu) = V(z;\mu) + \frac{\mu}{V_z(z;\mu) + V^2(z;\mu) + \frac{1}{2}z} \quad . \tag{3.6}$$

Then the B.T. (3.4), by the combined use of eq.(3.6) and (2.14), furnishes

$$\tilde{V}(z;1-\mu) = -V(z;\mu) \ . \tag{3.7}$$

The set of the two eqs.(3.6) and (3.7) is the B.T. for the solutions of the second Painlevé equation.

Eq.(3.6) is a new hidden symmetry property of the second Painlevé transcendents.

4. SUPERPOSITION PRINCIPLE AND SPECIAL SOLUTIONS

The reiterated use of the B.T. for the second Painlevé transcendents, defined by eqs.(3.6) and (3.7), allow us to calculate from any $V(z;\mu)$ all the succession $\tilde{V}(z;\pm(\mu-k))$ with $k=0,1,2,...$

Derivations and algebraic operations only are needed.

The B.T. for the potential similarity solutions can be written simply in terms of the V's:

$$\tilde{\phi}(\eta;(\mu-1)^2) = \phi(\eta;\mu^2) + 2^{1/3}V(z;\mu) \quad . \tag{4.1}$$

The n-th BT looks very beautiful and transparent

$$\tilde{\phi}(\eta;(\mu-n)^2) = \phi(\eta;\mu^2) + 2^{1/3}\sum_{k=o}^{n-1} V(z;\mu-k) \quad , \tag{4.2}$$

where the sequence $V(z;\mu-k)$ must be calculated, as indicated above, starting from the $V(z;\mu)$ defined in eq.(2.13).

It is well known that the second Painlevé equation is not integrable in terms of elementary or classical functions or transcendents defined by linear equations, except for special values of μ and of arbitrary constants contained in the solution.

The B.T. for the second Painlevé transcendents can be actually used for obtaining a 1-parameter family of explicit solutions for integer μ and a parameterless family for semi-integer μ.

The parent solutions of the two families satisfy respectively the Riccati equation

$$V_z(z;0) + V^2(z;0) + \frac{1}{2}z = 0 \tag{4.3}$$

and the equation

$$V_z(z;1/2) \equiv 0 \quad . \tag{4.4}$$

The fact that the second Painlevé equation has elementary solutions for both $\mu=0$ and $\mu=1/2$ is not casual, because these solutions are related. For more details see ref.[9].

5. ASYMPTOTIC BEHAVIOUR OF THE PAINLEVÉ TRANSCENDENTS.

We will try to show in this part of our lecture that the Painlevé transcendents y(x), via a Boutroux transformation $y = x^{\alpha}Y$ and $X = x^{\beta}$ with convenient α and β, become asymptotic to the elliptic functions in a peculiar sense that we will explain.

For simplicity we will confine ourselves to the first Painlevé transcendent. It is usually assumed in the literature that similar results can be obtained for all the Painlevé transcendents. But, as far as we know, there are not detailed results, apart from real second Painlevé transcendents and in a framework different from the general one of Boutroux.[12]

5.1 - *General Theorems*

The first Painlevé transcendent y(x) is a solution of the second-order ordinary differential equation

$$y'' = 6y^2 - 6x \tag{5.1.1}$$

It can be transformed, as suggested by Boutroux, to a function Y(X) defined by

$$\begin{cases} y(x) = x^{1/2}\, Y(X) \\ X = \frac{4}{5}\, x^{5/4} \end{cases} , \tag{5.1.2}$$

which is a solution of the differential equation

$$Y'' = 6Y^2 - 6 - \frac{Y'}{X} + \frac{4}{25}\frac{Y}{X^2} . \tag{5.1.3}$$

This equation is apparently asymptotic at great X to the equation

$$Y'' = 6Y^2 - 6 , \tag{5.1.4}$$

whose solutions are the biperiodic $\mathcal{P}$ Weierstrass elliptic functions.

More precisely one can state the following:

<u>Fundamental Theorem</u>. Fixed η and $\eta'(\neq 0)$ and a path L starting from η in the complex Y-plane, for any $\varepsilon>0$, there is a $X_\circ > 1/\varepsilon$ for which the solution Y(X) of eq.(5.1.3), with initial conditions

$$Y(X_\circ) = \eta \quad , \qquad Y'(X_\circ) = \eta' ,$$

satisfies all along L

$$|\mathscr{P}^{-1}(Y) - X| < \varepsilon^{\frac{1}{2}} \quad .$$

$X = \mathscr{P}^{-1}(Y)$ is the inverse function of $Y = \mathscr{P}(X)$, which satisfies the same initial conditions as $Y(X)$ and which is a solution of eq.(5.1.4) or, equivalently, of equation

$$Y'^2 = 4Y^3 - 12Y + D$$

with

$$D = \eta'^2 - 4\eta^3 + 12\eta \quad .$$

We will conventionally name 2ω and $2\omega'$ the periods of $\mathscr{P}$ and e_1, e_2, e_3 its three simple critical algebraic points and $\Delta = 27(8 - D)(8 + D)$ its discriminant.

From the fundamental theorem it follows that the critical points of the Painlevé transcendent $Y(X)$ (i.e. the branch points of Y^{-1}), for large X, can be collected in three series with limit points e_1, e_2, e_3. Moreover the Riemann surfaces of Y^{-1}(at large X) are topologically the same as those of $\mathscr{P}^{-1}$, which are drawn at the end of the paper.

Let us now choose in the complex Y-plane two closed curves Λ and Λ', both starting from η, but enclosing respectively e_2, e_3 and e_2, e_1. See Fig.1, where for definitness we have choosen $\Delta>0$.

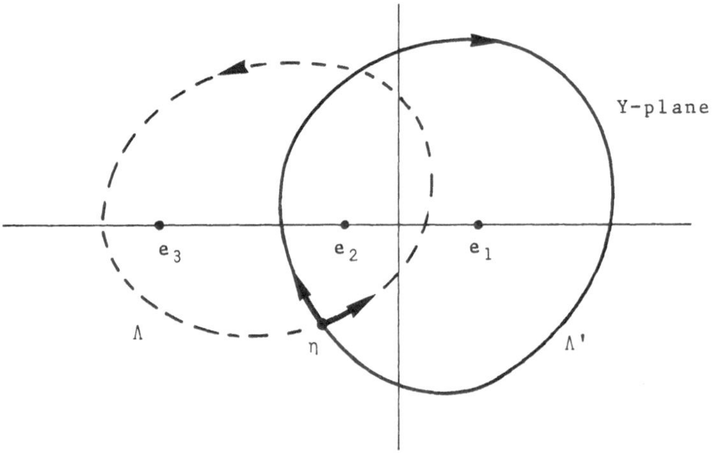

Fig.1

If one starts from $\eta=Y(X_o)$ on the Riemann surface of Y^{-1} and turns in the positive sense firstly round Λ and secondly round Λ', or if one, starting from the same point $\eta=Y(X_o)$, turns firstly round Λ' and secondly round Λ, one recovers at the end of both trips the same representative point of η on the Riemann surface. X_o must of course be sufficiently great.

Therefore the corresponding X point in the complex X-plane, during the two trips on the Riemann surface, that we described above, draws the curvilinear quadrilateral Q of Fig.2, where the full curves are images of Λ' and the dotted curves of Λ.

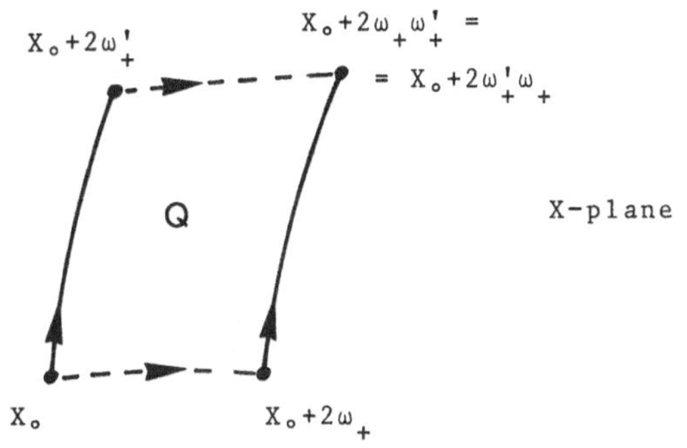

Fig. 2

$2\omega_+$ and $2\omega'_+$, defined in Fig.2, are called the "periods" of Y(X) at X_o. Because of the equality $2\omega_+\omega'_+ = 2\omega'_+\omega_+$ they are said to commute.

If Λ and Λ' are choosen to be respectively the images of the segments $(X_o, X_o+2\omega_+)$ and $(X_o, X_o+2\omega'_+)$ under the map $\mathscr{P}(X) = Y$, the curvilinear quadrilateral Q becomes a parallelogram as $X_o\to\infty$. This is our fixed choice in the following.

In every quadrilateral Q the transcendent Y has a double pole $\overline{X}_o$ and $|Y - \mathscr{P}|\to 0$ for $X_o\to\infty$ everywhere in Q (a neighbourhood of $\overline{X}_o$ excluded).

Turning around Λ (or Λ') n-times in the positive and negative sense the $X=Y^{-1}$ point in the complex X-plane describes a "periodic line", which for large X is asymptotically parallel to a straight line.

The end points of the curved segments image of Λ (or Λ') in the X-plane are named "vertices" of the periodic line.

If $X_0=\overline{X}_0$, i.e. $\eta=\infty$ the periodic line is called "line of infinities" or "of poles".

The $D = Y'^2 - 4\eta^3 + 12\eta$ value at the vertices of a periodic line tends as one moves toward infinity along the line to ±8, or to an imaginary value.

The paths Λ and Λ' can always be choosen in such a way that $\lim D = \pm 8$.

On these periodic lines the transcendent Y is therefore asymptotic (note that $D=\pm 8$ implies that the discriminant $\Delta=0$) to

$$1 - \frac{3}{\sin^2 i\sqrt{3}(X-\overline{X})} \tag{5.1.5}$$

or to

$$-1 + \frac{3}{\sin^2\sqrt{3}(X-\overline{X})} \tag{5.1.6}$$

One "period" of Y goes to infinity and the other to a finite value.

On the contrary on a line which is not periodic the limit of D is indetermined.

Therefore we must conclude that the Painlevé transendents are asymptotic to elliptic functions in a very peculiar sense, because a Painlevé transcendent is not asymptotic to a uniquely determinate elliptic function.

From the theorems on the asymptotic behaviour listed above one can also get global information on Painlevé transcendents. One really needs only a clever use of the continuity of solutions from initial conditions.

For instance if one moves the X_0 of a periodic line towards the origin of the X-plane, then also the periodic line moves with X_0 and finally it breaks into two semiperiodic lines, one to the left and the other to the right of the circle of radius $1/\varepsilon$ with centre at the origin, which is the region outside our asymptotic knowledge of Y(X).

5.2 - *Real solutions.*

The real solutions have a semi-periodic line on the real positive axis and on this line the D parameter tends to -8.

The solutions may be finite and in this case

$$Y(X) \simeq \mathcal{P}(X - \bar{X} + \omega') = e_3 + (e_2 - e_3)\,\mathrm{sn}^2(\lambda(X-\bar{X}),k) \qquad (5.2.1)$$

where $\lambda^2 = e_1 - e_3$, $k = \dfrac{e_2 - e_3}{e_1 - e_3}$ and $\bar{X}$ is real.

We stress once more the peculiar character of the asymptotic relation (5.2.1), because e_1, e_2, e_3 are not fixed parameters but change with D, which tends to -8 as $X \to +\infty$. In particular this means that $e_2 - e_3 \to 0$ as $X \to \infty$.

The Y(X) solution therefore does not stop oscillating with decreasing amplitude as X increases.

If the solution is singular

$$Y(X) \simeq \mathcal{P}(X-\bar{X}) = e_3 + \frac{e_1 - e_3}{\mathrm{sn}^2(\lambda(X-\bar{X}),k)} \qquad (5.2.2.)$$

The solution Y(X) has an infinite number of poles and the positive real axis is also a semi-line of infinities.

Because of the symmetry properties of eq.(5.1.3) these statements can be extended to the negative real axis and to the positive and negative imaginary axis.

5.3 - *Symmetric solutions*

There are two symmetric solutions meromorphic, whose Laurent expansions at the origin are

$$Y_1 = a_2X^2 + a_6X^6 + a_{10}X^{10} + \ldots \qquad (5.3.1)$$

$$Y_2 = \frac{16}{25X^2} + b_2X^2 + b_6X^6 + \ldots \qquad (5.3.2)$$

For these solutions one can easily get the behaviours of the lines of infinities in all the complex X-plane.

They are the dotted lines respectively in Fig.3 for Y_1 and in Fig.4 for Y_2. The full points are the poles. One must however be advised that these Figs. do not show the exact location of the poles, but the mere topological structure of the lattice they define in the plane, taking into account only the symmetry properties with respect to the origin.

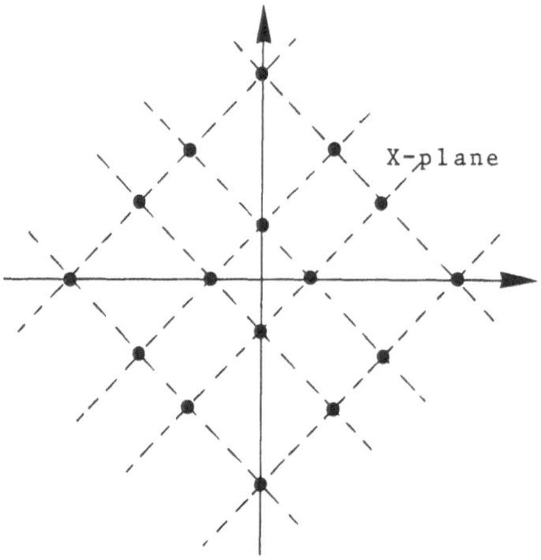

Fig. 3

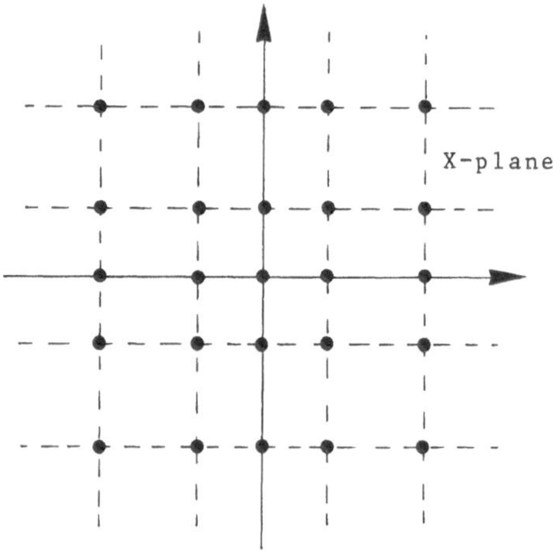

Fig. 4

The plane is then completely divided in quadrilaterals and in every quadrilateral Y_1 (or Y_2) is analytic and takes any value twice.

Because any solution Y(X) of eq.(5.1.3) can be obtained from Y_1 or Y_2 by changing continually the initial conditions η and η', the topological structure of the X-plane for any solution Y(X) is the same as that of Y_1 or Y_2 drawn respectively in Fig.3 and Fig.4. One can freely choose and actually these two structures are equivalent.

All these statements can be translated in the language of the first Painlevé transcendent y(x), which is connected to Y(X) by the Boutroux transformation $y = x^{1/2}\,Y$ and $X = \frac{4}{5}x^{5/4}$.

The complex x-plane must be divided in five equal angular sectors $A_i\hat{O}A_j$ and the lines of poles and the poles are displayed in Fig.5 and Fig.6 respectively for y_1 and y_2.

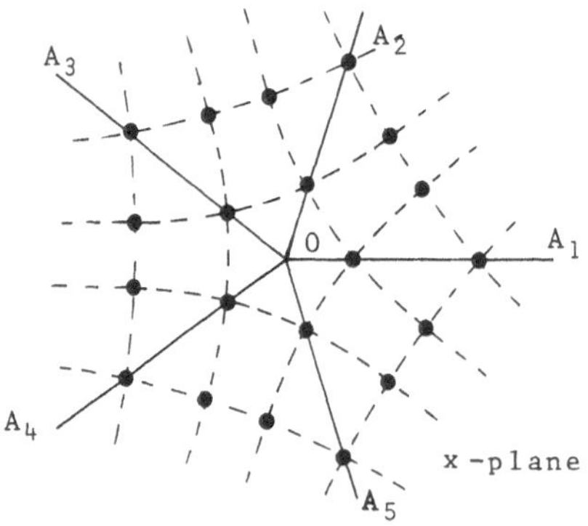

Fig.5

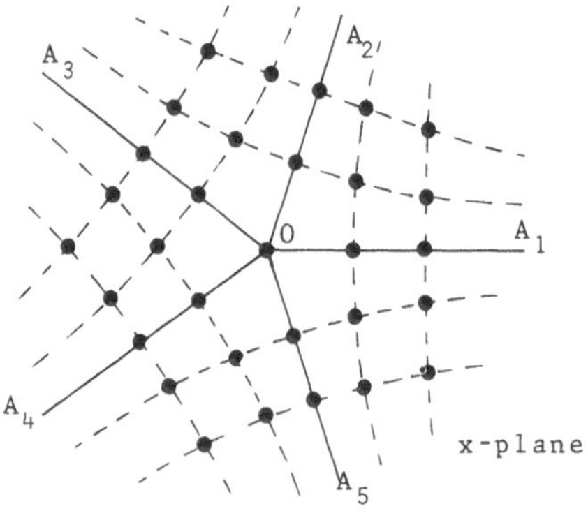

Fig.6

5.3 - *Truncated solutions*

In general we have seen that a line of poles has other lines of poles to its right and to its left.

It may happen however that, for instance to the left, all lines of poles are rejected to infinity.

These exceptional Y solutions are called truncated solutions.

One can show that a Y solution can be truncated in the direction of the positive (or negative) real axis and in the direction of the positive (or negative) imaginary axis. The solutions truncated in both directions of the real or imaginary axis are called bitruncated. To the best of our knowledge nothing is known about these exceptional solutions, not even their existence.

However one can show that there exist Y solutions truncated along any direction below, for instance, the real axis. They are called tritruncated. It is worthwhile to state the properties of the tritruncated solutions directly in the x-plane for the y(x) Painlevé transcendents. They are truncated in three consecutive directions OA_i (see Fig.5 or 6) and have poles (at large x) only in one angular sector. They are connected by simple symmetric relations. For instance

$$y_{215}(x) = e^{i\frac{6\pi}{5}} y_{154}(e^{i\frac{2\pi}{5}} x) \quad , \qquad (5.3.1)$$

where the subscripts of the y's indicate the OA_i directions on which they are truncated.

The tritruncates solutions are uniquely determined, when it is imposed the location of a pole.

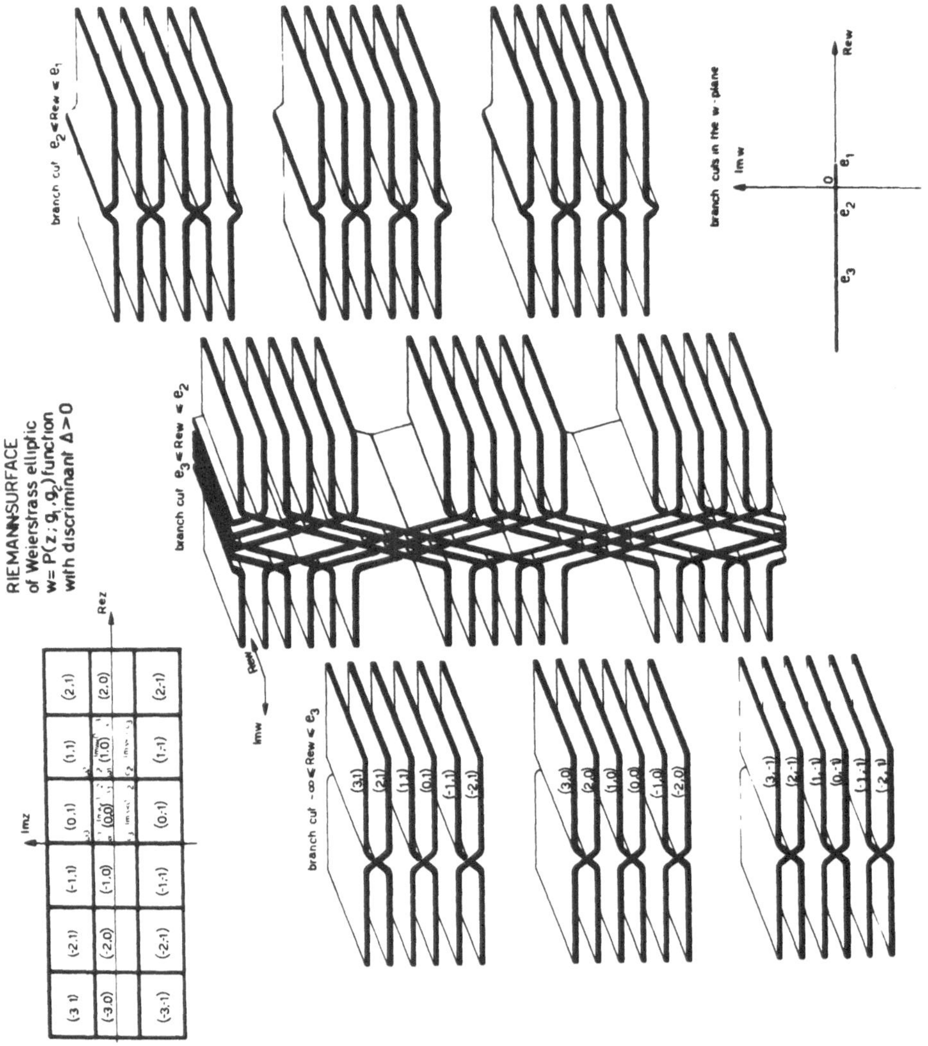
RIEMANN-SURFACE
of Weierstrass elliptic
w= P(z; g1, g2) function
with discriminant Δ>0
Imz
Rez
(-3,1)
(-2,1)
(-1,1)
(0,1)
(1,1)
(2,1)
(-3,0)
(-2,0)
(-1,0)
(0,0)
(1,0)
(2,0)
(-3,-1)
(-2,-1)
(-1,-1)
(0,-1)
(1,-1)
(2,-1)
branch cut e2 ≤ Rew ≤ e1
branch cut e3 ≤ Rew ≤ e2
branch cut -∞ ≤ Rew ≤ e3
Imw
Rew
(3,1)
(2,1)
(1,1)
(0,1)
(-1,1)
(-2,1)
(3,0)
(2,0)
(1,0)
(0,0)
(-1,0)
(-2,0)
(3,-1)
(2,-1)
(1,-1)
(0,-1)
(-1,-1)
(-2,-1)
branch cuts in the w-plane
Im w
0
e3
e2
e1
Rew

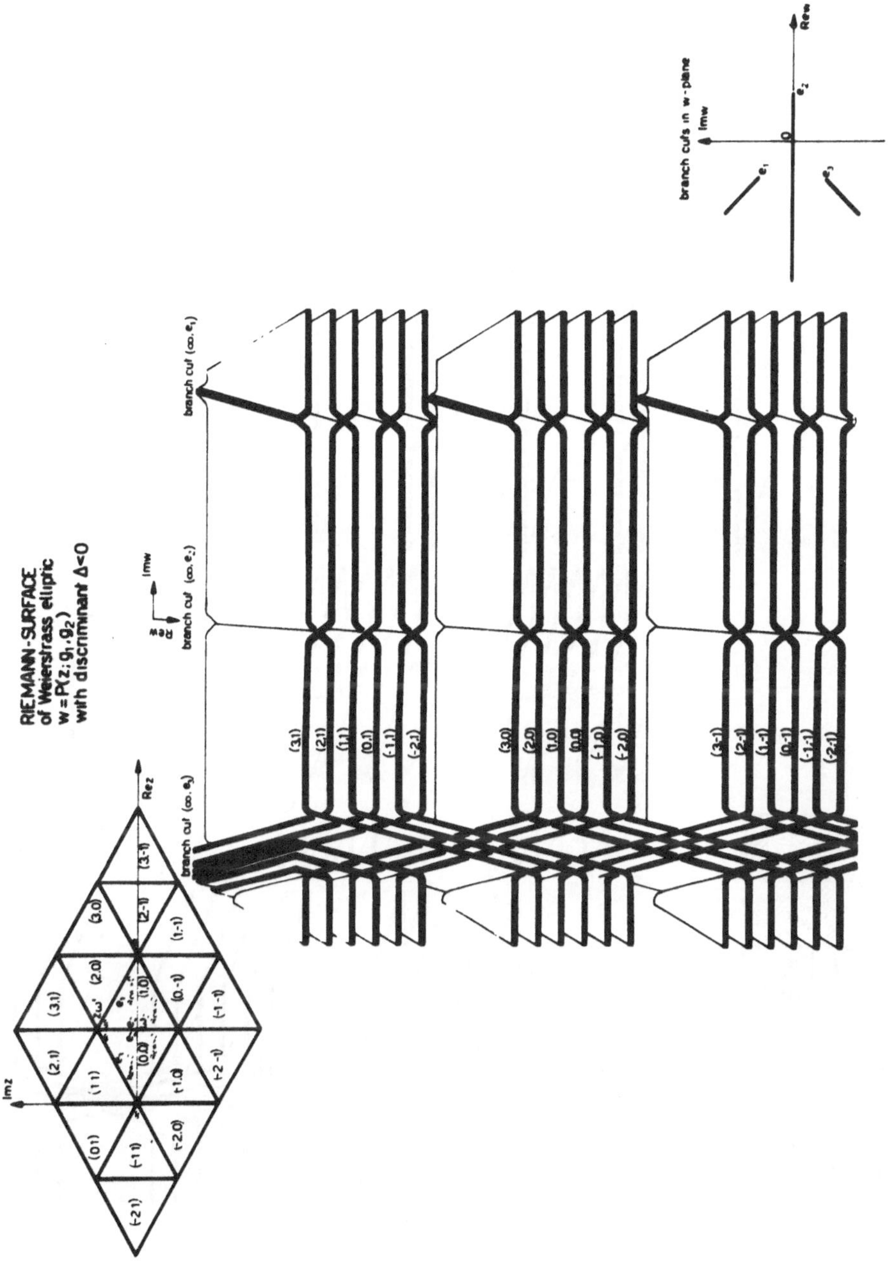

RIEMANN-SURFACE
of Weierstrass elliptic
w = P(z; g1, g2)
with discriminant Δ<0
branch cut (∞, e3)
branch cut (∞, e2)
branch cut (∞, e1)
branch cuts in w-plane
Imw
Rew
Imz
Rez

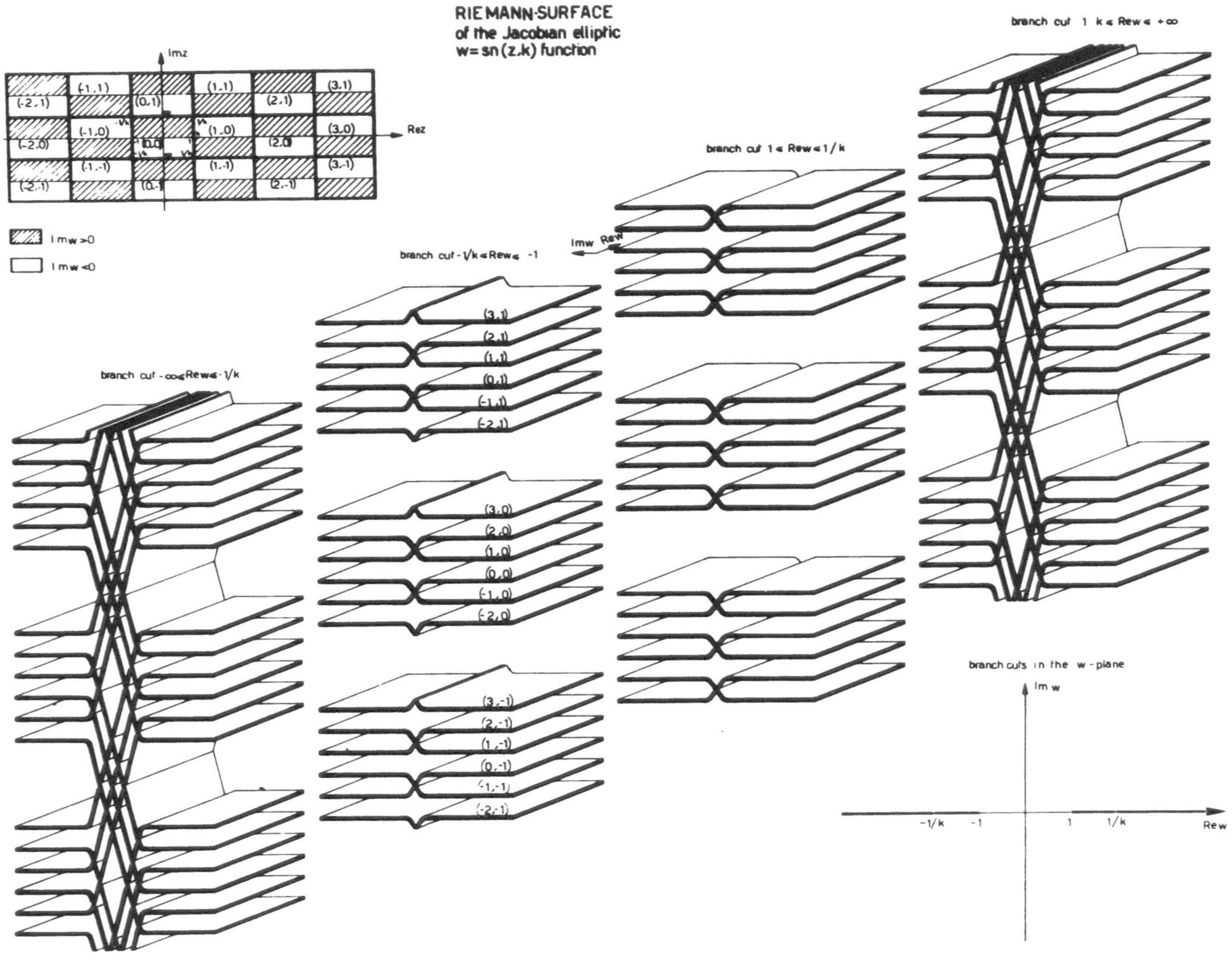
RIEMANN-SURFACE
of the Jacobian elliptic
w = sn (z,k) function
Imz
Rez
(-2,1)
(-1,1)
(0,1)
(1,1)
(2,1)
(3,1)
(-2,0)
(-1,0)
(0,0)
(1,0)
(2,0)
(3,0)
(-2,-1)
(-1,-1)
(0,-1)
(1,-1)
(2,-1)
(3,-1)
Im w > 0
Im w < 0
branch cut -∞ ≤ Re w ≤ -1/k
branch cut -1/k ≤ Re w ≤ -1
branch cut 1 ≤ Re w ≤ 1/k
branch cut 1/k ≤ Re w ≤ +∞
Im w
Re w
branch cuts in the w-plane
-1/k
-1
1
1/k

REFERENCES

1 D.J.Korteweg and G.de Vries: Phil.Mag. 39, 423 (1895).

2 A.C.Scott, F.Y.F.Chu and D.W.McLaughlin: Proc.IEEE, 61, 1443 (1973); R.K.Boullogh in these proceedings and references quoted therein.

3 C.S.Gardner, J.M.Green, M.D.Kruskal and M.Miura: Phys.Rev.Lett., 19, 1095 (1967); Comm.Pure Appl.Math., 27, 97 (1974).

4 F.Calogero and A.Degasperis in these proceedings and references quoted therein.

5 B.A.Dubrovin, V.B.Matveev and S.P.Novikov: Russian Math.Surveys, 31, 59 (1976).

6 H.D.Wahlquist and F.B.Estabrook: Phys.Rev.Lett., 31, 1386(1973).

7 P.Painlevé: Acta Math., 15, 1 (1902); B.Gambier: Acta Math., 33, 1 (1909).

8 E.L.Ince: Ordinary differential equations (New York, N.Y., 1956); H.T.Davis: Introduction to Nonlinear Differential and Integral Equations (New York, N.Y., 1962).

9 M.Boiti and F.Pempinelli: Nuovo Cimento, 51, 70 (1979).

10 A.C.Newell: The interrelation between Bäklund transformation and the inverse scattering transform, in Lecture Notes in Mathematics, N.515, edited by R.M.Miura (Berlin, 1974), p.227. M.J.Ablowitz: The inverse scattering transform - continuous and discrete, and its relationship with Painlevé transcendents, in Research Notes in Mathematics, N.26, edited by F.Calogero (London, 1977), p.9. M.J.Ablowitz, A.Ramani and H.Segur, Lett.Nuovo Cimento, 23, 333 (1978).

11 M.Boutroux, Ann.Ecole Norm.Sup., 30, 255 (1913); 31, 99 (1914).

12 R.R.Rosales, Proc.R.Soc.Lon., 361, 265 (1978); J.W.Miles, Proc.R.Soc.Lon., 361, 277 (1978); Stud.Appl.Math., 60, 59 (1979); R.Haberman, Stud.Appl.Math. 57, 247 (1977).

COADJOINT STRUCTURES, SOLITONS, AND INTEGRABILITY

Andrei Iacob
Mathematics Department
of the Weizmann Institute of Science,
Rehovoth, Israel

and

Shlomo Sternberg
Mathematics Department of the University of Tel Aviv, Israel

In this paper we explain some recent results relating the geometry of the dual space of a Lie algebra to the complete integrability of certain non-linear partial differential equations.

Our main purpose is to give a more or less self contained exposition of the recent results of Adler[1] and Lebedev-Manin[25] on the Poisson structures associated with non linear evolution equations. We briefly recall the background to this problem. By an evolution equation we mean a partial differential equation of the type

$$u_t = X(x,u,u_x,u_{xx},\ldots) ,$$

where u can be a vector valued function and X depends smoothly on x, u, and finitely many of its x derivatives. In general we know very little about the existence, uniqueness, or properties of solutions of such equations. However, for certain evolution equations such as the Korteweg-de Vries equation[18-2÷5-28÷29],

$$u_t = 6uu_x + u_{xxx} ,$$

certain families of solutions, the N-soliton solutions, were found. These solutions were first found numerically[4-5]; then verified analytically, cf [22÷24] or [5]. They have a number of truly remarkable properties. For large negative time a solution looks like the superposition of a number of isolated waves, spatially separated. In time these come together, interact and then for large positive time again emerge as spatially separated waves, moving with different velocities. For each fixed time, the set of functions of x can be parametrized by 2N real parameters (related to the locations and amplitudes of the waves) and the time evolution, in terms of these parameters, is given by the solution of a Hamiltonian system of ordinary differential equations. The manifold of solutions, at each time can be described as extremals of some "conserved quantity" of X, cf[22-23]. This last property, that a manifold of extremals of some F gives a class of solutions evolving according to some Hamiltonian systems can be understood in terms of the calculus of variations. Suppose that $u^t = u^t(x)$ is an actual solution of the evolution equation, with $u^\circ = u$ and F is some function of u and its x derivatives. Then

$$\frac{dF}{dt}(u^t, u_x^t, \ldots)\Big|_{t=0} = \frac{\partial F}{\partial u} u_t + \frac{\partial F}{\partial u_x} \frac{du_t}{dx} + \cdots$$

$$= \frac{\partial F}{\partial u} X + \frac{\partial F}{\partial u_x} \frac{dX}{dx} + \cdots \quad .$$

The operator

$$\hat{X} = X \frac{\partial}{\partial u} + \left(\frac{d}{dx} X\right) \frac{\partial}{\partial u_x} + \left(\frac{d^2}{dx^2} X\right) \frac{\partial}{\partial u_{xx}} + \cdots$$

can be applied to any F, whether or not we know about the solubility of the equation.

(Here

$$\frac{d}{dx} = \frac{\partial}{\partial x} + u_x \frac{\partial}{\partial u} + u_{xx} \frac{\partial}{\partial u_x} + \cdots$$

is the operation of total derivative with respect to x. See[6-26] or[31] for a more detailed and invariant description). We say that F is invariant under X if

$$\hat{X}F = 0 \; .$$

We can think of F as defining a "functional" on the space of functions, u, by assigning the "value"

$$\int F(u, u_x, \ldots)\, dx$$

to u. Of course, to get an actual functional we need to specify the domain of integration, and if infinite, worry about convergence. But the variational equations, i.e. the Euler-Lagrange equations associated to F are well-defined locally. If F satisfies the non-degeneracy condition

$$\frac{\partial^2 F}{\partial u_x^{(N)} \partial u_x^{(N)}} \neq 0 \qquad \text{anywhere}$$

where $u_x^{(N)}$ is the highest derivative of u occurring in F, then the set of extremals, i.e. of solutions of the Euler-Lagrange equations, form a manifold of dimension 2N, carrying a natural symplectic structure. If $\hat{X}F = 0$, then $\hat{X}$ is "tangent" in a reasonable sense to this manifold, and induces a Hamiltonian flow there, cf[6, 31] for example, for a short invariant proof of these facts.

Thus we are led to search for invariant F's. If also, we can find other $\hat{Y}$'s which commute with $\hat{X}$ and satisfying $\hat{Y}F = 0$, then this would help in the solution of the mechanical system. A method which has led to the discovery of such invariants and commuting vector fields, was the introduction of a Poisson bracket structure by Gardner[3], into the space of functions F, for the one dependent variable case of the Korteweg-de Vries equation. Roughly speaking, a Poisson structure is a rule which assigns to each function F a vector field $\hat{X}_F$, so that one has the Poisson bracket

$$\{F,G\} = -\hat{X}_F G$$

defining a Lie algebra operation on functions, and such that

$$[\hat{X}_F, \hat{X}_G] = \hat{X}_{\{F,G\}}$$

Gardner's Poisson bracket was given by assigning to F the evolutionary equation

$$u_t = X_F = \frac{d}{dx}\frac{\delta F}{\delta u}$$

where $\frac{\delta F}{\delta u}$ denotes the variational derivative (see below or[6, 16] or[31]). Then $\hat{X}_F$ is the associated operator on functions. If $\{F,H\} = -\hat{X}_F H = 0$, then $\hat{X}_H F = -\{H,F\} = +\{F,H\} = 0$. So if the original $X = \hat{X}_H$, the Poisson bracket is useful for the search for invariants and their commuting properties. The Gardner bracket was greatly generalized in an important series of papers by Gelfand and Dikii.[6-9] Their calculations, however, were extremely involved.

In the study of group representations and the foundations of mechanics in recent years, one had discovered that the dual space to the Lie algebra of a Lie group (or any invariant submanifold thereof) carries a natural Poisson structure. It was the remarkable discovery of Adler[1], and Lebedev-Manin[25] that the Gelfand-Dikii symplectic structures are obtained from the dual space of a certain infinite dimension Lie algebra, the algebra of formal pseudodifferential operators of negative degree. We wish to explain this circle of ideas.

We shall also describe some recent results of various authors leading to examples of complete integrability in group theoretical situations. We shall, finally, describe some examples in classical mechanics from the viewpoint of the dual of a Lie algebra.

1. POISSON STRUCTURES AND MOMENT MAPS

Let M be a manifold, and F a subspace of the space of smooth functions on M. We say that F carries a Poisson structure, if we are given a map

$$f \rightsquigarrow X_f$$

from F to vector fields on M such that

$$X_f \cdot F \subset F$$

$$X_{f_1} f_2 = -X_{f_2} f_1$$

and, setting

$$\{f_1, f_2\} = -X_{f_1} f_2$$

that

$$\{f_1,\{f_2,f_3\}\} = \{\{f_1,f_2\},f_3\} + \{f_2,\{f_1,f_3\}\} .$$

The fundamental example (Kirillov[15-16], Kostant[19], Souriau[30]) is the following:

Let g be a Lie algebra. Let g* denote the dual space of g -that is, the space of all linear functions on g. Let $F(g^*)$ denote the space of all differentiable functions on g^*. We shall show that $F(g^*)$ has a Poisson structure. Let <,> denote the pairing between g^* and g. In other words, if $\ell \in g^*$ and $X \in g$, then $\langle \ell, X\rangle$ denotes the value of the linear function ℓ on the Lie algebra element X. If f is any differentiable function on g^*, consider the mapping L_f of g^* into g given by

$$\langle m, L_f(\ell)\rangle = df_\ell(m) = \frac{d}{dt} f(\ell+tm)\Big|_{t=0} .$$

(The map L_f is just the Legendre transformation associated to the function f.)

If $X \in g$, we can think of X as a linear function on g^*, assigning to each ℓ the value $\langle \ell, X\rangle$. Then $\frac{d}{dt}\langle \ell+tm, X\rangle = \langle m, X\rangle$, so

$$L_X(\ell) \equiv X \quad \text{for} \quad X \in g .$$

Now define the Poisson bracket by

$$\{f_1, f_2\}(\ell) = \langle \ell, [L_{f_1}(\ell), L_{f_2}(\ell)]\rangle . \tag{1.1}$$

Notice that

$$L_{f_2 f_3}(\ell) = f_2(\ell) L_{f_3}(\ell) + f_3(\ell) L_{f_2}(\ell)$$

by Leibnitz's rule, so that $\{f_1, f_2 f_3\} = \{f_1, f_2\} f_3 + \{f_1, f_3\} f_2$. The bracket is clearly antisymmetric because the Lie bracket in g is. To prove the Jacobi identity it is convenient to introduce another expression for the Poisson bracket which will be useful for other purposes as well: Each X in the Lie algebra, g, defines a vector field $\tilde{X}$ on g^* by defining, for each $\ell \in g^*$, the vector $\tilde{X}(\ell) = X \cdot \ell$ by

$$\langle \tilde{X}(\ell), Y\rangle = -\langle \ell, [X, Y]\rangle \tag{1.2}$$

for every $Y \in g$. If f is any differentiable function on g^*, then the derivative $\tilde{X} f$ is defined by

$$(\tilde{X} f)(\ell) = \frac{d}{dt} f(\ell + t\tilde{X}(\ell))\Big|_{t=0} .$$

Thus

$$\tilde{X} f(\ell) = \langle \tilde{X}(\ell), L_f(\ell)\rangle = -\langle \ell, [X, L_f(\ell)]\rangle . \tag{1.3}$$

If f_1 is a function on g^*, then $L_{f_1}(\ell) \in g$ and hence $\widetilde{L_{f_1}(\ell)}$ is a vector field on g^* which we can evaluate at ℓ to get the vector $\widetilde{L_{f_1}(\ell)}(\ell)$. We let $\tilde{f}_1$ denote the vector field on g^* which assigns to each ℓ this vector in g^*, i.e.,

$$\tilde{f}_1(\ell) = \widetilde{L_{f_1}(\ell)}(\ell) \quad .$$

If $X \in g$ is thought of as a linear function on g^*, then this coincides with the previous definition of $\tilde{X}$. Observe that

$$-\tilde{f}_1 f_2(\ell) = \langle \ell, [L_{f_1}(\ell), L_{f_2}(\ell)] \rangle, \tag{1.4}$$

i.e.,

$$-\tilde{f}_1 f_2 = \{f_1, f_2\} \quad . \tag{1.5}$$

The bracket $\{f_2, f_3\}$ is clearly bilinear in f_2 and f_3, and therefore, for any vector field, V, on g^* we have, by Leibnitz's rule

$$V\{f_2, f_3\} = \{Vf_2, f_3\} + \{f_2, Vf_3\} \ .$$

Applying this to $V = -\tilde{f}_1$ gives Jacobi's identity.

Notice that linear functions on g^*, i.e., the element of g, form a subalgebra under Poisson bracket, for which Poisson bracket coincides with the original Lie bracket on g. It follows from the derivation property that the Poisson bracket of two polynomials is again a polynomial. We let $S(g) \subset F(g^*)$ denote the set of all polynomials on g^*. We have shown that $S(g)$ is a subalgebra not only for ordinary multiplication but also for Poisson bracket.

Let M be a submanifold of g^*. We say that M is invariant if $\tilde{X}(m)$ is tangent to M for each $m \in M$, $\forall X \in g$. Suppose that $F \in F(g^*)$ vanishes identically on M. Then for any function F_1 on g^*,

$$\{F_1, F\}_{|M} = (-\tilde{F}_1 \cdot F)_{|M} = 0$$

since $\tilde{F}_1$ is tangent to M. This has the following consequence: let f_2 be any function on M. Extend f_2 in any way so as to be a function, F_2 on g^*. Then

$$\{F_1, F_2\}_{|M}$$

is independent of the choice of extension. Similarly, the above Poisson bracket depends only on $F_{1|M}$. In other words, the Poisson structure on g^* induces a Poisson structure on any invariant submanifold, M, of g^*.

Suppose that g is the Lie algebra of some finite dimensional connected Lie group G. The adjoint representation of G on g, gives use to its contragradient representation, the coadjoint representation of G on g*. The condition that M be invariant then simply means that M is carried into itself by the action of G, i.e. that M is a union of G orbits. In particular, each G orbit is an invariant manifold, and hence carries its own Poisson structure. It is not hard to show, cf[15-16,19] or[30], that in this case the Poisson structure is non-degenerate, i.e. comes from an invariant symplectic structure . However, it is sometimes difficult for us to compute orbits, and it is then convenient to allow larger invariant manifolds.

In the next section we shall consider the case where g is a certain infinite dimensional (topological) Lie algebra. The elements of this algebra will be something like (actually sequences of) smooth functions. Then g* will be some space of (finite sums of) "generalized functions" or distributions. We shall then let $M = g^{\#}$ consist of those elements of g* which are actually smooth, and on $g^{\#}$ consider only very special kinds of functions (those which are local). These functions form a vector space $\mathcal{F}$ (which is not an algebra under multiplication). However the differentials of these functions extend in a natural way to be linear functions on all of g* and hence $L_f : g^* \to g$ will be well-defined, and gives a Poisson structure on $\mathcal{F}$.

Let M be a manifold and $\mathcal{F}$ a subspace of functions on M carrying a Poisson bracket. We say that we have a Poisson action of g on M if we are given a homomorphism of Lie algebras

$$g \longrightarrow \mathcal{F} .$$

Thus each $X \in g$ gives rise to a function f_X on M, which, in turn, gives rise to a vector field on M which we shall denote by $\tilde{X}$. It follows that $\widetilde{[X,Y]} = [\tilde{X},\tilde{Y}]$ so we get a Lie algebra action of g as vector fields on M. For each $m \in M$, the evaluation map

$$X \rightsquigarrow f_X(m)$$

depends linearly on X and can thus be thought of as a point in g*. We have thus defined a map

$$\Phi : M \to g^* \qquad \langle \Phi(m), X \rangle = f_X(m)$$

called the moment map.

For $Y \in g$

$$\langle Y \cdot \Phi(m), X\rangle = \langle \Phi(m), [X,Y]\rangle = f_{[X,Y]}(m)$$

$$= \tilde{Y} f_X(m) \qquad = \langle D_{\tilde{Y}} \Phi(m), X\rangle \quad .$$

In other words

$$D_{\tilde{Y}} \Phi = Y \cdot \Phi \quad ,$$

the map Φ is equivariant relative to the action of g. In case M is an invariant submanifold of g^* as before, the moment map is just the canonical injection.

Suppose g is the Lie algebra of G and that G acts on M. We say that we have a Hamiltonian action if F is invariant under the induced action of G and if the homomorphism $g \to F$ is equivariant under G. This amounts to the condition

$$\Phi(am) = a \cdot \Phi(m)$$

for all $a \in G$ and $m \in M$.

2. THE ALGEBRA OF FORMAL PSEUDODIFFERENTIAL OPERATORS

We begin by recalling the formula for the composition of two differential operators in $\mathbb{R}^n$. Let $x=(x_1,\ldots,x_n)$ and $\xi=(\xi_1,\ldots,\xi_n)$. For each $\alpha=(\alpha_1,\ldots,\alpha_n)$ with the α_i integers, we let ξ^α denote the monomial

$$\xi^\alpha = \xi_1^{\alpha_1} \ldots \xi_n^{\alpha_n} \quad .$$

We let

$$D_j = \frac{1}{i} \frac{\partial}{\partial x_j} \quad , \qquad i = \sqrt{-1}$$

and

$$D^\alpha = D_1^{\alpha_1} \ldots D_n^{\alpha_n} \quad .$$

We let

$$\xi \cdot x = \xi_1 x_1 + \cdots + \xi_n x_n \quad ,$$

so that

$$D^\alpha e^{i\xi \cdot x} = \xi^\alpha e^{i\xi \cdot x} \quad .$$

A differential operator of degree (at most) m is an operator of the form

$$P(x,D) = \sum_{|\alpha| \le m} a_\alpha(x) D^\alpha \quad , \; \alpha_j \ge 0$$

where the a_α are C^∞ functions of x and $|\alpha| = \alpha_1 + \cdots + \alpha_n$. Thus

$$P(x,D)e^{i\xi\cdot x} = e^{i\xi\cdot x} P(x,\xi)$$

where

$$P(x,\xi) = \sum a_\alpha(x)\xi^\alpha \quad .$$

If u is any C^∞ function which vanishes sufficiently rapidly at infinity we can write

$$u(x) = (2\pi)^{-n} \int e^{i\xi\cdot x}\, \hat{u}(\xi)\, d\xi$$

where the Fourier transform, $\hat{u}$, of u is given by

$$\hat{u}(\xi) = \int e^{-i\xi\cdot y}\, u(y)\, dy \quad .$$

If the a_α don't grow too fast at infinity, (for example if they are bounded), then we can pass the P under the integral sign and write

$$Pu(x) = (2\pi)^{-n} \int e^{i\eta\cdot(x-y)}\, P(x,\eta)u(y)\, dy d\eta \quad .$$

This formula allows us to compute the effect of conjugating P by multiplication by $e^{i\xi\cdot x}$. Let M_ξ denote this operation of multiplication, so that

$$(M_\xi u)(x) = e^{i\xi\cdot x}\, u(x) \quad .$$

Then

$$\begin{aligned}
\left[(M_\xi^{-1} P M_\xi)u\right](x) &= (2\pi)^{-n} e^{-i\xi\cdot x} \int e^{i\eta\cdot(x-y)} P(x,\eta) e^{i\xi\cdot y} u(y) dy d\eta \\
&= (2\pi)^{-n} \int e^{i(\eta-\xi)\cdot(x-y)} P(x,\eta) u(y) dy d\eta \\
&= (2\pi)^{-n} \int e^{i\eta\cdot(x-y)} P(x,\eta+\xi) u(y) dy d\eta \\
&= P(x,\xi+D)u \quad .
\end{aligned}$$

This last expression, or rather the next to the last integral is to be interpreted as follows: Take the Taylor expansion of $P(x,\cdot)$ about the point ξ:

$$P(x,\xi+\eta) = \sum \frac{1}{\alpha!}\, \partial_\xi^\alpha P(x,\xi)\eta^\alpha$$

which is a polynomial in η for each fixed x and ξ. Substitute D for η. Here $\alpha! = \alpha_1!\cdot\ldots\cdot\alpha_n!$ and

$$\partial_\xi^\alpha = \left(\frac{\partial}{\partial\xi_1}\right)^{\alpha_1}\cdots\left(\frac{\partial}{\partial\xi_n}\right)^{\alpha_n} .$$

Let P and Q be two differential operators. Then

$$\begin{aligned}(P\circ Q)e^{i\xi\cdot x} &= P\left[e^{i\xi\cdot x}\, Q(x,\xi)\right] = e^{i\xi\cdot x}(M_\xi^{-1}PM_\xi)Q(x,\xi) \\ &= e^{i\xi\cdot x}P(x,\xi+D)Q(x,\xi) \\ &= e^{i\xi\cdot x}\sum\frac{1}{\alpha!}\,\partial_\xi P(x,\xi)D^\alpha Q(x,\xi) \quad .\end{aligned}$$

In this last expression, the D^α acts as a differential operator on Q, thought of as a function of x for fixed ξ. This gives the formula for the composition of two differential operators. We have proved it under some restrictions on the growth of the coefficients of P and Q at infinity. But clearly the formula for the composition of two differential operators is purely local; and hence the above formula is valid without restriction. We can think of it as defining a new kind of multiplication on functions $P = P(x,\xi)$ which are polynomials in ξ with coefficients which are C^∞ functions of x:

$$(P\circ Q)(x,\xi) = \sum\frac{1}{\alpha!}\,\partial_\xi^\alpha P(x,\xi)D^\alpha Q(x,\xi) \quad . \tag{2.1}$$

In particular

$$(\xi^\beta\circ Q)(x,\xi) = (\xi+D)^\beta Q(x,\xi) \quad . \tag{2.2}$$

Now suppose that P and Q are formal series of the form

$$P = \sum_{-\infty}^{m} P_j(x,\xi)$$

$$Q = \sum_{-\infty}^{m'} Q_j(x,\xi)$$

where the P_j and Q_j are C^∞ functions of x and ξ defined for $\xi\neq 0$, which are homogeneous of degree j in ξ. We can now define the composition of P and Q by (2.1), where we apply (2.1) to P_j and Q_k and collect all terms of a fixed degree of homogeneity. Since any partial derivative with respect to ξ lowers the degree of homogeneity by one, there will be only a finite number of terms of each fixed degree of homogeneity, and hence $P \circ Q$ will be well defined as a formal sum. The operation, $\circ$, is associative. (In the case that P and Q are "symbols" of pseudodifferential operators, $P \circ Q$ is the symbol of their composition. The notion of pseudodifferential operators and the multiplication of their symbols was introduced in [17]. The ring of "formal pseudodifferential operators" described above, was introduced in [10]. See also [1-32]). We say that P is of degree m if the highest j such that $P_j \neq 0$ is $j=m$. If P is of degree m and Q is of degree m' then $P \circ Q$ is of degree m+m'. Also, the highest order term in $P \circ Q$ is just the ordinary product of P_m and $Q_{m'}$. Thus, the commutator

$$[P,Q] = P \circ Q - Q \circ P$$

has degree m+m'-1. In particular, the elements of negative degree form a Lie algebra which we shall denote by g. For future reference we record the following useful lemma: for any P and Q we can write

$$[P,Q] = \sum \partial_{\xi_i} A_i + \sum D_j B_j \qquad (2.3)$$

i.e. as a sum of ξ derivatives and of x derivatives. Indeed, for any E and F, we have

$$\partial_{\xi_j} E D_j F = \partial_{\xi_j} (E D_j F) - E \partial_{\xi_j} D_j F$$

and

$$D_j E \partial_{\xi_j} F = D_j (E \partial_{\xi_j} F) - E D_j \partial_{\xi_j} F \quad .$$

Since D_j and ∂_{ξ_j} commute, this implies that

$$\partial_{\xi_j} E D_j F - \partial_{\xi_j} F D_j E = \partial_{\xi_j} (E D_j F) - D_j (E \partial_{\xi_j} F) \quad .$$

This establishes (2.3) for the terms in $[P,Q]$ which involve only first derivatives in (2.1). Repeated application of this argument gives (2.3) for derivatives of any order.

Let us now restrict attention to the case of one independent variable, n=1. We shall also restrict ξ to be positive. We can write an element, X, of g as

$$X = \sum_{1}^{\infty} c_{-j}\xi^{-j} = \sum_{0}^{\infty} (\xi+D)^{-k-1} b_k$$

since we can use the binomial expansion of $(\xi+D)^{-k-1}$ to recursively solve for the b's in terms of the c's and vice versa. Thus

$$c_{-1} = b_0$$

$$c_{-2} = b_1 - Db_0$$

etc. We shall use the representation in terms of the b's to put a topology on g; each b_k is a C^∞ function, and we topologize the space of C^∞ functions on $\mathbb{R}$ in the standard way. Thus an element of g is a sequence of elements in a topological vector space, so we give g the weak product topology; thus an element A of g^*, i.e. a continuous linear function on g is a finite sequence $A=(a_0,\ldots,a_m)$ where each a_j is a distribution of compact support and

$$\langle A,X\rangle = \langle a_0,b_0\rangle + \ldots + \langle a_m,b_m\rangle \quad .$$

Let $g^{\#} \subset g^*$ denote the subspace of smooth distributions, i.e. those A for which all the a_j are smooth functions of compact support. Then we can write

$$\langle A,X\rangle = \int (a_0 b_0 + \cdots + a_m b_m)\, dx \quad .$$

We now make the key observation due to Adler. Let us write A as a differential operator

$$A = a_0 + a_1\xi + \cdots + a_m\xi^m \quad .$$

Then

$$A\circ X = (\sum a_k\xi^k) \circ (\sum(\xi+D)^{-j-1} b_j)$$

$$= \sum_{k,j} a_k(\xi+D)^k(\xi+D)^{-j-1} b_j = \sum_{k,j} a_k(\xi+D)^{k-j-1} b_j \quad .$$

Let us look for the coefficient of ξ^{-1} in this last expression. If $k>j$, then $(\xi+D)^{k-j-1}$ is a polynomial in ξ and D and so contains no ξ^{-1} terms. If $k<j$ then the expansion of $(\xi+D)^{k-j-1}$ starts with ξ^{k-j-1} and goes down in powers of ξ, and $k-j-1 \leq -2$. Thus only contributions come from $k=j$, and in these we have $(\xi+D)^{-1} = \xi^{-1}$ plus lower powers of ξ. Thus

$$(A\circ X)_{-1} = a_0b_0 + \cdots + a_mb_m$$

where $(\ )_{-1}$ denotes the coefficient of ξ^{-1}. We thus obtain Adler's formula

$$\langle A,X\rangle = \int (A\circ X)_{-1}dx \quad . \tag{2.4}$$

Next we have Adler's

Lemma. *For any* P *and* Q, $([P,Q])_{-1}$ *is always a total derivative, i.e.* $([P,Q])_{-1} = Db$ *for some function,* b. *Thus, if* P *or* Q *has compact support then* $\int([P,Q])_{-1}dx = 0$.

Proof. By (2.3) we know that $[P,Q] = \partial_\xi A + DB$. But ξ^{-1} is not the ξ derivative of any power of ξ, hence the first term can not contribute to the coefficient of ξ^{-1} and hence $([P,Q])_{-1} = (DB)_{-1} = DB_{-1}$, proving the lemma with $b=B_{-1}$.

For any P, let P_+ be the non-negative part of P, so that $P_+ = \sum_{j\geq 0} P_j$ if $P=\sum P_j$. We can now use the lemma to derive Adler's formula for the coadjoint action: For any X and Y in g,

$$\langle A,[X,Y]\rangle = \langle [A,X]_+,Y\rangle \quad . \tag{2.5}$$

Proof. Write $A\circ(X\circ Y-Y\circ X) = (A\circ X-X\circ A)\circ Y + (X\circ(A\circ Y) - (A\circ Y)\circ X)$. Since A has compact support, so does $A\circ Y$ and so, when taking the integral of the ξ^{-1} component, the second term vanishes by the lemma. This proves (2.5). Thus the coadjoint action of $X\in g$ acting on $A\in g^*$ is given by

$$X\circ A = [X,A]_+ \quad . \tag{2.6}$$

Remembering that X is of negative degree, we see that if $A=a_0+\ldots+a_m\xi^m$, the terms of degree m and m-1 in $[X,A]_+$ vanish. Thus the affine space given by

$$a_m = a \quad , \quad a_{m-1} = b$$

where a and b are fixed functions, is invariant under the coadjoint action.

Now let f be a function on $g^{\#}$. We wish to construct the Legendre transformation associated to f. We shall restrict attention only to those f which are *local*, that is to f which are of the form

$$f(A) = \int F(x,a_0(x),a_0'(x),\ldots,a_m(x),\ldots,a_m^{(k)}(x))\,dx \quad ,$$

where F is some smooth function of the variables $x,a_0,a_0',\ldots,a_0^{(k)},a_1,\ldots,a_1^{(k)},\ldots,a_m,\ldots,a_m^{(k)}$, compactly supported in x. Thus F is a function defined on the space of k-jets of the A's. Here each fixed F depends on only finitely many a_i and their derivatives up to some finite order. But the number, m, of the a_i and the number, k, of the derivatives may depend on the f. Now the expression for $df_A(B)$ will be an integral of a sum involving various partial derivatives of F, evaluated at the a's and their derivatives, multiplied by the values of the b's and their derivatives. Since the b's are of compact support, we can integrate by parts so as to eliminate all the expressions involving derivatives of the b's. The resulting expression will be

$$df_A(B) = \sum \int \frac{\delta F}{\delta a_i} b_i \, dx \quad .$$

The expression $\frac{\delta F}{\delta a_i}$ is called the variational derivatives of f with respect to a_i. It is to be evaluated, of course, at the points given by the values of the a_j and their derivatives. The important point about the above expression is that since the $\frac{\delta F}{\delta a_i}$ under the integral are smooth functions of x, the formula extends immediately to the case where the b's are allowed to be distributions. In other words, by the definition of $\langle B,X\rangle$ with $X=L_f(A)$ we see that

$$L_f(A) = (\xi+D)^{-1}\frac{\delta F}{\delta a_0} + (\xi+D)^{-2}\frac{\delta F}{\delta a_1} + \ldots \qquad (2.7)$$

$$X_f(A) = L_f(A)\cdot A = -[A,L_f(A)]_+ \qquad (2.8)$$

and

$$\{f,h\} = -X_f h = \langle A,[L_f(A),L_h(A)]\rangle = \int (A\circ[L_f(A),L_h(A)])_{-1}dx \ . \qquad (2.9)$$

The important point about formulas (2.7) and (2.8) is that they are local -the value of $L_f(A)$ at a point x, i.e. the values of each of the coefficients of the expansion of $L_f(A)$ at the point x, depends only on the values of the a's and their derivatives to sufficiently high order at x, and various partial derivatives of F (which enter into the expression for the variational derivatives of F) evaluated at these values. In particular, the right hand side of (2.7) makes sense without any compactness assumptions on either F or A, and we can take it as a definition of the left hand side. (Actually, in the absence of compactness assumptions on F, the "function" f is not defined since it involves an integration. So we should write $L_F(A)$ on the left of (2.7). On the other hand, if F and F' are two functions such that

$$\frac{\delta(F-F')}{\delta a_i} \equiv 0 \quad \text{for all } i, \tag{2.10}$$

then clearly the right hand side of (2.7) gives the same answer when evaluated on F or F'. We could then take (2.10) as defining an equivalence relation and let f denote the equivalence class of F. It can be shown, cf.[6], that (2.10) holds if and only if F-F' is a "total-derivative" and so, if F and F' have compact support, they define the same honest function, f.) Similarly, (2.8) is a local formula, and we can regard X_f as a vector field

$$X_f = (X_f)_0 \frac{\partial}{\partial a_0} + (X_f)_1 \frac{\partial}{\partial a_1} + \dots + (X_f)_{m-2} \frac{\partial}{\partial a_{m-2}} \tag{2.11}$$

where the coefficients $(X_f)_i$ depend on the a's and their derivatives up to some high order (the explicit expression involving the values of various derivatives of F at these points) and so again does not depend on whether or not A or F is compact. Now (2.9) does involve an integration, and so requires some slight reformulation. If $h(A) = \int H(\cdots)dx$, and X is a vector field like (2.11), then Xh is also given by a density, easily computable in terms of X and H. Rather than clutter up the formula with indices, we illustrate how this computation goes when m=2 and we write a for a_0, X for X_f and assume that H only depends on a and its first two derivatives, i.e. on a, a' and a". Then

$$\frac{\delta H}{\delta a} = \frac{\partial H}{\partial a} - \frac{d}{dx}\frac{\partial H}{\partial a'} + \left(\frac{d}{dx}\right)^2 \frac{\partial H}{\partial a''}$$

where

$$\frac{d}{dx} = \frac{\partial}{\partial x} + a'\frac{\partial}{\partial a} + a''\frac{\partial}{\partial a'} + a^{(3)}\frac{\partial}{\partial a''} + \dots$$

is the total derivative in the x direction. Writing $X = X_0 \frac{\partial}{\partial a}$ where X_0 can depend on a and its various derivatives, we have

$$(Xh)(A) = \int X_o \frac{\delta H}{\delta a}\, dx$$

$$= \int \left[X_o \frac{\partial H}{\partial a} + \left((\frac{d}{dx}) X_o \right) \frac{\partial H}{\partial a'} + \left((\frac{d}{dx})^2 X_o \right) \frac{\partial H}{\partial a''} \right] dx$$

$$= \int \hat{X} H \, dx$$

where $\hat{X}$ is the vector field given by

$$\hat{X} = X_o \frac{\partial}{\partial a} + \left(\frac{d}{dx} X_o \right) \frac{\partial}{\partial a'} + \left((\frac{d}{dx})^2 X_o \right) \frac{\partial}{\partial a''} + \cdots \qquad (2.12)$$

In general, it is clear that any vector field of the form (2.11) gives rise to a vector field $\hat{X}_f$ obtained from it by the formula (2.12) (applied to all the a_i), and the density corresponding to $X_f h$ is $\hat{X}_f H$.

The content of (2.9) is that

$$\int \hat{X}_f H \, dx = -\int (A \circ [L_f(A), L_h(A)])_{-1} \, dx \quad .$$

Now the integrands on both sides of this equation are local expressions in A, F and H and the equation is to hold identically in F and H when they have compact support. This can only happen if the reason that the two integrals are equal is because of some (possibly complicated) integration by parts; i.e. that the two integrands differ by a total derivative. So we can write

$$\hat{X}_f H \triangleq -(A \circ [L_f(A), L_h(A)])_{-1} \qquad (2.13)$$

where $\triangleq$ means equality up to a total derivative.

It is easy to check that the vector fields of the form (2.11) form a Lie algebra under Lie bracket and that they all commute with $\frac{d}{dx}$; in particular, they carry total derivatives into total derivatives. So, if h denotes the equivalence class of H, we can use (2.13) to define $\{f,h\}$; thus Poisson bracket is well defined on the equivalence classes and, from general considerations we know that

$$[\hat{X}_f, \hat{X}_g] = \hat{X}_{\{f,g\}} \quad . \qquad (2.14)$$

Let us illustrate the computation of this Poisson stucture for the simplest case, where m=2 and we fix $a_2=1$ and $a_1=0$. We write a for a_o so

$$A = a + \xi^2$$

is the time independent Schrodinger operator with potential a. Then

$$L_f(A) = (\xi+D)^{-1}\frac{\delta F}{\delta a} = \frac{\delta F}{\delta a}\xi^{-1} - D\frac{\delta F}{\delta a}\xi^{-2} + \ldots$$

$$L_h(A) = (\xi+D)^{-1}\frac{\delta H}{\delta a} = \frac{\delta H}{\delta a}\xi^{-1} - \ldots$$

$$[L_f(A), L_h(A)] = -\left(\frac{\delta F}{\delta a}D\frac{\delta H}{\delta a} - \frac{\delta H}{\delta a}D\frac{\delta F}{\delta a}\right)\xi^{-3} + \ldots$$

so

$$\{f,h\} = \int (A\circ[L_f(A), L_h(A)])_{-1}dx = \int -\left(\frac{\delta F}{\delta a}D\frac{\delta H}{\delta a} - \frac{\delta H}{\delta a}D\frac{\delta F}{\delta a}\right)dx$$

is the Gardner Poisson bracket. Also

$$X_f(A) = -[A, L_f(A)]_+ = -[\xi^2+a, L_f(A)]_+ = -[\xi^2, \frac{\delta F}{\delta a}\xi^{-1}]_+ = -2D\frac{\delta F}{\delta a},$$

all other terms being of negative degree. If $F=-i(a^3+\frac{1}{2}a_x^2)$ then

$$X_f = (6aa_x + a_{xxx})\frac{\partial}{\partial a}$$

is the evolution field of the Korteweg-de Vries equation, where we have written a_x instead of a' to conform to standard usage.

3. POISSON COMMUTATIVITY OF SPLIT AND TRANSLATED INVARIANTS

For many purposes, one would like to find non-trivial families of functions which Poisson commute among themselves. In this section we explain some recent techniques due to Kostant[20] and Symes[33] and another method due to Miscenko and Fomenko[27] for producing such families out of invariants. A function I on g^* is called an invariant, if $\{X,I\}=0$ for all $X\in g$. This means that $\tilde{X}(\ell)I=0$ for all $X\in g$ and $\ell\in g^*$ and hence that $\{F,I\}\equiv 0$ for all F. Roughly speaking, an invariant is (locally) constant on each orbit in g^*, and hence the corresponding vector field on g^* is identically zero. The invariants themselves are not interesting functions from the point of view of Poisson brackets (although they are of crucial importance in describing the orbits). However, they can be used to produce other functions as we now describe. We begin with the method of Kostant and Symes. Suppose that we have a vector space decomposition of the algebra g into a direct sum of two subalgebras:

$$g = a+b \qquad [a,a] \subset a, \quad [b,b] \subset b \quad . \tag{3.1}$$

For example we could let $g=g\ell(n)$ be the algebra of all n by n matrices, let a consist of all strictly lower triangular matrices and b consist of all upper triangular matrices. Or, we could let g be the algebra of *all* pseudodifferential operators in one variable, let b be the subalgebra of differential operators (pseudo differential operators of non-negative degree) and let a be the subalgebra of pseudo differential operators of negative degree. We studied this subalgebra a as an algebra in its own right in the preceding section.

The decomposition $g=a+b$ gives the corresponding decomposition

$$g^* = b^{\perp} + a^{\perp} \tag{3.2}$$

which allows an identification

$$b^{\perp} \sim a^* \quad , \quad a^{\perp} \sim b^* \quad . \tag{3.3}$$

The following Proposition was proved by Kostant and Symes[20-33] independently when $\lambda=0$, and by Kostant[20] with general λ.

Prop. 3.1 - *Suppose that* $\lambda \in g^*$ *satisfies*

$$<\lambda, [a,a]> = 0 \tag{3.4}$$

and

$$<\lambda, [b,b]> = 0 \quad . \tag{3.5}$$

For any function f, *on* g^*, *let* $f_{a,\lambda}$ *be the function on* $b^{\perp} \sim a^*$ *given by*

$$f_{a,\lambda}(\ell) = f(\ell+\lambda) \qquad \ell \in b^{\perp}$$

and similarly

$$f_{b,\lambda}(\ell) = f(\ell+\lambda) \qquad \ell \in a^{\perp} \quad .$$

Then, if f *and* h *are invariants, then*

$$\{f_{a,\lambda}, h_{a,\lambda}\} = 0.$$

Proof. For any $X \in g$, we let X^a and X^b describe its a and b components in the decomposition (3.1). Then for any $\ell \in b^{\perp}$

$$\begin{aligned}\{f_{a,\lambda}, h_{a,\lambda}\}(\ell) &= <\ell, [L_{f_{a,\lambda}}(\ell), L_{h_{a,\lambda}}(\ell)]> \\ &= <\ell, [L_f(\ell+\lambda)^a, L_h(\ell+\lambda)^a]> \\ &= <\ell+\lambda, [L_f(\ell+\lambda)^a, L_h(\ell+\lambda)^a]>\end{aligned}$$

by (3.4). Now f is an invariant so

$$<\ell+\lambda, [L_f(\ell+\lambda), Y]> = 0$$

for any $Y \in g$. Since $L_f(\ell+\lambda) = L_f(\ell+\lambda)^a + L_f(\ell+\lambda)^b$ we have

$$\begin{aligned} <\ell+\lambda, [L_f(\ell+\lambda)^a, L_h(\ell+\lambda)^a]> &= -<\ell+\lambda, [L_f(\ell+\lambda)^b, L_h(\ell+\lambda)^a]> \\ &= <\ell+\lambda, [L_f(\ell+\lambda)^b, L_h(\ell+\lambda)^b]> \end{aligned}$$

since h is also an invariant. But b is a subalgebra and $\ell \in b^{\perp}$ so $<\ell, [b,b]> = 0$, and $<\lambda, [b,b]> = 0$ by hypothesis. Hence the last expression above vanishes, proving the proposition.

This Proposition was used by Kostant[20] and Symes[33] to prove the complete integrability of the Toda[34-35] lattice and related systems. (For the Toda lattice one takes $g=s\ell(n)$ with the decomposition described above and $f(A)=hA^b$. One may use $\lambda=0$, or λ a principal nilpotent matrix,[13-14] or[20].) Let us give a heuristic sketch, following Adler[1], of how this Proposition also implies the Gelfand-Dikii construction of integrals for evolution equations. (Our discussion in only heuristic because we don't want to go into the technical problems of giving a precise definition of the complex powers of a pseudodifferential operator or the proof that the appropriate component of the complex power is a local function.) "Define"

$$f_\sigma(A) = \int (A^\sigma)_{-1}\, dx \quad .$$

Then

$$df_{\sigma A} = \int (\sigma A^{\sigma-1} \circ dA)_{-1}\, dx \; .$$

(Here we have to use Adler's lemma. For example $dA^2 = A \circ dA + (dA) \circ A$. But

$$\int (dA \circ A)_{-1}\, dx = \int (A \circ dA)_{-1}\, dx$$

so

$$\int (dA^2)_{-1}\, dx = 2\int (A \circ dA)_{-1}\, dx \quad .)$$

Thus, on the full algebra g we have

$$L_f(A) = \sigma A^{\sigma-1}$$

and the corresponding vector field (given by (2.8) without the subscript +) is $[A, \sigma A^{\sigma-1}] = 0$ since all powers of A commute. Thus f_σ is an invariant.

The restriction of these invariants to $b^{\perp}=a^{*}$ give the Gelfand-Dikii family of commuting integrals.

In the proof of Prop.3.1 we made use of both (3.4) and (3.5). However, it turns out that if b=0, condition (3.4) is unnecessary.

Prop. 3.2 - *(Miscenko and Fomenko[27]). Let* f *and* h *be invariant functions on* g^{*} *and define the functions* $f_{s\lambda}$, $h_{t\lambda}$ *by*

$$f_{s\lambda}(\ell) = f(\ell+s\lambda)$$

$$h_{t\lambda}(\ell) = h(\ell+t\lambda)$$

where s *and* t *are real numbers and* λ *is any element* g^{*}. *Then* $f_{s\lambda}$ *and* $h_{t\lambda}$ *Poisson commute.*

Proof. It suffices to prove the theorem when $s\neq t$ since it then follows for s=t by continuity. Set

$$\ell = a(\ell+s\lambda) + b(\ell+t\lambda) \qquad a = \frac{t}{t-s}, \quad b = \frac{s}{s-t} \quad .$$

Then

$$\begin{aligned}\{f_{s\lambda},h_{t\lambda}\}(\ell) &= \langle\ell,\left[L_f(\ell+s\lambda),L_h(\ell+t\lambda)\right]\rangle \\ &= a\langle\ell+s\lambda,\left[L_f(\ell+s\lambda),L_h(\ell+t\lambda)\right]\rangle + \\ &\quad + b\langle\ell+t\lambda,\left[L_f(\ell+s\lambda),L_h(\ell+t\lambda)\right]\rangle \ .\end{aligned}$$

The first term vanishes because f is an invariant and the second term vanishes because h is an invariant, proving the proposition.

In view of Prop.3.2 it would seem that (3.4) should not be necessary in Prop.3.1, or, in any event, the hypotheses of Prop.3.1 are too stringent. The proper setting is not clear to us at present.

There are some other points about Prop.3.1 which can use some clarification. For simplicity (although this is not relevant) assume that $\lambda=0$. If we take the simultaneous level surfaces of *all* the invariants of g, these are, locally, the maximal orbits of G acting on g^{*}. Let $\mathcal{O}$ be a G orbit in g^{*} passing through a point $\ell\in b^{\perp}$. Let A be the simply connected group associated to a. We can form the A orbit of ℓ in g^{*}, call it $A\cdot\ell$. We can also form the A orbit of ℓ in $b^{\perp}\approx a^{*}$, call it $\mathcal{O}^{a}$. Clearly $\mathcal{O}^{a}$ is the image of $A\cdot\ell$ under the projection of ℓ^{*} onto $b^{\perp}$ along $a^{\perp}$. Let $g_{\ell}\subset g$ denote the isotropy subalgebra of ℓ, i.e. the set of all $Z\in g$ such that $\tilde{Z}(\ell)=0$. (In case $\mathcal{O}$ is a generic orbit of maximal dimension,

g_ℓ will be spanned by $L_f(\ell)$ as f ranges over the invariants. For a lower dimensional orbit g_ℓ will be larger.) Write every $Z \in g_\ell$ as

$$Z = Z^a + Z^b \qquad Z^a \in a , \qquad Z^b \in b \quad .$$

Then $Z \cdot \ell = 0$ implies that $Z^a \cdot \ell = -Z^b \cdot \ell$. But since $Z^b \in b$, we have $Z^b b^\perp \subset b^\perp$ so $Z^a \cdot \ell \in b^\perp$. Thus $Z^a \cdot \ell$ is tangent to $\mathcal{O}^a$. If X and Y are two elements g_ℓ, then the symplectic scalar product of $X^a \cdot \ell$ $Y^a \cdot \ell$ is

$$\langle \ell, [X^a, Y^a] \rangle = \langle \ell, [X, Y^a] \rangle - \langle \ell, [X^b, Y^a] \rangle$$

$$= -\langle \ell, [X^b, Y] \rangle - \langle \ell, [X^b, Y^b] \rangle = 0$$

since $X \cdot \ell = Y \cdot \ell = 0$ and $\ell \perp [b, b]$. Thus the set

$$\{Z^a \cdot \ell, \ Z \in g_\ell\}$$

is an isotropic subspace of the tangent to $\mathcal{O}^a$ at ℓ. Moving it around by the action of A gives an isotropic distribution. In case $\mathcal{O}^a$ is a maximal orbit, this distribution is just the null foliation associated to the level surfaces of the f^a's and hence is completely integrable in the Frobenius sense. It would be nice to know whether this fact is true in general.

4. THE MOMENT MAP, PHASE SPACE AND SEMI-DIRECT PRODUCTS

Every symplectic manifold carries its canonical Poisson structure. In particular, the phase space $M = T^*N$ of some "configuration" manifold N carries a symplectic structure, and hence a Poisson bracket. Any diffeomorphism ϕ of N into N induces a transformation of T^*N into T^*N as follows: A point (q,p) of T^*N is determined by a point $q \in N$ and a linear function p on the tangent space, TN_q, to N at q. The differential of ϕ at q, which we denote by $d\phi_q$, maps TN_q into $TN_{\phi(q)}$. Its transpose, $d\phi_q^*$, will map a linear function on $TN_{\phi(q)}$ into a linear function on TN_q. Thus $d\phi_q^{*-1}$ maps a linear function on TN_q into a linear function on $TN_{\phi(q)}$ and the induced transformation, $\hat{\phi}$, on T^*N is given by

$$\hat{\phi}(q,p) = (\phi(q), \, d\phi_q^{*-1} p) \quad . \tag{4.1}$$

If Z_N is a vector field on N whose expression is $Z_N = Z(q)\partial/\partial q_1 + \cdots + Z^n(q)\partial/\partial q_n$ in terms of local coordinates, then the infinitesimal version of the above equation shows that the corresponding vector field $\hat{Z}_N$ on T^*N is given by

$$\hat{Z}_N(q) = Z^1(q)\frac{\partial}{\partial q_1} + \ldots + Z^n(q)\frac{\partial}{\partial q_n} - \frac{\partial Z_N \cdot p}{\partial q_1}\frac{\partial}{\partial p_1} - \ldots - \frac{\partial Z_N \cdot p}{\partial q_n}\frac{\partial}{\partial p_n} .$$

From this we see that we may choose the function corresponding to $\hat{Z}_N$ as

$$f_{Z_N} = Z_N \cdot p = Z_N^1 \cdot p_1 + \ldots + Z_N^n \cdot p_n . \tag{4.2}$$

It is easy to see that if ϕ and ψ are two diffeomorphisms of N, then

$$(\phi \circ \psi)\hat{} = \hat{\phi} \circ \hat{\psi}$$

as transformation of T^*N. From this it follows that if Y_N and Z_N are vector fields on N, then

$$[\hat{Z}_N, \hat{Y}_N] = [Z_N, Y_N]\hat{}$$

Also

$$\hat{Z}_N \cdot f_{Y_N} = [Z_N, Y_N] \cdot p = f_{[\hat{Z}_N, \hat{Y}_N]} .$$

If ϕ is a diffeomorphism of N, then

$$f_{Z_N}(\hat{\phi}^{-1}(q,p)) = f_{Z_N}(\phi^{-1}(q), d\phi^*_{\phi^{-1}q}\, p) = Z_N(\phi^{-1}(q)) \cdot d\phi^*_{\phi^{-1}q}\, p$$
$$= d\phi^*_{\phi^{-1}q}\left[Z_N(\phi^{-1}(q))\right] \cdot p .$$

Let $\psi_t = \exp tZ_N$ be the one parameter group generated by Z_N. Then $\phi \circ \psi_t \cdot \phi^{-1}$ is another one parameter group whose infinitesimal generator, which we shall denote by $Ad_\phi Z_N$, is clearly the vector field which assigns to the point q the tangent vector to the curve $\phi\psi_t(\phi^{-1}q)$, i.e., the tangent vector $d\phi_{\phi^{-1}q}\, Z_N(\phi^{-1}q)$. Thus we may write the last equation as

$$\rho(\hat{\phi}) f_{Z_N} = f_{Ad\phi\, Z_N} . \tag{4.3}$$

We have thus shown that the group, Diff(N), of all diffeomorphisms of N, whose "Lie algebra" can be thought of as the algebra, D(N), of all vector fields on N, has a Hamiltonian action on T^*N. An action of a Lie group G on N is a homomorphism of G into the group of all diffeomorphisms of N. Hence restricting the preceding discussion to the image of G, we get a Hamiltonian action of G on T^*N. For example, suppose that $N=\mathbb{R}^{3k}$ is the configuration space for k particles, with coordinates $q_{x_1}, q_{y_1}, q_{z_1}, q_{x_2}, \ldots, q_{z_n}$ and $G=\mathbb{R}^3$ is the group of translations in three dimensions, where G acts on N by simultaneously making the same translation of all three particles. Let $X=(1,0,0)\in\mathbb{R}^3$ be infinitesimal translation in the x-direction. Then the corresponding vector field on N is $X_N=\partial/\partial q_{x_1} + \partial/\partial q_{x_2} + \ldots + \partial/\partial q_{x_n}$, and (4.2) gives

$$f_{X_N}(q,p) = p_{x_1} + p_{x_2} + \ldots + p_{x_n} \qquad ,$$

the familiar expression for the total linear momentum in the x-direction. Thus $\Phi(q,p) = \vec{p}_1+\ldots+\vec{p}_n$ is the total linear momentum. Similarly, we may take $G=SO(3)$ to act on $N=\mathbb{R}^{3k}$ by simultaneous rotation on each particle. Then if we let X be infinitesimal rotation about some axis, the function f_{X_N} is easily seen to be the corresponding total angular momentum about that axis.

The functions f_{X_N} given by (4.2) are all (homogeneous) linear in p, with coefficients which are arbitrary functions of q. It is clear that the Poisson bracket of any two such functions is again of the same type. On the other hand, if f is a function of the q's alone, i.e. f is a function on N, then $\{f_{X_N}, f\}=X_N f$ is again a function on N. Thus we can consider the Lie algebra $D(N)\times F(N)$ (semi-direct product) with bracket relation

$$[(X_N^1, f_1), (X_N^2, f_2)] = ([X_N^1, X_N^2], X_N^2 f_1 - X_N^1 f_2)$$

and the map of $D(N)\times F(N)$ into $F(T^*N)$ sending (X_N, f) into $f_{X_N}+f$ is a homomorphism of Lie algebras. (Here by abuse of notation, we have used the same symbol, f, to denote a function on N and the function thought of as a function on T^*N). In what follows, it will be useful for us to describe the picture for the group whose "Lie algebra" is $D(N)\times F(N)$. This is the semi-direct group product

$$\mathrm{Diff}(N)\times F(N)$$

where

$$(\phi_1,f_1)\cdot(\phi_2,f_2) = (\phi_1\cdot\phi_2, f_1+\rho(\phi_1)f_2) \quad .$$

We wish to show that this group acts on T^*N. For this we observe that any function f on N determines a transformation t_f on T^*N defined by

$$t_f(q,p) = (q, p+df_q) \quad .$$

It is easy to see that this is a canonical transformation. If sf is a one parameter family of functions, the corresponding infinitesimal generator is $\frac{\partial f}{\partial q_1}\frac{\partial}{\partial p_1} + \ldots + \frac{\partial f}{\partial q_n}\frac{\partial}{\partial p_n}$, which corresponds to the function f on T^*N.

If $\phi \in \mathrm{Diff}(N)$, then

$$\hat\phi t_f \hat\phi^{-1}(q,p) = \hat\phi t_f(\phi^{-1}(q), d\phi^*_{\phi^{-1}q}p) = \hat\phi(\phi^{-1}(q), d\phi^*_{\phi^{-1}q}p + df_{\phi^{-1}q})$$

$$= (q, p+d\phi^{*-1}_{\phi^{-1}q}\ df_{\phi^{-1}q}) = (q, p+d(f\cdot\phi^{-1})_q) \quad ,$$

the last equation following from the chain rule. Thus

$$\hat\phi t_f \hat\phi^{-1} = t_{\rho(\phi)f} \tag{4.4}$$

and this implies that we have a Hamiltonian action of the semidirect product $\mathrm{Diff}(N)\times F(N)$ on T^*N.

We shall apply this result in the following form. Suppose we have a representation of the Lie group H on a vector space V. Suppose we have a map $v \to f_v$ of V into the space of functions, $F(N)$, on some manifold N. Suppose that we have an action $H\times N \rightsquigarrow N$ of H on N, and suppose that

$$f_v(a^{-1}n) = f_{av}(n) \tag{4.5}$$

for all $a\in H$, $v\in V$ and $n\in N$. Then we get a Hamiltonian action of the semidirect product $H\times V$ on T^*N where $a\in H$ acts by $\hat a$ and $v\in V$ acts by t_v. The Lie algebra of $H\times V$ is $h\times V$, where h is the Lie algebra of H. For $X\in g$, let X_N denote the corresponding vector field on N. Then

$$f_X = f_{X_N} \quad . \tag{4.6}$$

Let H be a Lie group with a representation on some vector space V. We shall let G denote the semi-direct product of H and V. Let p be some fixed element of V^*. We let ap denote a^{*-1}, that is

$$\langle ap, v\rangle = \langle p, a^{-1}v\rangle \qquad \text{for any } v\in V \ .$$

Define the functions f_v^p on G by

$$f_v^p(b) = \langle bp,v\rangle \qquad b\in G. \tag{4.7}$$

Then

$$f_v^p(a^{-1}b) = \langle a^{-1}bp,v\rangle = \langle bp,av\rangle = f_{av}(b) \quad .$$

The condition (4.5) is satisfied and hence we get a moment map, $\Phi^p: T^*H \to g^* = h^* + V^*$. To describe this map we shall identify T^*H with $H\times h^*$ via left multiplication. That is, left multiplication by any $c\in H$ gives an identification of $TH_e = h$ with TH_c and hence of T^*H_c with h^*. If $X\in h$, left multiplication by exp tX is a one parameter group of transformation on H whose infinitesimal generator we denote by X_H. Since

$$(\exp tX)\, c = c\,(c^{-1}\exp tXc) = c\exp tAd_{c^{-1}}X$$

we see that if we use left multiplication to identify TH_c with h, then

$$X_H(c) = Ad_{c^{-1}}X \ . \tag{4.8}$$

By definition, $\langle\Phi^p(c,\alpha),(X,0)\rangle = \langle\alpha,X_H(c)\rangle = \langle c\alpha,X\rangle$ for $X\in h$. From this we see that the h^* component of $\Phi^p(c,\alpha)$ is just $c\alpha$. On the other hand, $f_v^p(c) = \langle cp,v\rangle$ shows that the V^* component of $\Phi^p(c,\alpha)$ is cp. Thus

$$\Phi^p(c,\alpha) = (c\alpha,cp) \quad . \tag{4.9}$$

More generally, let N be some manifold on which H acts transitively. Let $H_p = \{a\in H \mid ap=p\}$ denote the isotropy group of $p\in V^*$ and let $H_n = \{a\in H \mid an=n\}$ denote the isotropy group of $n\in N$. Suppose that $H_p \supset H_n$. (In case N=H, then $H_n = \{e\}$. At the other extreme, we could take N to be the orbit through p of H in V^* and n=p, in which case $H_n = H_p$.) Then every point of N can be written as bn for some $b\in H$ and we define the function f_v^p on N by

$$f_v^p(bn) = \langle bp,v\rangle \quad . \tag{4.10}$$

This is well-defined since bn=b'n if and only if b'=ba with an=n, whence ap=p, so bp=b'p. Then as before, it is clear that (4.5) is satisfied and so we get a moment map

$$\Phi^p: T^*N \to g^* = h^*\oplus V^* \ .$$

To describe the moment map we note that we have a map of h onto TN_n sending $X\in h$ into $X_N(n)$. The transpose of this map gives an injection

$\tau: T^*N_n \to h^*$. We can use the action of $b \in H$ to identify TN_n with TN_{bn} and hence to identify T^*N_{bn} with T^*N_n. Then arguing much as before, we see that with this identification (which depends on b)

$$\Phi^p(bn,\alpha) = (b\tau\alpha, bp) \quad . \tag{4.11}$$

For example, we may take N=H and let H act on itself by *right* translation

$$R_b a = ab^{-1}$$

Then (4.11) becomes (remember that we identify T^*H with $H \times h^*$ using left translations)

$$\Phi^p(a,\alpha) = (-\alpha, a^{-1}p) \quad . \tag{4.12}$$

Let $T(\alpha)$ be some quadratic function of α (for example, if we consider α as matrices, $T(\alpha) = \frac{1}{2}\mathrm{tr}(\alpha Q^{-1}\alpha^t)$) and let $x \in V$. Define the function P on g^* by

$$P(\alpha,\gamma) = T(\alpha) - \langle\gamma, x\rangle \quad .$$

Then

$$H = P \circ \Phi^p$$

is a Hamiltonian system on T^*H consisting of a left invariant kinetic energy plus a "representative function", cf Kuperschmidt and Vinogradov[21].

Let L_b denote left multiplication by b,

$$L_b a = ba \quad .$$

Then

$$L_b(a,\alpha) = (ba,\alpha)$$

so

$$\Phi^p \circ L_b(a,\alpha) = (-\alpha, a^{-1}b^{-1}p)$$

and

$$\Phi^p \circ L_b = L_b \quad \text{if} \quad b_p = p \ ,$$

hence H is invariant under L_b if bp=p.
On other hand

$$\Phi^p \circ R_b = b \cdot \Phi^p \quad .$$

So, if both T and x are invariant under $b \in H$, then the Hamiltonian H is invariant under right multiplication by b.

Let us illustrate this discussion for the case where H=SO(3) and $V = \mathbb{R}^3$. The vector p will represent a constant force field, say the gravitational field, and $x \in \mathbb{R}^3$ the center of mass of a rigid body. The metric T is given by the inertia tensor, and H is the Hamiltonian of a rigid body with one point fixed in a uniform gravitational field. The algebra g in this case is the algebra $\mathscr{E}(3)$ of Euclidean motions in three space. We may identify $o(3)^*$ with $\Lambda^2\mathbb{R}^3$ and write the most general element of g^* as

$$(\nu,\gamma) \quad \text{where} \quad \nu\in\Lambda^2\mathbb{R}^3, \ \gamma\in\mathbb{R}^3 \ .$$

The functions $|\gamma|^2$ and $|\nu\Lambda\gamma|$ are invariants on g^* and the generic orbit is four dimensional. The function $|\gamma|^2$ measures the intensity of the gravitational field, while $|\nu\Lambda\gamma|$ pulls back to the angular momentum about the p axis, call it M_p. Then M_p and H Poisson commute. Suppose the inertia tensor has some axis of symmetry and the center of mass lies on this axis. Then if b is a rotation about this axis, b preserves both T and x and hence right multiplication by b preserves H. The infinitesimal generator of this group then gives rise to a third integral. This is the case of Lagrange's top.

It is instructive to look at the flow generated by P on g^*. Let e_1,e_2,e_3 be an orthonormal basis of $o(3)$ in terms of which T is diagonal, and f_1,f_2,f_3 the corresponding basis of $\mathbb{R}^3$ so that if we consider $e_1,\ldots,f_3$ as functions on g^*

$$P = \frac{1}{2}\sum\frac{e_j^2}{A_j} - \sum x_j f_j \quad .$$

Clearly

$$\{P,e_1\} = \frac{-e_2e_3}{A_2} + \frac{e_3e_2}{A_3} + x_2f_3 - x_3f_2$$

$$\vdots$$

$$\{P,f_1\} = \frac{e_3f_2}{A_3} - \frac{e_2f_3}{A_2}$$

$$\vdots$$

If we let $\omega_i=e_i/A_i$ and think of ω_i and f_i as coordinates on g^*, the corresponding differential equations are

$$A_1\frac{d\omega_1}{dt} = (A_2-A_3)\omega_2\omega_3+x_2f_3-x_3f_2$$

$$\vdots$$

$$\frac{df_1}{dt} = \omega_3f_2-\omega_2f_3$$

These are the Euler-Poinsot equations for the rigid body. Let ν_1,ν_2,ν_3, $\gamma_1,\gamma_2,\gamma_3$ be coordinates on g^* given be $e_1,\dots,f_3$. (That is, $\nu_1 = e_1$ considered as a function on g^* etc.; it is a bit more convenient to have a separate notation.)

Thus

$$P = \tfrac{1}{2}\sum_{1}^{3}\frac{\nu_i^2}{A_i} - \sum x_i\gamma_i \quad .$$

and the two invariants are

$$\Gamma = |\gamma|^2 = \gamma_1^2 + \gamma_2^2 + \gamma_3^2$$

and

$$M = |\nu\Lambda\gamma| = \nu_1\gamma_1 + \nu_2\gamma_2 + \nu_3\gamma_3 \quad .$$

For any function G, the Poisson bracket $\{P,G\}$ is given by

$$\{P,G\} = \sum \Big\{\left(\frac{\nu_i\nu_k}{A_i} - x_i\gamma_k\right)\frac{\partial G}{\partial\nu_j} - \left(\frac{\nu_j\nu_k}{A_j} - x_j\gamma_k\right)\frac{\partial G}{\partial\nu_i} + \frac{\nu_i\gamma_k}{A_i}\frac{\partial G}{\partial\gamma_j} - \frac{\nu_j\gamma_k}{A_j}\frac{\partial G}{\partial\gamma_i}\Big\}$$

where the sum is over cycle permutations of (1,2,3). We can now list the three cases of complete integrability.

I <u>The Euler-Poinsot case</u>

$$x_1 = x_2 = x_3 = 0 \qquad \text{(no potential!)}$$

Consider $J = \sum_{i=1}^{3}\nu_i^2$ (the squared total angular momentum)

Then

$$\{P,J\} = 2\sum_{(i,j,k)}\left(\frac{\nu_i}{A_i}\nu_j - \frac{\nu_j}{A_j}\nu_i\right)\nu_k = 0$$

and on every orbit J is 2 first integral for the Hamiltonian system induced by the Euler equation. It is clear that for $\{A_i$ not all equal$\}$ P and J are independent implying completely integrable system on every orbit.

II <u>The Lagrange-Poisson case</u>

$$A_1 = A_2 = \lambda\ , \qquad A_3 = \mu$$
$$x_1 = x_2 = 0\ , \qquad x_3 = \kappa$$

Let $N \in F(\mathscr{E}(3)^*)$, $N = \nu_3$. We have

$$\{P,N\} = \left(\nu_1\nu_2\left(\frac{1}{A_1} - \frac{1}{A_2}\right) + (\gamma_1 x_2 - \gamma_2 x_1)\right)\frac{\partial N}{\partial \nu_3} = 0$$

and on every orbit N is a first integral for the Hamiltonian system induced by the Euler equation. P and N are independent implying completely integrable system on every orbit.

III The Kovalevskaia case

$$A_1 = A_2 = 2A_3 = 2\lambda$$

$$x_1 = \kappa; \quad x_2 = x_3 = 0$$

Consider $K \in F(\mathscr{E}(3)^*)$

$$K(\nu,\gamma) = (\nu_1^2 - \nu_2^2 + \varepsilon\gamma_1)^2 + (2\nu_1\nu_2 + \varepsilon\gamma_2)^2 = \varphi_1^2 + \varphi_2^2$$

$$\frac{\partial K}{\partial \nu_1} = 4(\nu_1\varphi_1 + \nu_2\varphi_2), \qquad \frac{\partial K}{\partial \gamma_1} = 2\varepsilon\varphi_1, \qquad \frac{\partial K}{\partial \nu_3} = 0$$

$$\frac{\partial K}{\partial \nu_2} = 4(-\nu_2\varphi_1 + \nu_1\varphi_2), \qquad \frac{\partial K}{\partial \gamma_2} = 2\varepsilon\varphi_2, \qquad \frac{\partial K}{\partial \gamma_3} = 0$$

We claim that $\{P,K\} = 0$. Indeed

$$\{P,K\} = \left(\frac{\nu_1\nu_3}{A_1} - \kappa\gamma_3\right)\frac{\partial K}{\partial \nu_2} - \frac{\nu_2\nu_3}{A_2}\frac{\partial K}{\partial \nu_1} - \frac{\nu_3\nu_1}{A_3}\frac{\partial K}{\partial \nu_2} + \frac{\nu_3\nu_2}{A_3}\frac{\partial K}{\partial \nu_1} +$$

$$+ \frac{\nu_1\gamma_3}{A_1}\frac{\partial K}{\partial \gamma_2} - \frac{\nu_2\gamma_3}{A_2}\frac{\partial K}{\partial \gamma_1} - \frac{\nu_3\gamma_1}{A_3}\frac{\partial K}{\partial \gamma_2} + \frac{\nu_3\gamma_2}{A_3}\frac{\partial K}{\partial \gamma_1} =$$

$$= \nu_3\left[\frac{\nu_1}{2\lambda}\frac{\partial K}{\partial \nu_2} - \frac{\nu_2}{2\lambda}\frac{\partial K}{\partial \nu_1} - \frac{\nu_1}{\lambda}\frac{\partial K}{\partial \nu_2} + \frac{\nu_2}{\lambda}\frac{\partial K}{\partial \nu_1} - \frac{\gamma_1}{\lambda}\frac{\partial K}{\partial \gamma_2} + \frac{\gamma_2}{\lambda}\frac{\partial K}{\partial \gamma_1}\right] +$$

$$+ \gamma_3\left[-\kappa\frac{\partial K}{\partial \nu_2} + \frac{\nu_1}{2\lambda}\frac{\partial K}{\partial \gamma_2} - \frac{\nu_2}{2\lambda}\frac{\partial K}{\partial \gamma_1}\right] =$$

$$= \frac{1}{2\lambda}\nu_3\left[\nu_2\frac{\partial K}{\partial \nu_1} - \nu_1\frac{\partial K}{\partial \nu_2} + 2\gamma_2\frac{\partial K}{\partial \gamma_1} - 2\gamma_1\frac{\partial K}{\partial \gamma_2}\right] +$$

$$+ \frac{1}{2\lambda}\gamma_3\left[-2\lambda\kappa\frac{\partial K}{\partial \nu_2} + \nu_1\frac{\partial K}{\partial \gamma_2} - \nu_2\frac{\partial K}{\partial \gamma_1}\right] =$$

$$= \frac{2}{\lambda}\nu_3\left[\nu_2\nu_1\varphi_1 + \nu_2^2\varphi_2 + \nu_1\nu_2\varphi_1 - \nu_1^2\varphi_2 + \varepsilon\gamma_2\varphi_1 - \varepsilon\gamma_1\varphi_2\right] +$$

$$+ \frac{1}{2\lambda}\gamma_3\left[-8\lambda\kappa(-\nu_2\varphi_1 + \nu_1\varphi_2) + 2\varepsilon\nu_1\varphi_2 - 2\varepsilon\nu_2\varphi_1\right] =$$

$$= \frac{2}{\lambda}\nu_3[(2\nu_1\nu_2+\varepsilon\gamma_2)\varphi_1-(\nu_1^2-\nu_2^2+\varepsilon\gamma_1)\varphi_2] +$$

$$+ \frac{1}{2\lambda}\gamma_3[8\lambda\kappa\nu_2\varphi_1-8\lambda\kappa\nu_1\varphi_2+2\varepsilon\nu_1\varphi_2-2\varepsilon\nu_2\varphi_1] =$$

$$= \frac{2}{\lambda}\nu_3[\varphi_2\varphi_1-\varphi_1\varphi_2] + \frac{1}{2\lambda}\gamma_3[(8\lambda\kappa-2\varepsilon)\nu_2\varphi_1-(8\lambda\kappa-2\varepsilon)\nu_1\varphi_2] =$$

$$= 0 \quad \text{iff} \quad 8\lambda\kappa = 2\varepsilon \Leftrightarrow \boxed{\varepsilon = 4\lambda\kappa}$$

So $K=(\nu_1^2-\nu_2^2+4\lambda\kappa\gamma_1)^2+(2\nu_1\nu_2+4\lambda\kappa\gamma_2)^2$ is a first integral for P on every orbit. Clearly it is independent of P implying complete integrability on every orbit.

It would be nice to have a group theoretical explanation of this case.

Take now in (4.12) $p=f_3\in\mathbb{R}^3\simeq\mathbb{R}^{3*}$ and denote $\Phi^{f_3}=T^*SO(3)\to\mathscr{E}(3)^*$ by Φ^3. Then we have:

$$H(a,\alpha) = (P\circ\Phi^3)(a,\alpha) = \tfrac{1}{2}tr(\alpha Q^{-1}\alpha^t) - \langle a^t f_3, x\rangle$$

which is the Hamiltonian of the rigid body.

I E-P case: $J\circ\Phi^3 = \sum_{i=1}^{3}\alpha_i^2$ - the total angular momentum

II L-P case: $N\circ\Phi^3 = -\alpha_3$ - the first integral on $T^*SO(3)$

III K-case: $K\circ\Phi^3$ is the classical Kovalevskaia integral

In all the case:

$$\Gamma\circ\Phi^3 = \sum_{i=1}^{3}\gamma_i^2 = \|f_3\| = 1$$

$$M_3 = M\circ\Phi^3 = \left(\sum_{i=1}^{3}\gamma_i\nu_i\right)\circ\Phi^3 = (\gamma,\nu)\circ\Phi^3 = (a^t f_3,-\alpha) = -(f_3,a\cdot\alpha)$$

is the angular momentum relative to the vertical -the only first integral of $P\circ\Phi^3$ which is independent of the <u>Q&x</u> parameters.

<u>Obs.</u> In this model the Q&x parameters are still "outsiders".

The paper[12] suggests that in order to "include" the parameters x&Q as active elements of the model one must use the larger 12-dimensional Lie algebra $g=h\times V$, (semi-direct product) where h is again o(3), but V is now $\mathbb{R}^3\otimes\mathbb{R}^3 = \Lambda^2\mathbb{R}^3\oplus S^2\mathbb{R}^3$ -the 9-dimensional vector space which is the direct sum of skew-symmetric and symmetric tensor. If we fix a pair (x,Q), $(x,Q)\in\Lambda^2\mathbb{R}^3\oplus S^2\mathbb{R}^3=V\simeq V^*$ ($x\in\Lambda^2\mathbb{R}^3$ is the "center of mass" and $Q\in S^2\mathbb{R}^3$ is the "inertia tensor"), we obtain a moment map $\Phi^{(x,Q)}: T^*SO(3)\to g^*$. It is easy to see that the same function P as above (but considering now Q as variable in $S^2\mathbb{R}^3\simeq S^2\mathbb{R}^{3*}$) pulls back to the Hamiltonian of the rigid body, $H=P\circ\Phi^{(x,Q)}$. Now, to fix the physical parameters of the body, x&Q, means to fix an orbit in g^* so that the "cases" of the integration of the Euler-Poinsot equations of the motion of a rigid body are classified by the corresponding orbits in g^* and the Hamiltonian system induced on these orbits. In principle, one can use the integration procedure given in[12] in order to understand more about this classical problem (and in particular about the Kovalevskaia case).

REFERENCES

1 Adler, M., On a trace functional for formal pseudodifferential operators and the symplectic structure for the Korteweg-de Vries type equations, Inventiones math. 50, 219-248 (1979).

2 Faddeev, L., and Zakharov, V.E., Korteweg-de Vries equation as completely integrable Hamiltonian system, Funk.Anal.Priloz., 5 (1971), pp.18-27.(In Russian).

3 Gardner, C.S., Korteweg-de Vries equation and generalizations. IV: The Korteweg-de Vries equation as a Hamiltonian system, J.Math.Phys., 12 (1971) pp.1548-1551.

4 Gardner, C.S., Greene, J.M., Kruskal, M.D. and Miura, R.M., Method for solving the Korteweg-de Vries equation, Phys.Rev.Lett., 19 (1967), pp.1095-1097.

5 Gardner, C.S., Greene, J.M., Kruskal, M.D. and Miura, R.M., Korteweg-de Vries Equation and Generalizations VI, Methods of Exact Solution, Comm.Pure Appl.Math., 27, 97-133 (1974).

6 Gelfand, I.M., and Dikii, L.A., Asymptotic behavior of the resolvent of Sturm-Liouville equations and the algebra of the Korteweg-de Vries equations, Uspekhi Mat.Nauk 30:5 (1975) 67-100.

7 Gelfand, I.M., Dikii, L.A., Fractional Powers of Operators and Hamiltonian,Systems Funkcional'nyi Analiz i ego Prilozenija, 10, No.4 (1976).

8 Gelfand, I.M., Dikii, L.A., The Resolvent and Hamiltonian Systems, Funkcional'nyi Analiz i ego Prilozenija, 11, 11-27 (1977).

9 Gelfand, I.M., Manin, Yu.I., Shubin, M.A., Poisson Brackets and the Kernel of a Variational Derivative in Formal Variational Calculus. Funkt.Anal.Prilozen., 10, No.4 (1976).

10 Guillemin, V., Quillen, D., Sternberg, S.: The Integrability of Characteristics, C.P.A.M., 23, 39-77 (1970).

11 Guillemin, V., and Sternberg, S., Geometric Asymptotics, AMS Publ., 1977.

12 Guillemin, V., and Sternberg, S., The Moment Map and Collective Motion, to appear.

13 Hermann, R., Toda lattices, Cosymplectic Manifolds, Bäcklund Transformations, Kinks; Part A, 15, Part B, 18, Interdisciplinary Math., Math.Sci.Press.

14 Kazhdan, D., Kostant, B., Sternberg, S.: Hamiltonian Group Actions and Dynamical Systems of Calogero Type, C.P.A.M. (1978).

15 Kirillov, A., Unitary Representations of Nilpotent Lie Groups, Russ. Math.Surveys 17 (4) 57-101 (1962).

16 Kirillov, A.A.: Elements of the Theory of Representations. Berlin, Heidelberg, New York, Springer-Verlag pp.226-235,290-292 (1976).

17 Kohn, J.J., and Nirenberg, L., "Pseudodifferential Operators", Comm. in Pure and Appl.Math., vol.XVIII, 269-305 (1965).

18 Korteweg, D.J., and De Vries, G., On the change of form of long waves advancing in a rectangular canal, and on a new type of long stationary waves, Philos.Mag., 39 (1895), pp.422-443.

19 Kostant, B., Quantization and unitary representations, Lecture Notes in Math. 170, 87-208, Springer-Verlag (1970)

20 Kostant, B., "The solution to a generalized Toda lattice and representation theory", to appear.

21 Kuperschmidt, B., and Vinogradov, A.M., Uspehi Mat.Nauk 32 No.4(1977).

22 Lax, P., Periodic solutions of the KdV equations, Comm.Pure Appl.Math. 28 (1975), pp. 141-188.

23 Lax, P., Almost Periodic Solutions of the KdV Equation, SIAM Review, (1976).

24 Lax, P.D., Integrals of Nonlinear Equations and Solitary Waves. Comm.Pure Appl.Math., 21, 467-490 (1968).

25 Lebedev, D.R., and Manin, Yu.I., Gelfand-Dikii Hamiltonian Operator and Coadjoint Representation of Volterra Group, (1978) -to appear.

26 Manin, Yu.I., Algebraic Aspects of Nonlinear Differential Equations (Russian) Itogi Naukii Tekliniki, Sovremennye Problemy Matematiki, vol.11 (1978).

27 Miscenko, A.S., and Fomenko, A.T., Euler Equations of finite-dimensional Lie groups, Izvestija Akad.Nauk S.S.S.R., vol.42, No.2 (1978) Mathematics of the U.S.S.R., Izvestija vol.12, No.2 (1978).

28 Miura, R.M., The Korteweg-de Vries equation: A survey of results, SIAM Review (1975), pp. 412-458.

29 Miura, R., Gardner, C.S., and Kruskal, M.D., Korteweg-de Vries equation and generalizations, II: Existence of conservation laws and constants of motion. J.Math.Phys., 9 (1968),1204-1209.

30 Souriau, J.M., Structure des systemes dynamiques, Dunod, Paris (1970).

31 Sternberg, S., Some preliminary remarks on the formal variational calculus of Gelfand-Dikii Springer-Lecture Notes in Math., No.676 (1978) pp.399-407.

32 Symes, W., A Poisson Structure on Spaces of Symbols, MRC Technical Summary Report # 184, Univ.of Wisconsin-Madison (1978).

33 Symes, W., On systems of Toda type, Technical Summary Report, Univ. of Wisconsin, Madison Mathematics Research Center (1979).

34 Toda, M., Wave Propagation in Harmonic Lattices, J.Phys.Soc.Japan 23, 501-506 (1967).

35 Toda, M., Studies of a non-linear lattice, Physics Reports (sec.C of Physics Letters)18,1-125 (1975).

AROUND THE CLASSICAL STRING PROBLEM*+

Pierre C. SABATIER

Laboratoire de Physique Mathématique**

Université des Sciences et Techniques du Languedoc

34060 MONTPELLIER CEDEX FRANCE

Abstract : The problem of reconstructing the density $\rho(x)$ of a vibrating string given the N first eigenfrequencies of two vibrating configurations admits solutions that minimize certain weighted averages of the density. There exists a simple set of necessary conditions of these weights. In particular, it has been shown that the only weight functions $f(x)$ than can be consistent in all cases with the existence of an extremal density which is made up of a finite number of points masses are polynomials of degree two. In the present paper, it is shown that the weighted averages can be calculated exactly. Explicit formulas are given, which in certain cases depend only on the spectrum of one vibrating configuration. The results strongly suggest applications to the Earth inverse problem. They are also extended to other problems, which suggest applications to non-linear questions. In particular, a new non linear evolution equation is studied, and conservation laws are exhibited.

PM/79/14 June 1979

* Ce travail a été effectué dans le cadre de la R.C.P. n°264 : Etude interdisciplinaire des Problèmes inverses.

** Physique Mathématique et Théorique, Equipe de Recherche Associé au C.N.R.S.

+ This paper is presented in International Meeting on "Nonlinear Evolution Equations and Dynamical Systems" - June 20-23, 1979 - Lecce, ITALY - as an invited lecture.

Introduction

Inverse problems are often underdetermined, because information coming from experiments is always truncated. It is then important to class the equivalent solutions, and preferably to do it by means of properties which have a physical meaning. Looking for extremal properties is a seducing way to handle this problem. It works very well for constrained linear inverse problems, whose "ideal solutions", the solutions for which the upper bound is minimized, proved to be of a considerable physical interest. It non linear inverse problems, one may hope in addition that a solution which is characterized by its extremal properties can be constructed by means of a stable algorithm, following for instance the gradient of this extremal property. Besides, if the inverse problem yields an inverse method for solving a non-linear partial differential equation, the extremal property may characterize the corresponding solution of this non linear partial differential equation.

The present note is part of a series of papers in which these ideas are applied to the string problem and related problems. We first deal with the string problem.

The problem of reconstructing the density $\rho(x)$ of a vibrating string given the first N eigenfrequencies for two vibrating configurations admits an infinite number of solutions. Among all such strings compatible with the truncated data set, we look for solutions which achieve an extremum for a weighted average of the density :

$$M[\rho] = \int_0^1 f(x)\, \rho(x)\, dx \qquad (f(x) \geq 0) \tag{1}$$

The solutions were studied for $f(x) \equiv 1$ by Krein[1], then by Barcilon[2] who gave another precise example where $M[\rho]$ is a true minimum, the extremal solution is made up by a finite number of concentrated point masses, and $f(x)$ is a certain quadratic function. Besides,

Barcilon made plausible that, for large classes of weight functions $f(x)$, extremal solutions are made up of separated point masses (we shall call them degenerate strings). In a recent note, we wrote down necessary conditions for which $M[\rho]$ may be a local minimum in the class of degenerate strings and we showed that $f(x)$ cannot be chosen arbitrarily. If the extremal property is meant to be used to characterize a degenerate solution for any (allowed) truncated data set, $f(x)$ has to be a polynomial of degree 2. Here we first recall these results, then we explicitly calculate the extremal values of $M[\rho]$. We shall then generalize these results to certain other problems.

The general problem

We study the eigenvalue problems

$$w_n'' + \nu_n \rho w_n = 0 \qquad n = 1,2,\ldots \tag{2}$$

$$\left.\begin{aligned} w_n(0)\cos\alpha - w_n'(0)\sin\alpha &= 0 \\ w_n(1)\cos\gamma + w_n'(1)\sin\gamma &= 0 \end{aligned}\right\} \text{(even } n) \tag{3}$$

$$\left.\begin{aligned} w_n(0)\cos\beta - w_n'(0)\sin\beta &= 0 \\ w_n(1)\cos\gamma + w_n'(1)\sin\gamma &= 0 \end{aligned}\right\} \text{(odd } n) \tag{4}$$

where $0 \leq \alpha < \beta \leq \frac{\pi}{2}$ $\;0 \leq \gamma \leq \pi/2$, and $\rho(x)$ is made up by a finite number of point masses :

$$\rho(x) = \sum_{j=1}^{J} m_j\, \delta(x - x_j) \qquad (m_j \geq 0) \tag{5}$$

For a density of this form, the equation (2) is solved in such a way that a Dirac measure at x_i is included if the upper bound of the integral is x_i. This means that (2) is equivalent to the equations :

$$w(x) - w(0) - xw'(0) = -\nu \sum_i m_i w(x_i) \int_0^x \theta(t-x_i)dt \qquad (6)$$

$$w'(x) - w'(0) = -\nu \sum_i m_i w(x_i) \theta(x-x_i)$$

where θ is the "Heaviside function". These equations are consistent with the continuity conditions that should be fixed for w et w' across a point mass, i.e. w continuous and $w'(x_i^+) - w'(x_i^-) = -\nu m_i w(x_i)$.

This problem has exactly 2J different eigenvalues, which can be organized in an increasing sequence $\nu_1, \nu_2, \ldots, \nu_{2J}$. Let us be given an increasing sequence of 2N numbers $\nu_i^{(o)}$ $(i = 1,\ldots,2N)$, and let us call $\mathbf{C}_o$ the set of functions $\rho(x)$ which are of the form (5) and yield $\nu_i = \nu_i^{(o)}$ for $i = 1,2,\ldots,2N$. Since one can choose any $J \geq N$, and any value ν_n for $2N < n \leq J$, $\mathbf{C}_o$ is either void or contains an infinity of elements. It is convenient to define provisionnally a very large number J such that all densities we consider are of the form (5), with vanishing or non vanishing m_j's. Then one can define a distance between two densities, e.g. by $d(\rho,\rho') = \sup_{1\leq j\leq J} \{|x_j-x'_j|, |m_j-m'_j|\}$. In the metric space K_J of all functions like (5), the definition of $\mathbf{C}_o$ is achieved by imposing what we shall call the measurements constraint $\nu_i = \nu_i^{(o)}$ $(i=1,2,\ldots,2N)$, and the positivity constraint $(m_j \geq 0)$.

The extremal problem

We now recall the results we previously obtained[3]. Let us assume that a degenerate density ρ^o yields a local minimum of the function (1). Thus ρ^o is such that there exists $R > 0$ with $M[\rho] \geq M[\rho^o]$ for any ρ that belongs to the intersection of $\mathbf{C}_o$ and the ball $B(\rho^o,R)$. For the sake of convenience, we set $\mathbf{C}_o \subset K_J$,

$x_1^{(o)} \neq 0$, $x_J^{(o)} \neq 1$. Now, for $\rho \neq \rho^o$, if $d(\rho,\rho^o)$ is small enough, for each eigenvalue ν^o corresponding to ρ^o, it is possible to find an eigenvalue ν corresponding to ρ and such that $|\nu-\nu^o| \to 0$ as $d(\rho,\rho^o) \to 0$ [Hint : Transform (2) into integral equations by using the Green's function of $(w'' = 0)$ and conditions (3) or (4). For $\rho \in K_J$, these equations are algebraic and their characteristic polynomials, whose roots are eigenvalues, depend continuously on ρ for the distance in K_J]. The wronskian relation yields

$$\int_0^1 (\nu\rho - \nu^o\rho^o)\, w^o w \, dx = 0 \tag{8}$$

The measurements constraint imposes the value of ν_n for $n = 1,\ldots,2N$. Therefore, all ρ's in $C_o \cap B$ must satisfy $\int_0^1 (\rho-\rho^o)\, w_n^o\, w_n \, dx = 0$ for $n = 1,2,\ldots,2N$. Letting $d(\rho,\rho^o) \to 0$, and taking care of the continuity conditions, we obtain the first order "measurement constraints" on ρ :

$$0 = \sum_{i=1}^{J} \{\delta m_i\, [w_n(x_i)]^2 + m_i\, \delta x_i\, w_n(x_i)\, [w_n'(x_i^-) + w_n'(x_i^+)]\}$$

$$\text{for } n = 1,2,\ldots,2N \tag{9}$$

where δm_i can have any sign if $m_i \neq 0$, and can only be positive if $m_i = 0$. The ℓ.h.s in (9) define 2N linear forms on the space R_{2J} of vectors with components δm_i, δx_i. Equaling these forms to zero defines a manifold in R_{2J}. In the following, we call S^o the set of abscissas x_i with $m_i \neq 0$ in the definition (5) of ρ^o, with card $S^o = J'$, $N \leq J' \leq J$. It is convenient to define K_J in such a way that J' abscissas can vary freely, whereas the $J-J'$ ones are fixed in a set T, with $S^o \cap T = \emptyset$. For ρ in $B(\rho^o,R)$, $|\delta x_i|$ and $|\delta m_i|$ are all smaller than R. Hence, we can find R so small that for each couple (x_i^o, m_i^o), there exists for ρ a couple (x_i, m_i), such that the points of T stay apart from the intervals $|x_i - x_i^o| < R$. It is also convenient to arrange the abscissas of S^o, or the corresponding ones for ρ, in an increasing sequence labelled from 1 to J', $x_1 < x_2 < \ldots < x_{J'}$, and the ones which are in T also in an increasing sequence, labelled from $x_{J'+1}$ to x_J, and to set :

$$m_i w_n(x_i)\,[w_n'(x_i^+) + w_n'(x_i^-)] = \alpha_i(n) \qquad (i = 1,\ldots,J')$$

$$[w_n(x_i)]^2 = \alpha_{J'+i}(n) \qquad (i = 1,\ldots,J)$$

$$\delta x_i = \delta_i \quad (i = 1,\ldots,J') \qquad \delta m_i = \delta_{J'+i} \quad (i = 1,\ldots,J)$$

so that (9) reads

$$\sum_{i=1}^{2J'} \alpha_i(n)\,\delta_i + \sum_{2J'}^{J'+J} \alpha_i(n)\,\delta_i = 0 \qquad n = 1,2,\ldots,2N \tag{10}$$

Now we have been able to prove[3] the following results :

(a) The system $\alpha_i(n)$ for $n = 1,\ldots,2N$, and $i = 1,2,\ldots,2J'$ (ie for the abscissas of the "heavy" points) is of rank 2N (Hint : this follows from an expansion lemma that is proved in our previous paper). Let $\mathcal{U}_1$ be a set of indices i, with card $\mathcal{U}_1 = 2N$, such that $\alpha_i(n)$, with $i \in \mathcal{U}_I$, is invertable. Let $\mathcal{U}_2$ be the set of remaining indices that correspond to heavy points (card $\mathcal{U}_2 = 2J'-2N$), and $\mathcal{U}_3$ the set that correspond to points of T.

(b) In the limit $R \to 0$, the condition $M[\rho] - M[\rho^o] \geq 0$, when only linear terms are written, reduces to

$$\sum_{\mathcal{U}_1} f_i\,\delta_i + \sum_{\mathcal{U}_2} f_i\,\delta_i + \sum_{\mathcal{U}_3} f_i\,\delta_i \geq 0 \tag{11}$$

(c) The "linearized" problem is defined by imposing that the inequality (11) holds for any set δ_i such that $|\delta_i| < R$

for $i \in \mathcal{U}_I + \mathcal{U}_2$, $0 \leq \delta_i \leq R$ for $i \in \mathcal{U}_3$, and (10) is satisfied. This problem can be dealt with by a variant of the Lagrange's method. It follows from (a) that we can determine a set of 2N Lagrange multipliers such that

$$i \in \mathcal{U}_1 \qquad f_i = \sum_{n=1}^{2N} F_n \alpha_i(n) \tag{12}$$

and thus the extremal problem reduces to imposing that for any $|\delta_i| < R$ $(i \in \mathcal{U}_2)$, $0 \leq \delta_i \leq R$ $(i \in \mathcal{U}_3)$, the inequality holds :

$$\sum_{i\in\mathcal{U}_2} [f_i - \sum_{n=1}^{2N} F_n \alpha_i(n)]\delta_i + \sum_{i\in\mathcal{U}_3} [f_i - \sum_{n=1}^{2N} F_n \alpha_i(n)]\delta_i \geq 0 \tag{13}$$

(b) The non linear problem can be shown to reduce to the linear one in the limit $R \to 0$, in such a way that the necessary conditions obtained from the linear problem are exact necessary conditions.

(e) Hence the necessary conditions to be satisfied at a minimum are

$$\left.\begin{aligned} f(x_i) &= \sum_1^{2N} F_n w_n^2(x_i) \quad (i = 1,2,\ldots,J') \\ f'(x_i) &= \sum_1^{2N} F_n w_n(x_i) [w_n'(x_i^+) + w_n'(x_i^-)] \quad (i=1,2,\ldots,J') \end{aligned}\right\} \text{(14) Heavy Points}$$

$$f(x) \geq \sum_1^{2N} F_n w_n^2(x) \quad \text{for any } x \notin S_o$$

From these conditions it follows that unless $f(x)$ is a 2nd degree polynomial, with f and $f' > 0$, there exists minimum problems of the type we described that have no solution in the class (5).

Extremal values

We now introduce "Jost solutions" of the equation

$$w'' + \nu\rho w = 0 \tag{15}$$

which are defined to be solutions of (15) that satisfy in addition the following "initial" condition :

$$\Omega_1(\nu,0) = 1 \quad ; \quad \Omega_1'(\nu,0) = \text{cotg}\,\beta \tag{16}$$

$$\Omega_2(\nu,0) = 1 \quad ; \quad \Omega_2'(\nu,0) = \text{cotg}\,\alpha \tag{17}$$

$$F(\nu,1) = 1 \quad ; \quad F'(\nu,1) = -\,\text{cotg}\,\gamma \tag{18}$$

$$G(\nu,1) = \sin\gamma\cos\gamma \;;\; G'(\nu,1) = \sin^2\gamma \tag{19}$$

We introduce also the "Jost functions"

$$\mathcal{F}_1(\nu) = W(F,\Omega_1) = \Omega_1'(\nu,1) + \text{cotg}\,\gamma\,\Omega_1(\nu,1) = F(\nu,0)\,\text{cotg}\,\beta - F'(\nu,0) \tag{20}$$

$$\mathcal{F}_2(\nu) = W(F,\Omega_2) = \Omega_2'(\nu,1) + \text{cotg}\,\gamma\,\Omega_2(\nu,1) = F(\nu,0)\,\text{cotg}\,\alpha - F'(\nu,0) \tag{21}$$

$\mathcal{F}_1$ and $\mathcal{F}_2$ are also coefficients in the expansion of Ω_1 and Ω_2 along F and G :

$$\Omega_1 = \mathcal{G}_1 F + \mathcal{F}_1 G \quad ; \quad \Omega_2 = \mathcal{G}_2 F + \mathcal{F}_2 G \tag{22}$$

where

$$\begin{aligned} \mathcal{G}_1 &= [W(\Omega_1,G) = \sin\gamma\cos\gamma\,[\Omega_1(\nu,1)\text{tg}\,\gamma - \Omega_1'(\nu,1)]] \\ \mathcal{G}_2 &= [W(\Omega_2,G) = \sin\gamma\cos\gamma\,[\Omega_2(\nu,1)\text{tg}\,\gamma - \Omega_2'(\nu,1)]] \end{aligned} \tag{23}$$

Now it follows from (6) and (7) that $w(x)$ is a polynomial in ν of $d^o \leq k$ for $x_k \leq x \leq x_{k+1}$. Thus $\Omega_{1,2}(\nu,1)$ are polynomials of d^o J' and so are $\mathcal{F}_1(\nu)$, $\mathcal{F}_2(\nu)$, $\mathcal{G}_1(\nu)$, $\mathcal{G}_2(\nu)$. It follows from (22) that $\mathcal{F}_1$ and $\mathcal{G}_1$ do not vanish simultaneously and that the zeros of $\mathcal{F}_1(\nu)$ are $\nu_1, \nu_3, \ldots, \nu_{2J'-1}$, whereas the zeros of $\mathcal{F}_2(\nu)$ are $\nu_2, \nu_4, \ldots, \nu_{2J'}$. Since $\mathcal{F}_1(0)$ and $\mathcal{F}_2(0)$ are exactly calculated from (20) and (21), we obtain :

$$\mathcal{F}_1(\nu) = (\text{cotg}\,\beta + \text{cotg}\,\gamma + \text{cotg}\,\beta\,\text{cotg}\,\gamma) \prod_{j=1}^{j=J'} \left(1 - \frac{\nu}{\nu_{2j-1}}\right) \tag{24}$$

$$\mathcal{F}_2(\nu) = (\text{cotg}\,\alpha + \text{cotg}\,\gamma + \text{cotg}\,\alpha\,\text{cotg}\,\gamma) \prod_{j=1}^{j=J'} \left(1 - \frac{\nu}{\nu_{2j}}\right) \tag{25}$$

So as to calculate $\mathcal{G}_1(\nu)$ and $\mathcal{G}_2(\nu)$, we first derive from (22) the formula

$$W(\Omega_1,\Omega_2) = \text{cotg } \alpha - \text{cotg } \beta = (\mathcal{G}_1 \mathcal{F}_2 - \mathcal{G}_2 \mathcal{F}_1) \tag{26}$$

which uniquely determines the polynomials $\mathcal{G}_1$ and $\mathcal{G}_2$ (Bezout theorem). Indeed, for $\nu = \nu_{2n}$, $\mathcal{G}_2(\nu)$ is determined from (26), whereas $\mathcal{G}_2(0)$ follows from the definition. $\mathcal{G}_2(\nu)$ can then be written down as the interpolation polynomial :

$$\mathcal{G}_2(\nu) = \mathcal{G}_2(0) + \lambda \sum_{n=1}^{J'} \frac{\mathcal{G}_2(\nu_{2n})\, \Pi_2(\nu)}{(\nu-\nu_{2n})\, \nu_{2n}\, \Pi_2'(\nu_{2n})} \tag{27}$$

where

$$\begin{cases} \Pi_2(\nu) = \prod_{j=1}^{J'} (\nu-\nu_{2j}) \\ \mathcal{G}_2(\nu_{2n}) = -\,[\mathcal{F}_1(\nu_{2n})]^{-1}\, (\text{cotg } \alpha - \text{cotg } \beta) \\ \mathcal{G}_2(0) = \sin^2 \gamma\, [1 + \text{cotg } \alpha - \text{cotg } \alpha \text{ cotg } \gamma] \end{cases} \tag{28}$$

By the same token, we obtain $\mathcal{G}_1(\nu)$:

$$\mathcal{G}_1(\nu) = \mathcal{G}_1(0) + \nu \sum_{n=1}^{J'} \frac{\mathcal{G}_1(\nu_{2n-1})\, \Pi_1(\nu)}{(\nu-\nu_{2n-1})\, \nu_{2n-1}\, \Pi_1'(\nu_{2n-1})} \tag{29}$$

where

$$\begin{cases} \Pi_1(\nu) = \prod_{j=1}^{J'} (\nu-\nu_{2j-1}) \\ \mathcal{G}_1(\nu_{2n-1}) = (\text{cotg } \alpha - \text{cotg } \beta)\, [\mathcal{F}_2(\nu_{2n-1})]^{-1} \\ \mathcal{G}_1(0) = \sin^2 \gamma\, (1 - \text{cotg } \beta \text{ cotg } \gamma + \text{cotg } \beta) \end{cases} \tag{30}$$

We are now in position to calculate the extremal values of several quadratic moments of ρ that are of particular interest. We give

$$I_{11} = \int_0^1 (1 + x \operatorname{cotg} \beta)^2 \rho(x)\, dx \tag{32}$$

$$= \int_0^1 [\Omega_1 (0,x)]^2 dx = \lim_{\lambda \to 0} \{\mathcal{G}_1(\lambda)\, \mathfrak{F}_1(0) - \mathfrak{F}_1(\lambda)\, \mathcal{G}_1(0)\}$$

$$= \frac{\mathfrak{F}_1(0)\,(\operatorname{cotg}\alpha - \operatorname{cotg}\beta)}{\mathfrak{F}_2(0)} \sum_{n=1}^{J'} \frac{1}{\nu_{2n-1} \prod_{j=1}^{J'}{}^{*} \left(1 - \frac{\nu_{2n-1}}{\nu_{2j-1}}\right) \prod_{j=1}^{J'} \left(1 - \frac{\nu_{2n-1}}{\nu_{2j}}\right)}$$

where the * means that the zero factor is omitted.

$$I_{10} = \int_0^1 \rho(x)\, [1 + (1-x) \operatorname{cotg}\gamma]\, [1 + x \operatorname{cotg}\beta]\, dx$$

$$= \lim_{\lambda \to 0} \int_0^1 \Omega_1(0,x)\, F(\lambda,x)\, \rho(x)\, dx = - \mathfrak{F}_1'(0) \tag{33}$$

$$= [\operatorname{cotg}\beta + \operatorname{cotg}\gamma + \operatorname{cotg}\beta \operatorname{cotg}\gamma] \left[\frac{1}{\nu_1} + \frac{1}{\nu_3} + \ldots + \frac{1}{\nu_{2J'-1}}\right]$$

$$I_{00} = \int_0^1 \rho(x)\, [1 + (1-x) \operatorname{cotg}\gamma]^2 dx$$

$$= [\operatorname{cotg}\alpha - \operatorname{cotg}\beta]^{-1} \int_0^1 [\mathfrak{F}_2(0)\, \Omega_1(0,x) - \mathfrak{F}_1(0)\, \Omega_2(0,x)]\, F(0,x)\, dx$$

$$= [\operatorname{cotg}\alpha - \operatorname{cotg}\beta]^{-1} [\mathfrak{F}_1(0)\, \mathfrak{F}_2'(0) - \mathfrak{F}_2(0)\, \mathfrak{F}_1'(0)]$$

$$= \frac{[\operatorname{cotg}\alpha + \operatorname{cotg}\alpha \operatorname{cotg}\gamma + \operatorname{cotg}\gamma][\operatorname{cotg}\beta + \operatorname{cotg}\beta \operatorname{cotg}\gamma + \operatorname{cotg}\gamma]}{\operatorname{cotg}\alpha - \operatorname{cotg}\beta}$$

$$\left[\frac{1}{\nu_1} - \frac{1}{\nu_2} + \frac{1}{\nu_3} \quad \ldots\ldots\ldots\ldots \quad - \frac{1}{\nu_{2J'}}\right] \tag{34}$$

Needless to say, similar formulas also hold, in which α is to be replaced by β, 1 by 2, the odd sequence by the even sequence.

In all these formulas we see that the necessary conditions that we have imposed are also sufficient, and that making ∞ the eigenfrequencies for any $n > 2N$ indeed gives a smaller result.

Besides, we notice the very remarkable fact that I_{10} (and I_{20}, which has not be written) depend only on one spectrum. This strongly suggest generalisations to problems like the Earth Inverse Problem. In this wery important problem, the experimental information is given by the mode frequencies that correspond to only one set of boundary conditions. The equations can be considered as complicated generalizations of the string equation, and it is sound to look for extremal solutions, since they would yield upper or lower bounds of certain moments of physical quantities. However we shall not present results here.

The way we have introduced extremal properties suggest that they may hold only in the class of ρ's made of δ-functions. Actually, they hold in the class of all ρ's that are Stieltjes measures of the form

$$\rho(x) = dM(x) \qquad (35)$$

where $M(x)$ is a non decreasing bounded variation function. For such a ρ, the Jost solutions and the Jost functions (20), (21) are entire functions of ν, of order[4] not larger than $\frac{1}{2}$, so that they have a Hadamard's product of the form (24), (25), with $J' = \infty$, and the remaining part of the derivations hold.

Let us now study connected problems, which may have "non-linear" applications.

A curious one is the spectral problem which is defined by the equation

$$[(\frac{d^2}{dx^2} - a^2) + \nu\rho]w = 0 \qquad (36)$$

where a is positive, and ρ is a Stieltjes measure of compact support $\mathcal{B}$. The solutions of (36) that belong to $L_2(-\infty, +\infty)$ are asymptotically $\exp[-ax]$ for $x \to +\infty$ and $\exp[ax]$ for $x \to -\infty$.

Let I be an interval that contains $\mathcal{B}$, $I = [x_0, x_\infty]$. The boundary conditions at x_0 and x_∞ are

$$\begin{cases} w' = +a\, w & x = x_0 \\ w' = -a\, w & x = x_\infty \end{cases} \tag{37}$$

The spectral problem is that of solving the Fredholm equation, made with the Green's function of the $\rho = 0$ problem :

$$w(x) = \frac{\nu}{2a} \{e^{ax} \int_x^{\infty} e^{-at} \rho(t)\, w(t)\, dt + e^{-ax} \int_{-\infty}^{x} e^{at} \rho(t)\, w(t)\, dt\} \tag{38}$$

where the integral go in fact only to x_0 and x_∞. For a degenerate ρ, it reduces to the matrix-spectral problem :

$$w(x_i) - \frac{\nu}{2a} \sum_{j=0}^{J} e^{-a|x_i - x_j|} m_j\, w(x_j) = 0 \tag{39}$$

The Jost solutions $\Omega^{\pm}(\nu,x)$ being defined by their asymptotic behavior $e^{\mp ax}$ as $x \to \pm\infty$ respectively, we see that their wronskian $\mathcal{F}(\nu)$ is zero for the eigenvalues. Now, the Jost solutions are obviously given by solving the Voltera's equations

$$\Omega^{-}(\nu,x) = e^{ax} - \nu \int_{-\infty}^{x} \frac{\text{sh}\ a(x-t)}{a} \rho(t)\, \Omega^{-}(\nu,t)\, dt \tag{40}$$

$$\Omega^{+}(\nu,x) = e^{-ax} - \nu \int_{x}^{\infty} \frac{\text{sh}\ a(t-x)}{a} \rho(t)\, \Omega^{+}(\nu,t)\, dt \tag{41}$$

and their wronskian $W(\Omega^{+}, \Omega^{-})$ can be calculated anywhere, for instance as $x \to +\infty$, giving

$$\mathcal{F}(\nu) = 2a - \nu \int_{-\infty}^{+\infty} e^{-at} \rho(t)\, \Omega^{-}(\nu,t)\, dt \tag{42}$$

Solving (40) by iteration, we obtain a series $\sum_0^\infty \Omega_n^-$, with $\Omega_0^- = e^{ax}$ and

$$\Omega^-_{n+1}(\nu,x) = -\nu \int_{x_o}^{x} \frac{\text{sh } a(x-t)}{a} \rho(t)\, \Omega^-_n(\nu,t)\, dt \tag{43}$$

Thus, $|\Omega^-_{n+1}|$ satisfies the inequation

$$|\Omega^-_{n+1}(\nu,x)| < C|\nu| \int_{x_o}^{x} (x-t)\, d\, P(t)\, |\Omega^-_n(\nu,t)|\, dt$$

where $$P(x) = \int_{x_o}^{x} |\rho(t)|\, dt \tag{44}$$

so that $$|\Omega^-_{n+1}(\nu,x)| < \frac{[2(x-x_o)\, P(x)\, C|\nu|]^n}{(2n)!}$$

(<u>Hint</u> : the proof is made by induction, noticing that $(x-t)(t-x_o)^n$ is maximum for $t = (n+1)^{-1}(nx+x_o)$, and taking the maximum value out of the integral). Hence $\Omega^-(\nu,t)$ and $\mathfrak{F}(\nu)$ are entire functions of ν, of order less or equal to $\frac{1}{2}$. Besides, all the eigenvalues, ie the zeros of $\mathfrak{F}(\nu)$, are simple [if not, there would be two linearly independent solutions of the same equation (36) with the same homogeneous boundary conditions and thus with a zero wronskian]. Also, if ρ is positive, they are positive : if ν is a negative eigenvalue, one gets from (36), for the corresponding eigenfunction u, the impossible equality $a[u^2(x_\infty) + u^2(x_o)] = -\int_{x_o}^{x_\infty} [(a^2+|\nu^-|\rho)\, u^2+u'^2]\, dx$. Hence $\mathfrak{F}(\nu)$ has the representation

$$\mathfrak{F}(\nu) = 2a \prod_{j=1}^{\infty} \left(1 - \frac{\nu}{\nu_j}\right) \tag{45}$$

where we took into account the value of $\mathfrak{F}(0)$. Now, from (36) we can obtain

$$I = \lim_{\nu\to 0} \int_{-\infty}^{+\infty} \Omega^-(\nu,x)\, \Omega^+(0,x)\, \rho(x)\, dx = \int_{-\infty}^{+\infty} \rho(x)\, dx$$

$$= \lim_{\nu\to 0} \frac{\mathfrak{F}(\nu) - \mathfrak{F}(0)}{\nu} = -\mathfrak{F}'(0), \text{ and thus}$$

$$\int_{-\infty}^{+\infty} \rho(x)\, dx = 2a[\nu_1^{-1} + \nu_2^{-1} + \ldots + \nu_N^{-1} + \ldots] \tag{46}$$

Clearly, if the N first eigenvalues are fixed, the minimum mass distribution corresponds to infinite ν_j for $j > N$, and to a finite distribution of heavy points.

The analytic properties of $\mathfrak{J}(\nu)$ can be extended to cases where ρ has not a compact support. The proof which is given above holds for instance if a bounded variation function $P_\varepsilon(x)$ can be defined by

$$\int_{-\infty}^{x} e^{-\varepsilon t} |\rho(t)| \, dt = P_\varepsilon (x) \qquad (\varepsilon > 0) \qquad (47)$$

[Hint Replace in (43) $|\Omega_n^-(v,t)|$ by $4^n \varepsilon^{-n} [P_\varepsilon(x)]^n \exp[n\varepsilon+a]x \, [(2n)!]^{-1}$].

If $P_{-\varepsilon}(\infty) < \infty$, the corresponding results also hold for $\Omega^+(\nu,x)$, $\mathfrak{J}(\nu)$ is well defined by (42), and still is an entire function of order $\leq \frac{1}{2}$, with the representation (45). It follows that (46) holds for any ρ that is positive and has the properties we just defined. Hence, if we could write down an evolution equation for ρ such that the spectrum is invariant, the total mass would also be invariant. This remark leads us to study evolution equations that are associated to the string spectral problem (2) and to the special spectral problem (36). Unfortunately allowing ρ to be zero or to go to zero fast enough in the domain make these equations difficult to study, so that we shall limit here our study to a somewhat similar problem, but in which $\sigma = \rho^{-\frac{1}{2}}$ is the central quantity. Indeed, for $\rho > 0$, the spectral problem (36) is equivalent to the problem

$$S\,u + \nu\,u = 0 \qquad (48)$$

where $$S = \sigma\left(\frac{\partial^2}{\partial x^2} - a^2\right)\sigma$$

and $$u = \sigma^{-1} w = \rho^{\frac{1}{2}} w$$

Let us allow σ to depend on a parameter t. The Lax's method tells us that the spectrum is invariant if, for a certain antisymmetric operator B, the evolution of σ follows the law

$$[B,S] = \frac{\partial S}{\partial t} \equiv \sigma^{-1} \frac{\partial \sigma}{\partial t} S + S \sigma^{-1} \frac{\partial \sigma}{\partial t} \tag{49}$$

For $B = \frac{\partial}{\partial x}$, we readily get

$[B,S] = \frac{\partial \sigma}{\partial x} (\frac{\partial^2}{\partial x^2} - a^2) \sigma + \sigma(\frac{\partial^2}{\partial x^2} - a^2) \frac{\partial \sigma}{\partial x}$ and thus, the evolution equation simply yields the translations

$$\frac{\partial \sigma}{\partial t} = \frac{\partial \sigma}{\partial x} \tag{50}$$

For $B = AS + SA$, with A antisymmetric, so that B also is, we can write $[B,S] = [AS,S] + [SA,S] = (AS - SA)S + S(AS - SA)$.
Thus the evolution equation reads

$$AS - SA = \sigma^{-1} \frac{\partial \sigma}{\partial t} \tag{51}$$

with $A = \sigma \frac{\partial}{\partial x} + \frac{\partial}{\partial x} \sigma$, we obtain

$$\sigma^{-3} \frac{\partial \sigma}{\partial t} = - 4 a^2 \frac{\partial \sigma}{\partial x} + \frac{\partial^3 \sigma}{\partial x^3} = \frac{\partial}{\partial x} [\frac{\partial^2}{\partial x^2} - 4 a^2] \sigma \tag{52}$$

which is the simplest non trivial non linear evolution equation that keeps the spectrum invariant. It can also be written as

$$- \frac{1}{2} \frac{\partial \rho}{\partial t} = \frac{\partial}{\partial x} [\frac{\partial^2}{\partial x^2} - 4 a^2] \rho^{- \frac{1}{2}} \tag{53}$$

Clearly this equation is not very appropriate for a study of degenerate solutions. If ρ or ρ^{-1} is zero on an open set, one of the sides of the equation is infinite and the other is zero. If ρ tends to zero, ρ^{-1} tends to infinity so that even the smoothest case studied above does not fit. If it did, the fact that the total mass did depend only on eigenvalues would yield a simple conservation law. It is remarkable that this law remains true in the extremely opposite case where $\rho^{-\frac{1}{2}}$ is a Schwartz $\mathcal{D}$-function and, more generally, where $[\frac{\partial^2}{\partial x^2} - 4a^2]\rho^{-\frac{1}{2}}$ goes to the same constant as $x \to \pm\infty$.

Now, the equation (52) has a lot of interesting properties, which we shall tersely survey.

First, let us notice that "solitary wave" solutions of the form $\sigma = s(x-ct)$ can all be completely calculated, since s is then a solution of the solvable equation (in terms of hyperelliptic functions)

$$\frac{1}{2} s'^2 = -\frac{1}{2} c s^{-1} + \gamma + \beta s + 2a^2 s^2 \qquad (54)$$

where β and γ are arbitrary constants.

We can also find "separable" solutions of (52) or (53) of the form $\rho(x,t) = u(t)\, v(x)$, where u and v satisfy the equations

$$\begin{cases} -\frac{1}{2} u^{\frac{1}{2}}(t) \dfrac{du(t)}{dt} = \lambda \\ \dfrac{d}{dx}[\dfrac{d^2}{dx^2} - 4a^2]\, v^{-\frac{1}{2}}(x) = \lambda\, v(x) \end{cases} \qquad (55)$$

λ being the separation constant.

It is also sound to look for solutions that go to a constant as x goes to $\pm\infty$, so as to prove that (52) has non trivial smooth solutions.

Let $\sigma(x)$ be $b(1 + \varepsilon(x))$. If ε is always smaller than 1, we can hope in a "linearized form" of (52) such as :

$$\frac{\partial \varepsilon}{\partial t} = b^3 \frac{\partial}{\partial x} [\frac{\partial^2}{\partial x^2} - 4\, a^2]\, \varepsilon \qquad (56)$$

the true equation being obtained by adding to the left-hand side the (hopefully) small quantity $[(1-\varepsilon)^{-3} - 1]\frac{\partial\varepsilon}{\partial t}$. Needless to say, it is easy to study the iterated series by using the Riemann-Green's function of (56) and there are cases in which one can guarantee the convergence.

These equations (52)-(53) have a lot of other interesting properties. They can be transformed into integral equations by inverting the linear operator in the righthand side, and difference forms can be obtained in discrete cases. But the simplest way to study our original problem seems to forget about them and to study directly the spectral equation (39). In fact this equation readily furnishes all the constants of motion in an isospectral flow by simply writing that

$$\det [\lambda\ \delta_{ij} - \frac{1}{2a}\ e^{-a|x_i - x_j|}\ m_j]$$

is an invariant. This means for instance that the relative distances $|x_i - x_j|$ and the masses m_i are allowed to vary in such a way that the N symmetric functions of the roots of this characteristic polynomial are constants of motion. It is worth to notice that the first one yields the law of conservation of the total mass. The second one only refers to masses and to distances between neighbouring heavy points, etc.

The spectral problems we have studied are related to the Schrödinger problem by (complicated) transformations of functions and variables, so that these equations and problems are indeed related to the Korteweg de Vries equation.

Several points are still to be clarified in the connection between the discrete case and the continuous case, which are necessary to study the solutions neighbouring the extremal ones, a physically interesting problem. On the other hand, several results can easily be generalized and, in particular, it is clear that the equations we obtained are the simplest of a large class. But we shall not go further in the present lecture.

Acknowledgements

During the Lecce Meeting, Professor Calogero informed me that the $a = 0$ form of (53) has been introduced by Harry Dym and Kruskal some years ago, and that a research on the largest class of evolution equations associated with Sturm Liouville problems mixing the string form and the Schrödinger form has been recently undertaken by his own group.

It is also a pleasure to thank Doctor Matveev for very interesting discussions.

References :

(1) KREIN M.G. : On some case of effective determination of the density of an inhomogeneous cord from its spectral function. Dokl. Akad. Nauk SSSR 93, 617

(2) BARCILON V. : Ideal solution of an inverse normal mode problem with finite spectral data. Geophys. J. Roy. Soc. March 1979

(3) SABATIER P.C. : On extremal solutions of Sturm Liouville inverse problems in " Inverse and improlerly posed problems in differential equations " G. Anger Ed. Akademie-Verlag Berlin 1979

(4) KAC I. and KREIN M.G. : On the spectral functions of the string. Am. Math. Soc. Transl. II, 103, 19-102 (1974)

THE GENERALIZED RIEMANN-HILBERT PROBLEM AND THE SPECTRAL INTERPRETATION

D.V. Chudnovsky *

D.Ph.T.CEN-Saclay-91190

Gif-sur-Yvette, France

TABLE OF CONTENTS

---

* On leave from Dept. of Mathematics, Columbia University, New York, NY 10027, USA. This work was supported in part by Office of the Naval Research and National Science Foundations.

0. INTRODUCTION

This paper deals with several problems arising in the theory of evolutionary equations close to completely integrable. We consider two existing (and closely related) approaches to classical one, two and three-dimensional completely integrable systems. One, going back to Drach-Burchnall-Chaundy[1,15] and than Lax-Zakharov-Shabat[23,29], is based on isospectral deformation equations and uses well-developed for Schrödinger operator inverse scattering technique. Another method, used by L. and R. Fuchs[16] and Schlesinger[26] is the method of isomonodromy deformation for Fuchsian linear differential equations. The last method uses in its analytic part the Riemann boundary value problem. These two approaches are actually connected through the Riemann boundary value problem. The difference between them is actually the same as difference between equations with regular singularities (Fuchsian) and equations with irregular singularities. Indeed, the solution of inverse scattering problem (say, for Schrödinger operator) is the construction of Jost eigenfunction (asymptotic solution of the second order linear differential equation) with prescribed asymptotic behaviour.

We formulate below in §1 the isomonodromy deformation problem and present in §2 an effective solution of the Riemann-Hilbert problem (that belong to Lappo-Danilevski). We discuss then in §3 Schlesinger isomonodromy deformation equations (along with their derivation) and first approximation to the Schlesinger equations-system (A). System (A) is completely integrable Hamiltonian system and in §4 we show that any completely integrable one-dimensional isospectral deformation equation can be reduced to a system (A). In §§3-4 we defined also the special case of system (A): so-called Russian chain equivalent to stationary KdV equations.

In §5 we present, following scheme of Zakharov-Shabat, the most general two-dimensional isospectral deformation systems. The part 5.2 of §5 is devoted to operator generalizations of classical completely integrable systems. We study later special operator non-linear systems that are called to be of order two. These systems of order two are connected with inverse scattering problem for operator Schrödinger equation on a Hilbert space H. In §6 we study (following the results of Calogero-

Degasperis in the matrix case) scattering problem for operator Schrödinger equation. Then in §7 the most general class of operator isospectral equations of order two are dealt with. Examples (like operator KdV and mKdV equations) are given, and the most important is the operator non-linear Schrödinger equation. The stationary operator non-linear Schrödinger equation (operator Russian chain) introduced in 7.2 plays the key role in further studies. The §8 deals with first integrals of Operator Russian chain and decomposition theorems are proved. According to these theorems (x,t)-dimensional operator equations of order two for $U=U(x,t)$ are represented as common action of two commuting Hamiltonian one-dimensional flows for $(\Phi(\omega),\tilde{\Phi}(\omega))$, where $U=\int_\Omega \Phi(\omega)\tilde{\Phi}(\omega)\, d\sigma_\omega$. Commuting Hamiltonians appear from operator Russian chain $(\mathcal{OR})$ and Hamiltonian flows commuting with it. In §8 we prove also the representation $U=\int_\Omega \Phi(\omega)\tilde{\Phi}(\omega)\, d\sigma_\omega$ for a large class of operator potentials $U(x)$ and left and right operator eigenfunctions of the operator $(-d^2/dx^2 + U)$. In §9 we study both Dirac and Schrödinger operator equations (sometimes it's preferable to consider Dirac equation). Then in §10 we treat three-dimensional operator Zakharov-Shabat equations. E.g. for Kadomtsev-Petviashvili equation $3u_{yy} = \frac{\partial}{\partial x}(4u_t - 12uu_x + u_{xxx})$ we have $u=\int_\Omega \mathrm{tr}(\Psi_\lambda\tilde{\Psi}_\lambda)\, d\sigma_\lambda$ where Ψ_λ, $\tilde{\Psi}_\lambda$ satisfy operator Russian chain in x and commuting Hamiltonian flows in y and t.

I want to thank to Professors M.Boiti and G.Soliani for their kind invitation to this Conference and their warm hospitality.

1. MONODROMY GROUP OF FUCHSIAN LINEAR DIFFERENTIAL EQUATIONS AND THE RIEMANN PROBLEM IN ONE DIMENSIONAL CASE

We consider the Riemann monodromy problem in $\mathbb{C}\cup\{\infty\}$ only for Fuchsian linear differential equations (i.e. for linear differential equations with regular singularities).

The simplest example of Fuchsian matrix linear differential equation is the following system having simplest singularities:

$$\frac{d\bar{y}}{dx} = \sum_{i=1}^{n} \bar{y}\, \frac{A_i}{x-a_i} \quad , \tag{1.1}$$

where $\bar{y}=(y_1,\dots,y_m)$ and A_i are $m\times m$ constant matrices, $i=1,\dots,n$.

The system (1.1) is called normal (or canonical) form of Fuchsian linear differential equation. It is preferable to work with systems in normal form.

However we give the definition of the monodromy group for an arbitrary Fuchsian linear differential equation (1.2).

Let

$$\frac{d\bar{y}}{dx} = \bar{y}\,A(x) \quad , \quad \bar{y} = (y_1,\dots,y_m) \tag{1.2}$$

be a Fuchsian matrix linear differential equation for a matrix A(x) with rational elements. Let $\{a_1,\dots,a_n\}$ be the set of poles of A(x) in $\mathbb{C}\cup\{\infty\}$ and let Y=Y(x) be a fundamental matrix solution of (1.2). In general the fundamental matrix Y(x) is a multi-valued function having $a_1,\dots,a_n$ as its branch points and

$$Y(x) \longmapsto M(\gamma)\,Y(x)$$

when prolonged along closed curve γ. Here $M(\gamma)\in GL(m,\mathbb{C})$ and $M(\gamma)$ depends only on the homotopy class of γ in $\mathbb{P}\mathbb{C}^1\setminus\{a_1,\dots,a_n\}$. Then, of course,

$$M(\gamma_1\gamma_2) = M(\gamma_2)M(\gamma_1)$$

so the set $\mathfrak{M}=\{M(\gamma)\}$ is a group-monodromy group of the equation (1.2). This group is generated by the matrices $M_i=M(\gamma_i):i=1,\dots,n$, where γ_i is a clockwise circuit around a_i, which does not contain other singular points inside, and we have

$$M_1\dots M_n = I \quad . \tag{1.3}$$

Classical Riemann problem. Given branch points $a_1,\dots,a_n\in\mathbb{C}$ and matrices $M_1,\dots,M_n\in GL(m,\mathbb{C})$ find a linear differential equation (1.2) or, better, (1.1), whose monodromy group $\mathfrak{M}$ coincides with the group generated by $M_1,\dots,M_n$.

We also naturally demand in the *classical* Riemann problem for the fundamental matrix Y(x) of (1.2) to be at most regularly singular at the points $a_1,\dots,a_n$. This means that

$$Y(x) = (x-a_i)^{-L_i}\,H_i(x) \quad : i=1,\dots,n \quad , \tag{1.4}$$

(natural parameter near ∞ is $1/x$). In (1.4) $H_i(x)$ is an invertible holomorphic matrix at $x=a_i$ and L_i (the exponent) is a constant matrix such that

$$e^{2\pi i L_j} = M_j \quad : \quad j=1,\dots,n \ . \tag{1.5}$$

The reason that this problem appears as the "Riemann-Hilbert" problem is very simple: Riemann in 1857[30] was working with the system of functions satisfying the linear differential equation with a given monodromy group $\mathfrak{M}$. The only thing that was not proved by Riemann was the existence of the system of m functions analytic outside of $\{a_1,\dots,a_n\}$ and having a given monodromy group $\mathfrak{M}$.

It is unclear whether Riemann knew about the problem of existence or not. In any case, Riemann computed the dimensions of the set of Fuchsian linear differential equations with given singularities and the number of possible groups (more precisely the dimension of the possible monodromy matrices). On the basis of such computations he claimed that there are infinitely many systems (of functions) belonging to a given class (i.e. having a given monodromy group).

Hilbert proposed the proof of this statement as problem n.21 in his list and in 1912 gave[20] the first serious solution for m=2 on the basis of Riemann boundary value problem.

Since Riemann and Hilbert this problem was solved in different ways by Plemelj, Birkhoff (reduction to the singular integral equations); by Röhrl (for an arbitrary Riemann surface using fibre bundles); by Lappo-Danilevský[22] (using series expansions in hyperlogarithms).

From our present point of view the most important was the reduction of the linear Riemann problem to the non-linear "completely integrable" system of equations: Schlesinger's equations[26].

"Schlesinger's Theorem"[26]. Let $Y=Y(x_0;x)$ be the fundamental matrix solution of (1.1); $Y(x_0;x_0)=I$. The necessary and sufficient conditions for A_i : $i=1,\dots,n$ as the function of the parameters $a_1,\dots,a_n,x_0$ to have the fixed monodromy group of $Y(x)$ is the following completely integrable system of total differential equations

$$dA_j = -\sum_{i\neq j} \left[A_j,A_i\right] \, d\log\frac{a_j-a_i}{x_0-a_i} \quad : \quad j=1,\dots,n \ . \tag{1.6}$$

This Schlesinger system can be written in a classical form[26,25,18] :

$$\frac{\partial A_j}{\partial a_i} = [A_j, A_i]\left(\frac{1}{a_j - a_i} - \frac{1}{x_0 - a_i}\right) , \quad j \neq i$$

$$\frac{\partial A_i}{\partial a_i} = - \sum_{j \neq i} [A_i, A_j] \frac{1}{a_i - a_j} ,$$

$$\frac{\partial A_i}{\partial x_0} = \sum_{j \neq i} [A_i, A_j] \frac{1}{x_0 - a_j} \quad : \quad i=1,\ldots,n \quad . \tag{1.7}$$

This system of equations is the source of "classical" completely integrable systems like, e.g. stationary KdV equation.

There is, of course, a natural temptation to identify the deformation equations of the types (1.7) with the Lax type of isospectral deformation. This is not the case at all! The deformation equations (1.7) are reduced in the most interesting cases to the equations, differential with respect to a spectral parameter as it will be explained below.

Formal solution of the Riemann problem in terms of matrix exponents. Let now $a_1,\ldots,a_n$ and x_0 be distinct points on $\mathbb{P}\mathbb{C}^1$ and let $L_1,\ldots,L_n$ be m×m matrices satisfying the natural condition

$$e^{2\pi i L_1} \cdots e^{2\pi i L_n} = 1 \quad . \tag{1.8}$$

We consider the following precise version of the Riemann problem: Find a matrix $Y(x)$ with the properties

a) $Y(x)$ is a multivalued analytic matrix on $\mathbb{P}\mathbb{C}^1 \setminus \{a_1,\ldots,a_n\}$;

b) $Y(x) = (x-a_i)^{-L_i} H_i(x)$ at $x=a_i$ $(i=1,\ldots,n)$, where $H_i(x)$ is an invertible holomorphic matrix at $x=a_i$;

c) $\det Y(x) \neq 0$ for $x=a_1,\ldots,a_n$;

d) $Y(x_0) = 1$.

Such a matrix $Y(x)$ is unique (but exists not always). Lappo-Danilevski[22] have proved that for sufficiently small $|L_i|$: $i=1,\ldots,n$ (excluding $i=\infty$) such a matrix

$$Y = Y(x_0; x; {a_1,\ldots,a_n \atop L_1,\ldots,L_n})$$

exists and can be written as a series in L_i : $i=1,\ldots,n$.

Let us suppose now that $a_i \neq \infty$: $i=1,\ldots,n$, then the function $Y(y;x;{a_1,\ldots,a_n \atop L_1,\ldots,L_n})$ gives us simultaneously: 1) the Fuchsian linear differential system with the monodromy group, generated by $\{e^{2\pi i L_1},\ldots,e^{2\pi i L_n}\}$

and 2) the Schlesinger system of the equations (1.7) for the coefficients $A_1,\dots,A_n$ of (1.1).

We can write this for a function

$$Y = Y(y;x;{a_1,\dots,a_n \atop L_1,\dots,L_n})$$

as a linear total differential equation

$$dY = Y\Omega, \qquad (1.9)$$

$$\Omega = \sum_{i=1}^{n} A_i \, d\log\frac{x-a_i}{y-a_i} = \sum_{i=1}^{n} A_i \left(\frac{d(x-a_i)}{x-a_i} - \frac{d(y-a_i)}{y-a_i}\right)$$

where

$$A_i = A_i(y;{a_1\dots a_n \atop L_1\dots L_n}) = -H_i(a_i)^{-1}L_iH_i(a_i) \; : \; i=1,\dots,n \qquad (1.10)$$

are matrices independent of x satisfying

$$\sum_{i=1}^{n} A_i = 0 \quad .$$

The equation (1.9) means that Y as a function of x satisfies the Fuchsian system of linear ordinary differential equations

$$\frac{dY}{dx} = \sum_{i=1}^{n} Y \frac{A_i}{x-a_i} \quad .$$

Now the coefficients A_i : $i=1,\dots,n$ as functions of x_0 and $\bar{a}=(a_1,\dots,a_n)$ satisfy the Schlesinger's equations (1.6):

$$dA_j = -\sum_{j\neq i} \left[A_j,A_i\right] d\log\frac{a_i-a_j}{a_i-y} \; : \; j=1,\dots,n \; .$$

Recently Sato, Miwa, Jimbo[25] in a series of papers describe a quantum field theory approach of the construction of $Y(y;x;{a_1\dots a_n \atop L_1\dots L_n})$ and the solution of the Riemann problem. They represent Y and A_i in the terms of classical field theory operators and then apply such representations to the explicit expressions for the n-th correlation functions in the two-dimensional Ising model. It should be noted, however, that the series expansions for $Y(y;x;{a_1\dots a_n \atop L_1\dots L_n})$ proposed by Sato, Miwa, Jimbo are basically the same as in the papers of Lappo-Danilevsky[22].

2. LAPPO-DANILEVSKI SOLUTION OF THE RIEMANN PROBLEM

There exists an effective method for solving the Riemann problem and for the explicit construction of the fundamental solution in terms of the monodromy group and the singularities. This method, belonging to Poincaré and Lappo-Danilevski, uses matrix series in terms of polylogarithmic functions.

These polylogarithmic functions were introduced by Poincaré and studied by Lappo-Danilevski in the connection with the Fuchsian equation. However these functions were known already for a long time (especially dilogarithms were known to Euler and then studied e.g. by Abel and Lobachevsky).

In the *notation of Lappo-Danilevski* for a fixed x_0 we define these polylogarithmic functions by induction:

$$L_{x_0}(a_{j_1}|x) = \int_{x_0}^{x} \frac{dx}{x-a_{j_1}} = \log \frac{x-a_{j_1}}{x_0-a_{j_1}} ;$$

$$L_{x_0}(a_{j_1},\ldots,a_{j_\nu}|x) = \int_{x_0}^{x} \frac{L_{x_0}(a_{j_1},\ldots,a_{j_{\nu-1}}|x)}{x-a_{j_\nu}} dx .$$

Then we can define a fundamental matrix $Y_{x_0}(x)$ as follows:

$$\begin{aligned} Y_{x_0}(x) &= \Phi_{x_0}\left\{\begin{matrix} A_1,\ldots,A_n \\ a_1,\ldots,a_n \end{matrix}\middle| x\right\} \\ &= I + \sum_{\nu=1}^{\infty} \sum_{(j_1,\ldots,j_\nu)}^{1,\ldots,n} A_{j_1}\ldots A_{j_\nu} L_{x_0}(a_{j_1},\ldots,a_{j_\nu}|x) . \end{aligned} \tag{2.1}$$

The series defines an entire function of A_i and is uniformly convergent with respect to x in any finite domain $\mathcal{D}$ (of Riemann surface corresponding to cuts $(a_1,\infty);\ldots;(a_n,\infty)$) having no points a_j in $\mathcal{D}$ or on $\partial\mathcal{D}$-boundary of $\mathcal{D}$.

The series (2.1) is chosen in such a way that $Y_{x_0}(x)$ satisfies a linear differential equation with regular singularities at a_i and coefficients A_i:

$$\frac{dY_{x_0}}{dx} = \sum_{j=1}^{n} \frac{Y_{x_0} A_j}{x-a_j} \quad \text{and} \quad Y_{x_0}(x_0) = I . \tag{2.2}$$

It is now possible to rewrite the monodromy matrices M_i (or integral substitutions in the terminology of Lappo-Danilevski) in terms of matrix series of polylogarithms. For this let us consider a closed contour γ_i from x_0 traveling around a_i and returning back to x_0. We put

$$P_j(a_{j_1}|x_0) = \int_{\gamma_i} \frac{dx}{x-a_{j_1}} = \begin{cases} 2\pi i & : j=j_1 \\ 0 & : j\neq j_1 \end{cases}$$

$$P_j(a_{j_1},\ldots,a_{j_\nu}|x_0) = \int_{\gamma_i} \frac{L_{x_0}(a_{j_1},\ldots,a_{j_{\nu-1}}|x)}{x-a_{j_\nu}} dx .$$

An "integral substitution" M_i corresponding to a_i of a regular matrix Y normed at x_0 and having the coefficients of the differential equations A_j at singularities a_j : $j=1,\ldots,n$ is represented as an entire function of the coefficients $A_1,\ldots,A_n$:

$$M_j = I + \sum_{\nu=1}^{\infty} \sum_{(j_1,\ldots,j_\nu)}^{1,\ldots,n} A_{j_1}\ldots A_{j_\nu} P_j(a_{j_1},\ldots,a_{j_\nu}|x_0). \qquad (2.3)$$

Now $\{M_j\}$ give us the monodromy group of the equation (2.2). In order to look on the behaviour of the fundamental matrix $Y_{x_0}(x)$ of (2.2) defined in (2.1) it is better instead of monodromy matrices M_i to consider the corresponding exponents L_i : $i=1,\ldots,n$.

For this, following Lappo-Danilevski, suppose first that A_i : $i=1,\ldots,n$ are in the neighbourhood of a zero matrix.

Then the corresponding monodromy matrices $M_1,\ldots,M_n$ from (2.3) belong to the neighbourhood of unit matrix I and we can introduce "exponents" $L_1,\ldots,L_n$ as

$$L_j = \ell n\, M_j = \frac{1}{2\pi i} \sum_{\nu=1}^{\infty} \frac{(-1)^{\nu-1}}{\nu} (M_j-I)^\nu \qquad (2.4)$$

and by (2.3) they are holomorphic functions of $A_1,\ldots,A_n$ in that neighbourhood of a zero matrix : $j=1,\ldots,n$.

Moreover, Lappo-Danilevski have shown how to present L_j as an entire function of $A_1,\ldots,A_n$ divided by another entire function of $A_1,\ldots,A_n$, i.e. a meromorphic function. Moreover, it follows that L_j as a function of A_i has a singularity iff among the eigenvalues of A_ν there are such two that their difference is a non-zero integer.

Now in terms of the "exponents" L_j : $j=1,\ldots,n$ (2.4) we can present precisely the monodromy properties of the fundamental solution $Y_{x_0}(x)$.

We have for $Y_{x_0}(x)$ from (2.1):

$$Y_{x_0}(x) = \Phi_{x_0}\begin{Bmatrix} A_1,\ldots,A_n \\ a_1,\ldots,a_n \end{Bmatrix} x = \tag{2.5}$$

$$= \left(\frac{x-a_j}{x_0-a_j}\right)^{L_j} \tilde{\Phi}_{x_0}^{(j)}\begin{Bmatrix} A_1,\ldots,A_n \\ a_1,\ldots,a_n \end{Bmatrix} x \quad : \; j=1,\ldots,n,$$

where

$$\tilde{Y}_{x_0}^{(j)}(x) = \tilde{\Phi}_{x_0}^{(j)}\begin{Bmatrix} A_1,\ldots,A_n \\ a_1,\ldots,a_n \end{Bmatrix} x \tag{2.6}$$

and $\left[\tilde{Y}_{x_0}^{(j)}(x)\right]^{-1}$ are holomorphic with respect to x at $x=a_j$: $j=1,\ldots,n$. If now $A_1,\ldots,A_n$ are in the neighbourhood of the zero matrix, then there exist *no other* exponents L_j, different from those given by formula (2.4) with the property that in (2.5)-(2.6) both

$$\left[\tilde{Y}_{x_0}^{(j)}(x)\right] \text{ and } \left[\tilde{Y}_{x_0}^{(j)}(x)\right]^{-1}$$

are holomorphic at $a_j=x$: $j=1,\ldots,n$.

In other words, the fundamental matrix $Y_{x_0}(x)$ defined in (2.1) has indeed regular singularities at $x=a_j$: $j=1,\ldots,n$ and exponents $L_1,\ldots,L_n$ from (2.4) and is, what we called before

$$Y(x_0;x;{a_1,\ldots,a_n \atop L_1,\ldots,L_n}) \quad .$$

We now observe that the formalism of Lappo-Danilevski also solves the Riemann monodromy problem. Instead of constructing the fundamental solution $Y_{x_0}(x)$ and "exponents" L_j as meromorphic function of the coefficients $A_1,\ldots,A_n$ and $a_1,\ldots,a_n$; we can invert the problem and represent the coefficients $A_1,\ldots,A_n$ of the differential equation (2.2) and the fundamental solution $Y_{x_0}(x)$ of (2.2) in terms of composition (matrix) series in exponents $L_1,\ldots,L_n$ and $a_1,\ldots,a_n$. (So-called "effective" solution of Riemann-Hilbert problem).

This problem is solved by an inversion of the previous series. We start with the exponent matrices $L_1,\ldots,L_n$ in the neighbourhood of zero matrix. Then inverting the previous series we obtain the following representations for the coefficients $A_1,\ldots,A_n$ (again close to a zero matrix):

$$A_j(x_0) = \sum_{\nu=1}^{\infty} \sum_{(j_1,\ldots,j_\nu)}^{(1,\ldots,n)} L_{j_1}\ldots L_{j_\nu} R_j(a_{j_1},\ldots,a_{j_\nu}|x_0) \quad . \tag{2.7}$$

Here the coefficients $R_j(a_{j_1},\dots,a_{j_\nu}|x_0)$ are defined by induction in terms of the values of the polylogarithmic functions $P_j(\dots|\dots)$:

$$R_j(a_{j_1}|b) = \delta_{jj_1},$$

$$R_j(a_{j_1},\dots,a_{j_\nu}|x_0) = -\sum_{\mu=2}^{\infty}\ \sum_{(h_1,\dots,h_\mu)}^{(1,\dots,n)}\ \sum_{1\le k_1<\dots<k_{\mu-1}<\nu} \cdot$$

$$\cdot R_{h_1}(a_{j_1},\dots,a_{j_{k_1}}|x_0)\, R_{h_2}(a_{j_{k_1+1}},\dots,a_{j_{k_2}}|x_0)\dots$$

$$\dots R_{h_\mu}(a_{j_{k_{\mu-1}+1}},\dots,a_{j_\nu}|x_0)\, Q_j(a_{h_1},\dots,a_{h_\mu}|x_0),$$

for

$$Q_j(a_{j_1},\dots,a_{j_\nu}|x_0) = \frac{1}{2\pi i}\sum_{\mu=1}^{\nu}\ \sum_{0<k_1<\dots<k_{\mu-1}<\nu} \frac{(-1)^{\mu-1}}{\mu}\cdot$$

$$\cdot P_j(a_{j_1},\dots,a_{j_{k_1}}|x_0)\dots P_j(a_{j_{k_{\mu-1}+1}},\dots,a_{j_\nu}|x_0).$$

The series (2.7) are holomorphic functions of the exponents $L_1,\dots,L_n$ in the neighbourhood of zero matrix. Substituting (2.7) into (2.1), one now obtains effectively the whole fundamental solution

$$Y(x_0;x;{}^{a_1,\dots,a_n}_{L_1,\dots,L_n})$$

of the Riemann-Hilbert monodromy problem (together with the corresponding equations (2.2)).

3. SCHLESINGER AND SIMPLIFIED SCHLESINGER ISOMONODROMY DEFORMATION EQUATIONS

Isospectral deformation equations are the condition of consistency of two or more differential equations that are alebraic in spectral parameter λ. This is so called Zakharov-Shabat-Mikhailov general scheme (see below), which has been given by Burchnall and Chaundy[1] in one dimensional case.

The isomonodromy equations also arise as a consistency condition of differential equations but being now differential in spectral parameter λ as well. This is more general approach than the isospectral one.

We rigorously show below in §4 that in one dimensional case isomonodromy equations contain as a very particular case all the isospectral. We present the deduction of Schlesinger and simplified Schlesinger systems (A). These systems (A) are, in fact, equivalent to one-dimensional isospectral deformation equations.

Let us start as in §1 with the system of linear differential equations of the order m with n+2 singularities $t_1,\dots,t_{n+2}$ (where we can assume $t_{n+1}=0$, $t_{n+2}=1$):

$$\frac{dy_k}{dx} = \sum_{j=1}^{m} y_j \sum_{i=1}^{n+2} \frac{A^i_{jk}}{x-t_i} \quad : \quad k=1,\dots,m \ . \tag{3.1}$$

We assume that the equation (3.1) is Fuchsian and the monodromy group $\mathfrak{M}$ of (3.1) is independent on t_i $(i=1,\dots,n)$. Then the quantities A^i_{jk} satisfy the Schlesinger system of nonlinear non-autonomous equations[26]:

$$\frac{\partial A^j_{hk}}{\partial t_i} = \sum_{\ell=1}^{m} \frac{A^i_{h\ell}A^j_{\ell k} - A^j_{h\ell}A^i_{\ell k}}{t_j - t_i} \qquad j\neq i;\ h,k=1,\dots,m \tag{3.2}$$

$$\sum_{j=1}^{n+2} \frac{\partial A^j_{hk}}{\partial t_i} = 0 \quad : \quad i=1,\dots,n \ . \tag{3.3}$$

In order to obtain the system (3.2)-(3.3) we consider simply the evolution of y in t_i as

$$\frac{\partial y}{\partial t_i} = y\,\beta_i \quad : \quad i=1,\dots,n \ .$$

Together with the original equation (3.1)

$$\frac{\partial y}{\partial x} = ya = y\beta_0 \ ,$$

for $t_0=x$. Then the condition of consistency gives us the following equations

$$\frac{\partial \beta_i}{\partial t_j} - \frac{\partial \beta_j}{\partial t_i} = \left[\beta_i,\beta_j\right] \ . \tag{3.4}$$

Now let us look at the behaviour of y near t_j:

$$y = \left[(x-t_j)^r\right] z \ ,$$

where z is holomorphic near $x=t_j$. Then

$$\beta_i = z^{-1} \frac{\partial z}{\partial t_i} ,$$

and we can find that $\beta_i + a$ is a holomorphic function everywhere, i.e.

$$\beta_i = - \frac{A^i}{x-t_i} + \gamma_i \quad : \quad i=1,\ldots,n+2 ,$$

where γ_i are now independent of x. We can also make in (3.1) a transformation

$$y = Yc$$

for c independent on x in such a way that

$$\frac{\partial c}{\partial t_i} = c\gamma_i \quad \text{or} \quad \frac{\partial \gamma_i}{\partial t_j} - \frac{\partial \gamma_j}{\partial t_i} = [\gamma_i, \gamma_j] .$$

In other words we can assume

$$\beta_i = - \frac{A^i}{x-t_i} ,$$

and the system (4.4) is reduced to the form

$$\frac{\partial A^j}{\partial t_i} = [A^i, A^j] / (t_j - t_i) \quad : \quad i \neq j \; ; \quad \sum_{j=1}^{m} \frac{\partial A^j}{\partial t_i} = 0 \tag{3.5}$$

and the system (3.5) is naturally equivalent to (3.2)-(3.3).

We still do not know too much about such systems: we do not know whether all such systems are Hamiltonian, whether they are indeed completely integrable, etc.

There is, however, one very special case where we know a rather simple and nice answer to all such questions. This special case corresponds to the situation when we treat the Sclesinger system (3.5) in *the first approximation* in t_i. Such simplified systems were first introduced by Painlevé[24], who showed that Painlevé transcendents under a transformation like

$$t_i = a_i + \varepsilon t'_i , \qquad w_i = \varepsilon^{-1} w'_i ,$$

transform into a classical elliptic function. Then these investigations were continued by R.Garnier[18] (1916,1919), who considered many examples (especially m=2 or n=1).

In the Schlesinger system (3.5) we make the following transformation

$$t_i \to a_i + t_i \; ; \qquad A^i \to \varepsilon^{-1} A^i \quad : \quad i=1,\ldots,n+2 ,$$

for constants a_i, where, in particular, $a_{n+1}=a_{n+2}=1$. If we put $\varepsilon\to 0$ then the system (3.5) takes the following form

$$\frac{\partial A^j}{\partial t_i} = \frac{[A^i, A^j]}{a_j - a_i} \quad : \quad j\neq i$$
$$\sum_{j=1}^{n+2} \frac{\partial A^j}{\partial t_i} = 0 \quad : \quad i=1,\ldots,n \quad . \tag{A}$$

As we shall see the system (A) is indeed completely integrable and, moreover, can be solved in Abelian integrals (of the first and second kind) corresponding to a certain algebraic curve.

Moreover, *any* completely integrable one-dimensional system connected with the algebra of commuting differential operators can be represented in a form (A) (see the next chapter). We call the system (A) "simplified Schlesinger equations".

Let us investigate the system (A). In (3.1) we have

$$\frac{dy_k}{dx} = \sum_{k=1}^{m} a_{hk}y_h \text{ for } a_{hk} = \sum_{i=1}^{n+2} \frac{A^i_{hk}}{x-t_i} \tag{3.6}$$

Now in the simplified system (A) we put

$$a^o_{hk} = \sum_{i=1}^{n+2} \frac{A^i_{hk}}{x-a_i} \quad : \ h,k=1,\ldots,m \quad .$$

Let

$$\varphi(x) = \prod_{i=1}^{n+2} (x-a_i) = x(x-1) \prod_{i=1}^{n} (x-a_i) \ ,$$

and we define

$$a^o_{hk} = \frac{b_{hk}(x)}{\varphi(x)} \ , \tag{3.7}$$

for some polynomials $b_{hk}(x)$: $h,k=1,\ldots,m$ of degree $\leq n+1$ in x and $b_{hk}(x)$ depending on t_i (because A^j depend on t_i).

If we put

$$b = (b_{hk}(x)) \qquad h,k=1,\ldots,m$$

then the system of the equations (A) is equivalent to

$$\frac{\partial b}{\partial t_i} = \frac{[A^i, b]}{x - a_i} \quad : \quad i=1,\ldots,n. \tag{3.8}$$

One of the main results concerning the system (A) has the following form:

Theorem 3.1 - Let us define the following algebraic curve

$$f(x,y) = \det (b_{hk}(x) + y\delta_{hk}) = 0 \qquad h,k=1,\dots,m . \qquad (3.9)$$

Then the coefficients of $f(x,y)$ are the first integrals of the system (A), i.e. for any solution A^i of (A) the coefficients of $f(x,y)$ are constants.

On other hand, for an algebraic curve $f(x,y)=0$ we can always find a $b_{hk}(x)$ such that $f(x,y)=0$ can be represented in a form (3.9).

Proposition 3.2 [18] - For any $f(x,y)$ having the form

$$f(x,y) = y^m + \sum_{i=1}^{m} y^{m-i} P_{i(n+1)}(x) \qquad (3.10)$$

for polynomials $P_{i(n+1)}(x)$ of degree $\leq i(n+1)$ there are $b_{hk}(x)$ of degree $\leq n+1$ such that

$$f(x,y) = \det (b_{hk}(x) + y\delta_{hk}) \quad , \qquad h,k=1,\dots,m .$$

If the theorem (3.1) enables us to show that the system (A) is completely integrable ant to integrate it in terms of Abelian functions, then proposition (3.2) gives us the possibility to reduce any classical one-dimensional completely integrable system to a simplified Schlesinger equation (A).

Example 3.3 (Garnier) - Let us consider an arbitrary hyperelliptic curve of the genus $m-1$: $m\geq 2$. We can present such a curve in the most general form

$$f(x,y) = x^2 P(y) + x Q(y) + R(y) \qquad (3.11)$$

for polynomials $P(y)$, $Q(y)$, $R(y)$ of degrees $m-2$, $m-1$ and m, respectively. Here we put $n=1$ (i.e. singularities at $0,1,t$ but, possibly, some apparent singularities). Now we can put

$$b_{11}(x)=x^2+C_{11}; \; b_{hk}(x)=A_{hk}\cdot x+C_{hk} \quad : \quad (h,k) \neq (1,1) .$$

We can always transform b in such a way that $A_{hk}=0$ for $h>1$ or $k>1$. Looking at $b_{hk}(x)$ corresponding to the equation (3.11) we deduce from the equation (3.8), or the system (A), that

$$a_1(a_1-1)\frac{dA_{1k}}{dt_1} = -(A_{1k}a_1 + C_{1k})$$

$$a_1(a_1-1)\frac{dA_{h1}}{dt} = A_{h1}\,a_1 + C_{h1}$$

where putting

$$\tau = \frac{t_1}{a_1(a_1-1)} \quad ,$$

we see that

$$A_{1k} = f_k e^{-a_1\tau}, \qquad A_{h1} = g_h e^{a_1\tau}$$

$$C_{1k} = -f_k' e^{-a_1\tau}, \qquad C_{h1} = g_h' e^{a_1\tau}$$

(' is the derivative on τ). Next

$$a_1(a_1-1)\frac{dC_{hk}}{dt_1} = C_{h1}A_{1k} - A_{h1}C_{1k}$$

or for $h>1$ or $k>1$

$$C_{hk}=f_k g_h + C_{hk} \quad \text{and} \quad -C_{11} = f_2g_2+\dots+f_mg_m \quad .$$

From C'_{1h} and C'_{k1} we obtain

$$\begin{cases} f_i'' = \lambda_i f_i + 2f_i \sum\limits_{j=2}^{m} f_j g_j \\ g_i'' = \lambda_i g_i + 2g_i \sum\limits_{j=2}^{m} f_j g_j \qquad : \quad i=2,\dots,m \ . \end{cases} \qquad (R)$$

Here $\lambda_i=c_{ii}$ and we can consider (c_{kk}) to be reduced to a diagonal form.

This system (especially in the case of m=3) appears in many different works. E.g. in 1870 (m=3) it was considered by K.Neumann. Since 1880 it was been the subject (for m=3 again) of investigation for Russian mathematicians: Kowalevskaya, Tchapligin, Kolossoff, Goriachov, so we decided to call the system (R) the "Russian chain").

4. ALGEBRA OF COMMUTING DIFFERENTIAL OPERATORS AND SIMPLIFIED SCHLESINGER EQUATIONS

Let us describe results on the algebra of commuting differential operators and corresponding completely integrable systems. We start with two linear differential operators in d/dx:

$$L_1 = \sum_{i=0}^{n} u_i \frac{d^i}{dx^i} , \qquad L_2 = \sum_{j=0}^{m} v_j \frac{d^j}{dx^j} ,$$

where $u_n=v_m=1$, $u_{n-1}=v_{m-1}=0$. Then we can consider the following non-linear system of differential equations on u_i, v_j :

$$[L_1, L_2] = 0 \tag{4.1}$$

being equivalent to a condition of consistency of a system

$$L_1\psi = \lambda\psi , \qquad L_2\psi = \mu\psi . \tag{4.2}$$

The basic result here is

Theorem 4.1 (Burchnall-Chaundy[1]). An equation (4.1) is equivalent to an algebraic relation

$$Q(L_1, L_2) = 0$$

for a polynomial $Q(x,y)$. Here

1. For eigenfunctions ψ in (4.2) corresponding to the solution of (4.1), $Q(\lambda,\mu)=0$;
2. The coefficients of $Q(x,y)$ are the first integrals of the system (4.1); they can be expressed as a differential polynomials in u_i, v_j and are constant for any given solution of (4.1).

The case of relatively prime m and n is especially simple. In this case $Q(x,y)$ has the form $Q(x,y)=x^m-y^n+\dots$ and the space $\mathscr{L}_\lambda$ of ψ satisfying (4.2) is one-dimensional. In this case we have again

Theorem 4.2 (Burchnall-Chaundy[1], 1928). In the case of a one-dimensional $\mathscr{L}_\lambda$ (the space of ψ in (4.2)), in particular, for $(n,m)=1$ the system (4.1) is completely integrable. In this case the solutions of (4.1) (the coefficients u_i, v_j of L_1, L_2) are expressed in terms of a θ-function on $Q(x,y)=0$.

For any fixed curve $Q(x,y)=x^m-y^n+\dots$ the algebras $\mathscr{A}$ of commuting differential operators $\{L_1,L_2\}$ satisfying $Q(L_1,L_2)=0$ is naturally isomorphic to the Jacobian Jac(Q) of $Q(x,y)=0$.

For not relatively prime m and n or $\dim\mathscr{L}_\lambda=r>1$ the description of L_1, L_2 can be effectively given in terms of algebraic vector bundles of dimension r over $Q(x,y)$.

The case $m=2$, $n=2k+1$-odd number is especially interesting. The equations

$$\left[L_2, L_{2k+1}\right] = 0$$

for $L_2 = -\frac{d^2}{dx^2} + u$ have the form

$$Q(u, u_x, \ldots, u_{\underbrace{x\ldots x}_{2k+1}}) = 0$$

and are called stationary higher Korteweg-de Vries equations (k-th KdV) (Ref.7 and 19). The reason for such a name is very simple. If we consider the Korteweg-de Vries equation

$$u_t = 6uu_x - u_{xxx}$$

then it can be written in the form of a Lax representation

$$\frac{dL_2}{dt} = \left[L_2, L_3\right]$$

for $L_3 = -4\frac{d^3}{dx^3} + 3(u\frac{d}{dx} + \frac{d}{dx}u)$. Now the KdV have infinitely many first integrals (conserved quantities)

$$I_n = \int P_n(u, u', \ldots)dx$$

and the corresponding commuting Hamiltonian flows

$$u_t = \frac{\partial}{\partial x} \cdot \frac{\delta I_n}{\delta u}$$

Each of these Hamiltonian systems is equivalent to a Lax representation (Ref 17 and 19):

$$\frac{dL_2}{dx} = \left[L_2, A_{2n+1}\right] \quad .$$

Now, the stationary system

$$\left[L_2, A_{2n+1}\right] = 0$$

is

$$\frac{\delta I_n}{\delta u} = c$$

and is an n-th stationary KdV equation for the function u(x).

We showed[8-11] that a general n-th stationary KdV equation is equivalent to the following completely integrable Hamiltonian system on n parameters $\lambda_1, \ldots, \lambda_n$

$$f''_i = \lambda_i f_i + 2f_i \sum_{j=1}^{n} f_j^2 \quad : \quad i=1,\ldots,n . \tag{4.3}$$

But the system (4.3) is simply "Russian chain" corresponding to a simplified Schlesinger equation (A) with 3 singularities: 0,1,x and of the order n+1. In this case for L_1, L_2 in (4.1) we have a hyperelliptic curve $Q(x,y)$:

$$Q(x,y) = x^2 - P_{2n+1}(y) \quad ,$$

which can be reduced to a f(x,y) above.

This particular result can be generalized to a:

Theorem 4.3 - For any relatively prime m and n any system (4.1) with $\mathrm{ord}(L_1)=n$, $\mathrm{ord}(L_2)=m$ can be represented as the system (A):-"simplified Schlesinger equations". In this case the curves $Q(x,y)=0$ and $f(x,y)=0$ are birationally equivalent.

Moreover we can prove certain results in the case $(m,n)\neq 1$ and in the case of matrix differential operators as well.

Theorem 4.4 - For any m and n the system (4.1) can be represented as the system (A).

Theorem 4.5 - Let $\hat{L}_1, \hat{L}_2$ be matrix differential operators of size k×k:

$$\hat{L}_1 = \sum_{i=0}^{n} U_i \frac{d^i}{dx^i} \quad , \qquad \hat{L}_2 = \sum_{j=0}^{m} V_j \frac{d^j}{dx^j}$$

where $U_n=\mathrm{diag}(a_1,\ldots,a_k)$; $V_m=\mathrm{diag}(b_1,\ldots,b_k)$ and $U_{n-1,\alpha\beta}=0$ if $a_\alpha=a_\beta$; $V_{m-1,\alpha\beta}=0$ if $b_\alpha=b_\beta$.

Then the non-linear equation

$$[\hat{L}_1, \hat{L}_2] = 0$$

can be reduced to a form (A).

The order of (A) and the number of singularities of the system (A) are determined now by a genus and degree of $Q(x,y)=0$.

Of course, the Schlesinger approach is more efficient than the classical Burchnall-Chaundy-Lax-Novikōv-... approach[1,23,30], since it can be generalized to great extent to an arbitrary dimension (cf.14).

We should notice that the additional variable x in the Schlesinger system is not a variable t (the position of the singularity). The variable t is a "real" variable in a system (A) but now plays the role of a *spectral* parameter. Such an analogy can be continued further and we can indeed show that the system

$$\frac{d\bar{y}}{dx} = \sum_{i=1}^{n+2} \bar{y} \frac{A^i}{x-t^i}$$

can be considered as to be some auxiliary system of differential linear equations, where x *is* a spectral parameter for the Lax representation

$$L_1\psi = x\psi \quad ; \quad L_2\psi = y\psi \quad ; \quad Q(x,y) = 0$$

of the simplified Schlesinger system (A).

5. TWO-DIMENSIONAL COMPLETELY INTEGRABLE ISOSPECTRAL DEFORMATION SYSTEMS AS CONDITION OF COMMUTATIVITY.

5.1 - Scalar case.

The most general approach to the completely integrable system is based on the representation of the system of non linear equation as condition of commutativity of linear differential operators in partial derivatives. The first such representation belongs to Burchnall and Chaundy (Ref.1). We present the corresponding idea along the lines of Zakharov-Shabat-Mikhailov[28,29] approach. In (x,t)-space-time dimensional case, $x_1=x$, $x_2=t$ we introduce light-cone variables $2\xi=t-x$, $2\eta=t+x$.

The most general commutativity condition in (x,t) have the following form

$$i\psi_\xi = U\psi \ , \quad i\psi_\eta = V\psi \tag{5.1}$$

for (complex) operators U,V and ψ depending on ξ,η. Then (5.1) is equivalent to

$$U_\eta - V_\xi + i[U,V] = 0 \tag{5.2}$$

The system (5.1)-(5.2) is gauge-invariant under the transformation

$$U \to \tilde{U} = f\,U\,f^{-1} + i\,f_\xi\,f^{-1}$$

$$V \to \tilde{V} = f\,U\,f^{-1} + i\,f_\eta\,f^{-1} \ , \quad \psi \to \tilde{\psi} = f\psi \ ,$$

for an arbitrary nonsingular operator function f.

The system (5.2) *itself* is trivially solvable if you don't put restrictions on U, V.

The system (5.2) becomes non-trivial if you introduce spectral parameter λ. For this we demand U and V to be rational (or, in general meromorphic) functions in λ.

We assume according to the Zakharov-Mikhailov-Shabat[28,14,29] scheme that U and V, as function of λ, have poles of fixed orders $m_1,\dots,m_n$ at fixed points $\lambda_1^0,\dots,\lambda_n^0$ of a λ-plane.

Then the system (5.2) becomes a "*completely*" *integrable system of non-linear equations on* the coefficients of U and V: i.e. on *the residues of* U *and* V *at* $\lambda = \lambda_j^0$. The classical Lax representation is imbedded into this scheme, if you consider both U and V to have poles at one fixed λ_0, with orders of poles at $\lambda=\lambda_0$ for U and V being equal, respectively, to the orders of operators L and A.

In the simplest case, when $U(\lambda)$, $V(\lambda)$ have each only one pole, we put

$$U = U_0 + \frac{U_1}{\lambda + 1}\ , \qquad V = V_0 + \frac{V_1}{\lambda - 1} \tag{5.3}$$

where U_0, U_1, V_0, V_1 are now independent of λ. If we now choose the gauge f in which

$$\tilde{U}_0 = 0\ , \quad \tilde{V}_0 = 0\ , \quad \tilde{U}_1 = A\ , \quad \tilde{V}_1 = -B$$

we obtain very important for applications system of equations[12,28]:

$$A_\eta = \frac{i}{2}\,[A,B]\ , \qquad B_\xi = -\frac{i}{2}\,[A,B]\ . \tag{5.4}$$

Equation (5.4) is equivalent to the condition of consistency for the system

$$i\psi_\xi = \frac{A}{\lambda+1}\psi\ , \qquad i\psi_\eta = -\frac{B}{\lambda-1}\psi\ . \tag{5.5}$$

The system (5.4) leads to field theories, e.g. to σ models, connected with Lie Groups[28]. Let us suppose that we have at any point (ξ,η) some element $g(\xi,\eta)$ of the Lie group G, where G is considered as subgroup of group of complex matrices.

We discuss the equations of the motion

$$g_{\xi\eta} = \frac{1}{2}\,(g_\xi g^{-1} g_\eta + g_\eta g^{-1} g_\xi) \tag{5.6}$$

and the corresponding action

$$S = \int d\xi d\eta \frac{1}{2} T_2\left(\frac{\partial}{\partial\xi} g \frac{\partial}{\partial\eta} g^{-1}\right) .$$

The equation (5.6) can be reduced to the form (5.4). We define:

$$A = ig_\xi g^{-1} , \qquad B = ig_\eta g^{-1} . \tag{5.7}$$

Then in the notation (5.7), our system (5.6) is equivalent to the system (5.4):

$$A_\eta - B_\xi - i[A,B] = 0 , \qquad A_\eta + B_\xi = 0.$$

From a physical (and mathematical) point of view it is reasonable to restrict ourselves to the two most important cases

$$G = SO(N) \quad \text{and} \quad G = SU(N) .$$

General commutativity conditions can also be used to write three-dimensional equations which are suspicious for "complete integrability".

Three-dimensional systems that are of Zakharov-Shabat type. They can be written as

$$\frac{\partial}{\partial y} L_n - \frac{\partial}{\partial t} L_m = [L_m, L_n]$$

for matrix linear differential operators L_n, and L_m with coefficients being functions of x,y,t. Such a Zakharov-Shabat system is the condition of commutativity

$$L_m\psi = \frac{\partial\psi}{\partial y} , \qquad L_n\psi = \frac{\partial\psi}{\partial t} .$$

In the scalar case ($n=2$, $L_2=-d^2/dx^2+u$, L_m is of the third order) a so-called Kadomtsev-Petviashvili equation is obtained

$$3u_{yy} = \frac{\partial}{\partial x}(4u_t + u_{xxx} - 12uu_x) .$$

This will be treated later in this paper.

5.2 - Operator generalizations of isospectral deformation equations.

One can put in correspondence with any equation of Zakharov- Mikhailov-Shabat type an operator one. The most general of which is the condition of compatibility of two linear problems

$$\frac{\partial\psi}{\partial\xi} = U(\lambda)\psi , \qquad \frac{\partial\psi}{\partial\eta} = V(\lambda)\psi ,$$

where $U(\lambda)$ and $V(\lambda)$ are rational functions in the λ-plane. In the classical cases $U(\lambda)$ and $V(\lambda)$ are $N\times N$ matrices and all such systems are characterized by: i) the position λ_j^0 of the poles of $U(\lambda)$ and $V(\lambda)$ that are fixed; and ii) the orders of the poles of $U(\lambda)$ and $V(\lambda)$ at $\lambda=\lambda_j^0$.

In order to replace the classical system by an operator one $U(\lambda)$ and $V(\lambda)$ must be changed to be operators keeping the same poles λ_j^0 and the order of the poles to be fixed.

In general, in the quantum case there also appears the necessity to rearrange the order of terms in $U(\lambda)$ and in $V(\lambda)$ and in the equations themselves. Hence, in each concrete case this must be done separately.

5.3 - *Two-dimensional operator systems of order two.*

Here operator systems associated with isospectral deformation in the simplest case of *order two* are considered.

There are two equivalent definitions:

1) The class of equations of order two consist of non-linear operator systems having Lax representation

$$\gamma\frac{dL}{dt} = [A,L] \ ,$$

where A and L are linear differential operators in d/dz with operator coefficients such that one of the two following situations holds

i) $L=a\cdot\frac{d}{dx}+u$, where a and u are operators and $a^2=b$ is a projection operator, identical on its range; and

ii) $L=b\frac{d^2}{dx^2}+v\frac{d}{dx}+u$, where again b is either $I(=id)$ or is a projection operator, identical on its range.

2) If two-dimensional non linear operator equations arising from Zakharov-Mikhailov-Shabat scheme

$$\psi_{x_1}(x_1,x_2,\lambda) = U(\lambda)\psi \quad ;$$

$$\psi_{x_2}(x_1,x_2,\lambda) = V(\lambda)\psi \qquad \text{are given.}$$

Then the corresponding non-linear equation

$$U_{x_2} - V_{x_1} + [U,V] = 0$$

is considered to be of order two if each of the operators $U(\lambda)$, $V(\lambda)$ have only *one* pole in the λ-plane, the orders of which poles can be arbitrary.

In this class of equations is the majority of publicly known (non classified) two dimensional completely integrable systems: KdV, mKdV, non linear Schrödinger, sine-Gordon, chiral fields, σ-models, Gross-Neveu model, Thirring model, etc.

The theory of matrix Lax systems of order two was developed by Gelfand-Dikij[19] who introduced the Hamiltonian two dimensional structure on these kinds of systems and gave recurrent formulae for conservation laws. However, some simple way to present both equations of order two and to directly connect them with the evolution of scattering data is necessary.

In order to develop this simple method, it is best to use a certain single integro-differential operator associated with potential U(x). This integro-differential operator was introduced by Ablowitz, Kaup, Newell and Segur[35] and by Calogero and Degasperis[2,3] for the matrix case. This integro-differential operator denoted $\mathscr{L}$, is closely connected to the resolvent of the operator Schrödinger equation

$$\frac{d^2\psi}{dx^2} = U\psi - k^2\psi \ ;$$

$$\frac{d^2\tilde{\psi}}{dx^2} = \tilde{\psi}U - k^2\tilde{\psi} \ ;$$

moreover $\mathscr{L}$ is characterized by a very simple property:

$$\mathscr{L}\psi\tilde{\psi} = - k^2\psi\tilde{\psi} \quad ,$$

i.e. $\mathscr{L}$ has "squares" of eigenfunctions of Schrödinger as its own eigen functions. Using this single operator it is possible to write down all isospectral equations of order two.

6. SCATTERING DATA FOR AN OPERATOR SCHRÖDINGER EQUATION

In the scheme of inverse scattering for the operator Schrödinger equation on H , there is little different from an ordinary case. The classical operator approach in the arbitrary Banach space belongs to Krein, Berezanskij, Nizhnik, Levitan[31,4,32] and in the matrix case from the point of view of inverse scattering to Calogero, Degasperis, Wadati, Kamijo[2,3].

Assuming that U(x) is an operator on H with $\|U(x)\|$ vanishing asymptotically at an exponential rate or faster (as $|x|\to\infty$).

Consider the Jost functions: operator solutions of (two) Schrödinger equations with the potential U(x) corresponding to the continuous spectrum:

$$\psi_{xx}(x,k) = U(x)\psi(x,k) - k^2\psi(x,k) \tag{6.1}$$

and

$$\tilde{\psi}_{xx}(x,k) = \tilde{\psi}(x,k)U(x) - k^2\tilde{\psi}(x,k) \quad ; \quad k\geq 0 \tag{6.2}$$

together with the boundary conditions:

$$\psi(x,k) \to T(k)\exp(-ikx) \quad : \quad x\to-\infty , \tag{6.3}$$

$$\psi(x,k) \to \exp(-ikx) + R(k)\exp(ikx) \quad : \quad x\to+\infty , \tag{6.4}$$

and

$$\begin{aligned}&\tilde{\psi}(x,k) \to \tilde{T}(k)\exp(-ikx) \quad : x\to-\infty ;\\ &\tilde{\psi}(x,k) \to \exp(-ikx) + \tilde{R}(k)\exp(ikx) : x\to+\infty .\end{aligned} \tag{6.5}$$

Here R(k) is the reflection operator and T(k) is the transmission operator and, of course, it follows

$$\tilde{R}(k) = R(k) . \tag{6.6}$$

The most important property of scattering data is the analytic continuation of R(k), T(k) in k-plane. Under the conditions of the fast asymptotic vanishing of U(x), both R(k), T(k) are meromorphic in the whole k-plane.

In this case R(k) may have N simple poles at the values x_j : j=1, ...,N. Then x_j^2:j=1,...,N are exactly discrete eigenvalues of the problem (6.1). Of course, x_j^2:j=1,...,N are also eigenvalues of (6.2) and the corresponding eigenfunctions follow

$$\begin{aligned}&\psi^{(j)}_{xx} = U(x)\psi^{(j)} - x_j^2\psi^{(j)}\\ &\overline{\psi}^{(j)}_{xx} = \overline{\psi}^{(j)}U(x) - x_j^2\overline{\psi}^{(j)} \quad : j=1,\ldots,N.\end{aligned} \tag{6.7}$$

Scattering data corresponding to the eigenvalue x_j^2 have the following form:

$$\lim_{k\to x_j} \{[k-x_j]\,R(k)\} = P_j \quad : \quad j=1,\ldots,N,$$

and the system

$$\mathscr{S} = \{R(k); x_j, P_j : j=1,\dots,N\}$$

is called the system of scattering data associated to the potential U(x).

Additionally in the case of nondegenerate eigenvalues x_j^2 all P_j are projector operators of rank one.

Now an inverse scattering method consisting of reconstruction of the potential U(x) from the scattering data $\mathscr{S}$ can be generalized in a normal manner.

The best tool for this is the Gelfand-Levitan equation.

First, the construction of the spectral operator (initial data associated with $\mathscr{S}$) is necessary:

$$F(y) = \frac{1}{2\pi} \int_{-\infty}^{\infty} R(k) \exp(iky)\, dk + \sum_{i=1}^{N} P_i \exp(-x_i y) \quad .$$

Then the Gelfand-Levitan equation as the Fredholm operator integral equation is written as

$$K(x,x_1) + F(x+x_1) + \int_{x}^{+\infty} K(x,z)F(x+x_1)\,dz = 0 \quad , \quad x_1 \leq x.$$

Now the potential U(x) is reconstructed from $K(x,x_1)$ in a very simple way:

$$U(x) = -2 \frac{d}{dx} K(x,x) \quad .$$

7. OPERATOR TWO-DIMENSIONAL NON-LINEAR SYSTEMS OF ORDER TWO

Consider now the most general two-dimensional operator evolutionary equation for U(x,t) associated with the operator Schrödinger equation:

$$\frac{d^2\psi}{dx^2} = U\psi - k^2\psi \; ; \quad \psi = \psi(x,t,k)$$
$$\frac{d^2\tilde{\psi}}{dx^2} = \tilde{\psi}U - k^2\tilde{\psi} \; ; \quad \tilde{\psi} = \tilde{\psi}(x,t,k) \quad . \tag{S}$$

Considering only those equations of order two for which evolution in t of the scattering coefficient R(k,t) of (S) *is linear*, now we introduce the integro-differential operator $\mathscr{L}^{2,3}$

$$4\mathscr{L}F(x) = F_{xx}(x) - 2\{U(x,t),F(x)\} + G\int_x^{+\infty} dx'F(x') \quad ,$$

$$GF(x) = \{U_x(x,t),F(x)\} + \left[U(x,t), \int_x^{+\infty} dx'\left[U(x',t),F(x')\right]\right] \quad .$$

The class of the equations above can be described by iterative applications of the operator $\mathscr{L}$, keeping in mind that

$$[A,B] = AB - BA \ , \qquad \{A,B\} = AB + BA \quad .$$

Theorem 7.1 – [2,3,35] For fixed entire functions $\alpha(z)$, $\beta(z)$ and fixed constant operators M,N, the following non linear operator evolutionary equation for U(x,t)

$$U_t(x,t) = \alpha(\mathscr{L})\left[N,U(x,t)\right] + \beta(\mathscr{L})\cdot G\cdot M$$

is equivalent to a linear differential equation for the scattering coefficient R(k,t) of (S):

$$R_t(k,t) = \alpha(-k^2)\left[N,R(k,t)\right] + 2ik\beta(-k^2)\{M,R(k,t)\} \quad .$$

Of course, this is not a complete picture of evolution for scattering data, since the evolution for discrete part of the spectrum is still unknown. The evolution of discrete part of the spectrum can be obtained from the evolution of the continuous spectrum and we have

$$x_{j,t} = 0 \quad : \quad j=1,\dots,N$$

$$P_{j,t} = \alpha(-x_j^2)\left[N,P_j\right] + 2ix_j\beta(-x_j^2)\{M,P_j\} \ .$$

Examples. 1) The operator KdV equation:

$$\Phi_t = \Phi_{xxx} + 3(\Phi_x\Phi + \Phi\Phi_x) \ .$$

Here M=I, $\beta(\mathscr{L})=\mathscr{L}$; $\alpha(\mathscr{L})=0$.

2) The operator modified KdV equation

$$\psi_t = \psi_{xxx} - 3(\psi_x\psi^2 + \psi^2\psi_x)$$

Here

$$U = \begin{pmatrix} \psi^2 & \psi_x \\ \psi_x & \psi^2 \end{pmatrix} \quad , \quad M = \begin{pmatrix} I & 0 \\ 0 & I \end{pmatrix} \ , \ \beta(\mathscr{L}) = \mathscr{L} \ ; \quad \alpha(\mathscr{L}) = 0 \ .$$

3) The operator nonlinear Schrödinger equation:

$$i\psi_t = \psi_{xx} + \psi\psi^\dagger\psi \quad \text{(suggested by A.Neveu)} .$$

Here

$$U = \begin{pmatrix} \psi\psi^\dagger & \psi \\ \psi^\dagger & \psi^\dagger\psi \end{pmatrix} , \quad N = \begin{pmatrix} I & 0 \\ 0 & -I \end{pmatrix} , \quad \beta(\mathscr{L}) = 0 , \quad \alpha(\mathscr{L}) = \mathscr{L} .$$

Operator equations 2) and 3) (but not 1)) give us multi(infinite-) component equations if you consider operators ψ of *rank one*.

We have equations of the following form:

consider H as $L^2(\Omega,d\mu)$ with the scalar product $(a,b)=\int_\Omega a(\omega)b(\omega)^* d\mu$. Then, as a particular case of the operator non-linear Schrödinger equation the following system is obtained

$$i\overline{\varphi}_{j,t} = \overline{\varphi}_{j,xx} + \sum_{i=1}^{k} c_i\overline{\varphi}_i\cdot(\overline{\varphi}_j,\overline{\varphi}_i) \quad : \quad j=1,\dots,k$$

for the arbitrary constants $c_1,\dots,c_k$ and the vectors $\overline{\varphi}_1(x,t),\dots,\overline{\varphi}_k(x,t)$ from H.

In particular, k=1 corresponds only to a multicomponent nonlinear Schrödinger equation:

$$i\overline{\varphi}_t = \overline{\varphi}_{xx} + \overline{\varphi}\cdot(\overline{\varphi},\overline{\varphi}) \quad .$$

In complete analogy we get a multicomponent modified KdV equation:

$$\overline{\varphi}_t = \overline{\varphi}_{xxx} - 3\overline{\varphi}_x\cdot(\overline{\varphi},\overline{\varphi}) - 3\overline{\varphi}\cdot(\overline{\varphi}_x,\overline{\varphi})$$

for $\overline{\varphi}=\overline{\varphi}(x,t)$ from $H=L^2(\Omega,d\mu)$.

The combination of these two equations gives us Kadomtsev-Petviashvili equation in (x,y,t)-dimensions (two-dimensional KdV).

In the case of finite-dimensional H, dim(H)=n, an n-component nonlinear Schrödinger equation is obtained:

$$i\varphi_{\ell,t} = \sum_{j=1}^{n} \varphi_j\psi_j\cdot\varphi_\ell - \varphi_{\ell,xx}$$

$$-i\psi_{\ell,t} = \psi_\ell\cdot\sum_{j=1}^{n}\varphi_j\psi_j - \psi_{\ell,xx} \quad : \quad \ell=1,\dots,n$$

or

$$i\varphi_{\ell,t} = \sum_{j=1}^{n} |\varphi_j|^2\cdot\varphi_\ell - \varphi_{\ell,xx} \quad : \quad \ell=1,\dots,n .$$

The most interesting is the case of the stationary equation, where

$$\varphi_\ell = \varphi_\ell(x)\, e^{i\lambda_\ell t} \quad , \quad \psi_\ell = \psi_\ell(x)\, e^{-i\lambda_\ell t}$$

Then

$$\begin{cases} \varphi_{\ell,xx} = \sum_{j=1}^{n} \varphi_j\psi_j\cdot\varphi_\ell + \lambda_\ell\varphi_\ell , \\ \psi_{\ell,xx} = \sum_{j=1}^{n} \varphi_j\psi_j\cdot\psi_\ell + \lambda_\ell\psi_\ell \quad : \ell=1,\dots,n \end{cases}$$

with arbitrary $\lambda_1,\dots,\lambda_n$.

This system was called in §3 the Russian Chain ($\mathcal{R}$).

The Operator Russian Chain as a particular case of the Stationary Operator NonLinear Schrödinger equation.

Consider now the coupled Operator Non-linear Schrödinger

$$i\Phi_t = \Phi\psi\Phi - \Phi_{xx}$$

$$-i\psi_t = \psi\Phi\psi - \psi_{xx} \quad ,$$

and its stationary solutions

$$\Phi(x,t) = \Phi(x)e^{i\Lambda t}$$

$$\psi(x,t) = e^{-i\Lambda t}\,\psi(x)$$

for a constant Λ. Then the coupled stationary non-linear Schrödinger is:

$$-\Phi\cdot\Lambda = \Phi\psi\Phi - \Phi_{xx}$$

$$-\Lambda\cdot\psi = \psi\Phi\psi - \psi_{xx} \quad .$$

Now consider H as $L^2(\Omega,d\sigma)$-extension of H_0: $H= \oplus_{L^2(\Omega,d\sigma)} H_0$ and elements $\alpha(\omega)$, $\beta(\omega)\in L^2(\Omega,d\sigma)$, such that $\int_\Omega \alpha(\omega)\beta(\omega)d\sigma=1$. Taking $\lambda(\omega)$ as a measurable function and defining operators $\Phi(\omega)$, $\psi(\omega)$ over H_0, the operators Φ and ψ over H can be taken in the form:

$$\text{for } \varphi\in H, \quad (\Phi\cdot\varphi)(\omega) = \alpha(\omega)\cdot\int_\Omega \Phi(\omega_1)\varphi(\omega_1)d\sigma \ ;$$

$$(\psi\cdot\varphi)(\omega) = \psi(\omega)\cdot\int_\Omega \beta(\omega_1)\varphi(\omega_1)d\sigma \ ;$$

$$(\Lambda\cdot\varphi)(\omega) = \lambda(\omega)\varphi(\omega) \quad .$$

I.e. Φ and ψ in H are of rank one over H_0. Thus the stationary non-linear Schrödinger equation can be reduced from H to H_0 in the following multi-component form:

$$\Phi(\omega)_{xx} = \int_\Omega \Phi(\omega_1)\psi(\omega_1)d\sigma_1\cdot\Phi(\omega) + \lambda(\omega)\cdot\Phi(\omega) \ ;$$

$$\psi(\omega)_{xx} = \psi(\omega)\cdot\int_\Omega \Phi(\omega_1)\psi(\omega_1)d\sigma_1 + \lambda(\omega)\cdot\psi(\omega) \ .$$

In the particular case $\Omega\subseteq\mathbb{R}$, $\lambda(\omega)=-\omega^2$, the result is an Operator Russian Chain.

8. OPERATOR STATIONARY NONLINEAR SCHRÖDINGER EQUATION (OPERATOR RUSSIAN CHAIN)

Consider now the following monster: an infinite component operator Stationary Nonlinear Schrödinger equation (operator Russian Chain) derived in the previous page:

$$\begin{aligned} \Phi_{\lambda,xx} &= \int \Phi_\mu \tilde{\Phi}_\mu d\sigma_\mu \cdot \Phi_\lambda - \lambda^2 \Phi_\lambda \\ \tilde{\Phi}_{\lambda,xx} &= \tilde{\Phi}_\lambda \cdot \int \Phi_\mu \tilde{\Phi}_\mu d\sigma_\mu - \lambda^2 \tilde{\Phi}_\lambda \ . \end{aligned} \qquad (OR)$$

In other words, Φ_λ and $\tilde{\Phi}_\lambda$ are (left and right) eigenfunctions corresponding to the eigenvalue λ^2 of the operator Schrödinger equation with the potential

$$U(x) = \int \Phi_\mu(x) \cdot \tilde{\Phi}_\mu(x) d\sigma_\mu \ .$$

However from now on (OR) is considered to be (an infinite-dimensional) Hamiltonian system.

Assuming first that Φ_λ belongs to (a Banach) algebra of operators, and secondly that $d\sigma_\mu$ is a measure on $\mathbb{R}$ (or $\mathbb{C}$). Moreover, rather restrictive conditions on U and Φ_λ shall be imposed such that

$$U = \int \Phi_\mu \tilde{\Phi}_\mu d\sigma_\mu$$

is to be a bounded operator (for example, in analogy with the classical inverse scattering U(x) can vanish exponentially on infinity). Furthemore, all Φ_μ, $\tilde{\Phi}_\mu$ and

$$\int \Phi_\mu^2 \, d\sigma_\mu \quad , \qquad \int \tilde{\Phi}_\mu^2 \, d\sigma_\mu$$

are demanded to be bounded operators.

Now we are able to supply you with the conservation laws for (OR):

Theorem 8.1 - For the system (OR) we have the following first integrals:

1) $K[\lambda] = \tilde{\Phi}_{\lambda x} \Phi_\lambda - \tilde{\Phi}_\lambda \Phi_{\lambda x}$,

2) $$C[\lambda] = \tilde{\Phi}_{\lambda x} \Phi_{\lambda x} + \lambda^2 \tilde{\Phi}_\lambda \Phi_\lambda - \frac{1}{2} \tilde{\Phi}_\lambda \cdot \int \Phi_\mu \tilde{\Phi}_\mu d\sigma_\mu \cdot \Phi_\lambda + \frac{1}{2} \int \frac{1}{\lambda^2 - \eta^2} [\tilde{\Phi}_{\lambda x} \Phi_\eta - \tilde{\Phi}_\lambda \Phi_{\eta x}][\tilde{\Phi}_{\eta x} \Phi_\lambda - \tilde{\Phi}_\eta \Phi_{\lambda x}] d\sigma_\eta \ .$$

Moreover, all the first integrals $C[\lambda]$ are in involution.

Here we consider elements of $\Phi_{\lambda x}$ as the conjugate variables to the corresponding elements of Φ_λ (if one views Φ, $\tilde{\Phi}$ as infinite matrices).

In particular we have first integrals of the system $(\mathcal{OR})$ in a traditional form; for an arbitrary constant operator $S(\lambda)$ the following Hamiltonian is in involution with $(\mathcal{OR})$:

$$\mathcal{H}_S = \int tr_H\{S(\lambda)C[\lambda]\}dS_\lambda \quad .$$

In particular, on the solutions of $(\mathcal{OR})$,

$$\mathcal{H}_S = \text{const.}$$

Also the Hamiltonian of $(\mathcal{OR})$ has the form

$$\mathcal{H} = \frac{1}{2}\int tr_H C[\lambda]d\sigma_\lambda \quad .$$

The most important class of Hamiltonians $\mathcal{H}_S$ is the class of "momental" Hamiltonians defined as

$$\mathcal{H}_{n,M} = \int tr\{\lambda^n MC[\lambda]\}d\sigma_\lambda$$

for a constant M and integer n.

In general Hamiltonians $\mathcal{H}_S$ define a rather complicated evolution for $(\Phi_\lambda, \Phi_{\lambda x}; \tilde{\Phi}_\lambda, \tilde{\Phi}_{\lambda x})$, and indeed the evolution of $U=\int\Phi_\mu\tilde{\Phi}_\mu d\sigma_\mu$ under the Hamiltonian flow $\mathcal{H}_S$ may be non-local in U.

However, the evolution of U under the Hamiltonian $\mathcal{H}_{n,M}$ is always *local* in the sense that the equation of the evolution of U under $\mathcal{H}_{n,M}$ takes the form

$$\mathcal{R}_{n,M}\left[U_t, U_{tt}, \ldots; U, U_x, U_{xx}, \ldots\right] = 0$$

where $\mathcal{R}_{n,M}(x_1, \ldots; y_1, \ldots)$ is a polynomial with the coefficients depending on M and n.

In particular, for n-non-negative and even integer, $n=2m\geq 0$, the evolution of U under the action of $\mathcal{H}_{2m,M}$ takes the form:

$$U_{t_{2m}} = P_{2m,M}\left[U, U_x, \ldots, U_{\underbrace{x\ldots x}_{2m+1}}\right]$$

In particular for M=I we obtain simply the m-th Operator KdV equation. E.g.:

$$U_{t_2} = 6(UU_x + U_x U) - U_{xxx} \quad \text{etc.}$$

Now we come to our main result concerning the *(OR)*. In fact, *all* operator nonlinear equations of order two connected with the operator Schrödinger equations can be represented as the action of one of the Hamiltonians $\mathcal{H}_S$.

More precisely, any equation of order two presented before and corresponding to $\alpha(\mathcal{L})$ and $\beta(\mathcal{L})$ for the potential

$$U(x,t) = \int \Phi_\mu(x,t)\cdot\tilde{\Phi}_\mu(x,t)\,d\sigma_\mu$$

is equivalent to the evolution of

$$(\Phi_\mu(x,t),\ \tilde{\Phi}_\mu(x,t))$$

according to

$$\mathcal{H} = \frac{1}{2}\int \mathrm{tr}C[\lambda]\,d\sigma_\lambda$$

in the x-direction and according to the Hamiltonian of the form

$$\mathcal{H}_{f,M} = \int \mathrm{tr}\{f(\lambda)MC[\lambda]\}\,d\sigma_\lambda$$

in the t-direction.

E.g. in the case $\alpha(\mathcal{L})=0$, $\beta(\mathcal{L})$-arbitrary, $M=I$, we have

$$\mathcal{H}_{f,M} = \int \mathrm{tr}\{\beta(-\lambda^2)C[\lambda]\}\,d\sigma_\lambda$$

and

$$U_t = \frac{\partial}{\partial x}\left\{\int \beta(-\lambda^2)\Phi_\lambda\tilde{\Phi}_\lambda\,d\sigma_\lambda\right\}$$

together with

$$U_t = \mathcal{R}_n[U, U_x, \ldots, U_{\underbrace{x\ldots x}_{2n+1}}]$$

for $\beta(\lambda)=\lambda^n$.

The scalar case of the Russian Chain.

In the scalar case the Russian chain is considered on the Hilbert space $L^2(\Omega, d\mu_k)$ for $\Omega\subset\mathbb{C}$. Then we have the following Hamiltonian system for $f=f(x,k)$, $g=g(x,k)$:

$$\begin{cases} f_{xx} = \int_\Omega f\cdot g\, d\mu_k \cdot f - k^2 f \\ g_{xx} = \int_\Omega f\cdot g\, d\mu_k \cdot g - k^2 g\ . \end{cases} \tag{$\mathcal{O}$}$$

This system has several internal symmetries described by the Bäcklund transformation.

Theorem 8.2 - We have the following first integral of the system ($\mathcal{O}$):

$$K[k] = f_x(k)g(k) - f(k)g_x(k) \quad ,$$

$$C[k] = f_x(k)g_x(k) + k^2 f(k)g(k) - \frac{1}{2}f(k)g(k)\int_\Omega f\cdot g d\mu_\lambda +$$
$$+ \frac{1}{2}\int \frac{d\mu_\lambda}{\lambda^2-k^2}\,(f_x(\lambda)g(k)-f(\lambda)g_x(k))(g_x(\lambda)f(k)-g(\lambda)f_x(k)) \ .$$

The first integrals $K[k]$, $C[k]$ are in involution, and independent. There are additional first integrals:

$$C_2[k] = 2f_x(k)^2 + 2k^2 f(k)^2 - f(k)^2\int_\Omega fg\, d\mu_\lambda +$$
$$+ \int \frac{d\mu_\lambda}{\lambda^2-k^2}(f_x(\lambda)f(k)-f(\lambda)f_x(k))(g_x(\lambda)f(k)-g(\lambda)f_x(k)) \ ,$$

$$C_3[k] = 2g_x(k)^2 + 2k^2 g(k)^2 - g(k)^2\int_\Omega fg\, d\mu_\lambda +$$
$$+ \int \frac{d\mu_\lambda}{\lambda^2-k^2}(f_x(\lambda)g(k)-f(\lambda)g_x(k))(g_x(\lambda)g(k)-g(\lambda)g_x(k)) \ .$$

New relations between $K[k]$, $C[k]$, $C_2[k]$, $C_3[k]$ after quantization give rise to nice commutation relations between Hamiltonians and different scattering coefficients $b(\xi)$, $b^*(\xi)$ for: nonlinear Schrödinger, sine-Gordon, and the massive Thirring model obtained by Honerkamp[34], Thacker[33], Faddeev, Sclianin, Tahtadzan and others.

In particular

$$[K(k'),C_j(k)] = 0 \quad : \quad k'\neq k \ ; \quad [K(k),K(k')] = 0 \ ;$$
$$[K(k),C_j(k)] = 2C_j(k) \quad : \ j=2,3 \ .$$

Appendix.

One of questions that arises in the connection of the Russian Chain ($\mathcal{OR}$) is the possibility of the representation

$$U(x) = \int \Phi_\mu \tilde{\Phi}_\mu d\sigma_\mu$$

of the potential U(x) in terms of (left and right) eigenfunctions Φ_μ, $\tilde{\Phi}_\mu$ corresponding to an eigenvalue $-\mu^2$ of the operator Schrödinger equation with the potential U(x).

Another important problem is: how the measure $d\sigma_\mu$ is connected with the spectral measure naturally associated to U(x).

The answer to these questions have been developed previously for the scalar case, so now the answers for the general case can be proposed and in certain cases proven.

The first case: U(x) is hermitian, $U=U^\dagger$ and U(x) vanishes asymptotically exponentially or faster, i.e. for some $\varepsilon>0$,

$$\lim_{|x|\to\infty} \left[\exp(\varepsilon|x|)U(x)\right] = 0 \ .$$

We assume also that $-d^2/dx^2 + U(x)$ does not have degenerate or negative eigenvalues (this restriction, however, may be removed).

Theorem 8.3 - Under the assumptions above the representation for U(x) becomes

$$U(x) = \frac{2i}{\pi}\int_{-\infty}^{\infty} k\Phi_1(x,k)\tilde{\Phi}_1(x,k)dk \ ,$$

where $\Phi_1(x,k)$, $\tilde{\Phi}_1(x,k)$ are eigenfunctions corresponding to U(x):

$$\Phi_{1,xx}(x,k) = U\cdot\Phi_1(x,k) - k^2\Phi_1(x,k) \ ,$$

$$\tilde{\Phi}_{1,xx}(x,k) = \tilde{\Phi}_1(x,k)\cdot U - k^2\tilde{\Phi}_1(x,k) \ .$$

These functions $\Phi_1(x,k)$, $\tilde{\Phi}_1(x,k)$ are closely connected with Jost eigenfunctions (cf. ref.27):

$$\Phi_1(x,k) \longrightarrow A(k)\, e^{ikx} \qquad : x\to+\infty \ ,$$

$$\tilde{\Phi}_1(x,k) \longrightarrow \tilde{A}(k)\, e^{ikx} \qquad : x\to+\infty$$

and

$$A(k)\tilde{A}(k) = R(k) \ ,$$

where R(k) is the reflection coefficient.

The same type of result is true for periodic potentials U(x),

$$U(x+T) \equiv U(x) \quad : \quad T\neq 0 \ .$$

However, $d\alpha_\mu$ should be a singular measure with the support at the set of ends of the lacunae of the Bloch spectrum.

E.g. if in the spectrum of

$$\left(-\frac{d^2}{dx^2}+U-k^2\right)\Phi = 0$$

only n forbidden zonae, $U=U^\dagger$ are given and U is a periodic, then for any n eigenvalues $\lambda_1^2,\ldots,\lambda_n^2$ from 2n+1 ends of the forbidden zonae, the representation becomes (see refs.8 and 11)

$$U(x) = \sum_{i=1}^{n} \Phi_i\tilde{\Phi}_i + C$$

for a constant operator and Φ_i, $\tilde{\Phi}_i$-eigenfunctions corresponding to eigenvalues λ_i^2:

$$\Phi_{i,xx} = U\Phi_i - \lambda_i^2\Phi_i \quad ,$$

$$\tilde{\Phi}_{i,xx} = \tilde{\Phi}_i U - \lambda_i^2\tilde{\Phi}_i \quad : \quad i=1,\ldots,n.$$

Of course, for $U=U^\dagger$ and real k, k≥0, $\tilde{\Phi}_k = \Phi_k^\dagger$, i.e., for $U=U^\dagger$,

$$U = \sum_{i=1}^{n} \Phi_i\Phi_i^\dagger + C \qquad \text{etc.}$$

can be written.

9. CONNECTION BETWEEN SYSTEMS ARISING FROM AN OPERATOR DIRAC EQUATION AND OPERATOR SCHRÖDINGER EQUATION

Starting from a Hilbert space H over $\mathbb{C}$ with defined involution $\cdot^\dagger$ and considering operators R, Q, ϕ, ψ,... from H to H, in the matrix (2×2 case) over H the Hilbert space H_2=H×H is acheived.

According to non-commutativity of operator multiplication consider two operator Dirac equations (where potential acts on l.h.s. and r.h.s.):

$$\begin{cases} \check{\phi}_{1,x} = Q\check{\phi}_2 - i\zeta\check{\phi}_1 \\ \check{\phi}_{2,x} = R\check{\phi}_1 + i\zeta\check{\phi}_2 \end{cases} \qquad (9.1)$$

and

$$\begin{cases} \hat{\phi}_{1,x} = \hat{\phi}_2 Q - i\zeta\hat{\phi}_1 \\ \hat{\phi}_{2,x} = \hat{\phi}_1 R + i\zeta\hat{\phi}_2 \end{cases} \quad . \qquad (9.2)$$

Next, consider matrix (over H×H) solutions of two Dirac equations (9.1) and (9.2):

$$\check{\Phi} = \begin{pmatrix} \check{\phi}_1 & \check{\bar{\phi}}_1 \\ \check{\phi}_2 & \check{\bar{\phi}}_2 \end{pmatrix} \qquad \text{for (9.1)}$$

and

$$\hat{\Phi} = \begin{pmatrix} \hat{\phi}_2 & \hat{\phi}_1 \\ \hat{\bar{\phi}}_2 & \hat{\bar{\phi}}_1 \end{pmatrix} \qquad \text{for solutions } (\hat{\phi}_1, \hat{\phi}_2) \text{ and } (\hat{\bar{\phi}}_1, \hat{\bar{\phi}}_2) \text{ of (9.2).}$$

In the case of $\|R\|$, $\|Q\|$ exponentially decreasing as $|x| \to \infty$ there exists a natural choise of $\hat{\Phi}$, $\check{\Phi}$ as Jost functions.

The basic result from the spectral theory of an operator Dirac equation that is needed is the representation of "potentials" R, Q as an integral of the products $\check{\phi}_1\hat{\phi}_1$, $\check{\phi}_2\hat{\phi}_2$,... of the eigenfunctions of "left" and "right" Dirac equations (9.1) and (9.2) over the spectral measure $d\mu_\zeta$.

Such representation for the usual (scalar) Dirac equation was written down explicitly by D.Kaup[27] and also follows from the corresponding representation for Schrödinger operator.

The result is the most general expression for "potentials" R, Q in terms of product of eigenfunctions $\check{\phi}_i\hat{\phi}_i$, $\check{\bar{\phi}}_i\hat{\bar{\phi}}_i$, ...

Theorem 9.1 - There exist eigenfunctions $(\check{\phi}_1, \check{\phi}_2)$, $(\check{\bar{\phi}}_1, \check{\bar{\phi}}_2)$ of (9.1) and eigenfunctions $(\hat{\phi}_1, \hat{\phi}_2)$, $(\hat{\bar{\phi}}_1, \hat{\bar{\phi}}_2)$ of (9.2) such that

$$R = \int \check{\phi}_2 \hat{\phi}_2 d\mu_\zeta + \int \check{\bar{\phi}}_2 \hat{\bar{\phi}}_2 d\mu_\zeta \quad ,$$

$$Q = -\int \check{\phi}_1 \hat{\phi}_1 d\mu_\zeta - \int \check{\bar{\phi}}_1 \hat{\bar{\phi}}_1 d\mu_\zeta \quad .$$

In the case of $\|R\|$, $\|Q\|$ rapidly decreasing as $|x| \to \infty$ the eigenfunctions $(\check{\phi}_1, \check{\phi}_2)$, $(\check{\bar{\phi}}_1, \check{\bar{\phi}}_2)$ and $(\hat{\phi}_1, \hat{\phi}_2)$, $(\hat{\bar{\phi}}_1, \hat{\bar{\phi}}_2)$ are proportional up to (scattering) factors to the canonical Jost functions.

There is another set of useful formulae for the higher momentum of the products $\check{\phi}_i\hat{\phi}_j + \check{\bar{\phi}}_i\hat{\bar{\phi}}_j$. We define:

$$r_n = \int (2i)^n \zeta^n \check{\phi}_2 \hat{\phi}_2 d\mu_\zeta + \int (2i)^n \zeta^n \check{\bar{\phi}}_2 \hat{\bar{\phi}}_2 d\mu_\zeta \quad ;$$

$$q_n = \int (2i)^n \zeta^n \check{\phi}_1 \hat{\phi}_1 d\mu_\zeta + \int (2i)^n \zeta^n \check{\bar{\phi}}_1 \hat{\bar{\phi}}_1 d\mu_\zeta \quad ;$$

$$s_n^{12} = \int (2i)^n \zeta^n \check{\phi}_1 \hat{\phi}_2 d\mu_\zeta + \int (2i)^n \zeta^n \check{\bar{\phi}}_1 \hat{\bar{\phi}}_2 d\mu_\zeta \quad ;$$

$$s_n^{21} = \int (2i)^n \zeta^n \check{\phi}_2 \hat{\phi}_1 d\mu_\zeta + \int (2i)^n \zeta^n \check{\bar{\phi}}_2 \hat{\bar{\phi}}_1 d\mu_\zeta \quad .$$

This leads to:

$$R = r_0 \ , \ Q = -q_0 \ , \ s_0^{12} = s_0^{21} = 0 \ ;$$

$$Q_x = q_1 \ , \ R_x = r_1 \ , \ QR = s_1^{12} \ , \ RQ = s_1^{21} \ ;$$

$$2QRQ - Q_{xx} = q_2 \ , \ R_{xx} - 2RQR = r_2 \ ,$$

$$QR_x - Q_xR = s_2^{12} \ , \ R_xQ - RQ_x = s_2^{21} \ ; \ \ldots$$

$$Q_{xxx} - 3Q_xRQ - 3QRQ_x = q_3 \ ,$$

$$R_{xxx} - 3R_xQR - 3RQR_x = r_3 \ , \ \ldots$$

Now let's come from Dirac equations (9.1) (9.2) to an operator Schrödinger equation.

Defining the potential U in H×H by the formula

$$U = \begin{pmatrix} QR & Q_x \\ R_x & RQ \end{pmatrix} = V^2 + V_x$$

where

$$V = \begin{pmatrix} 0 & Q \\ R & 0 \end{pmatrix}$$

and introducing the matrix $\sigma_3 = \begin{pmatrix} I & 0 \\ 0 & -I \end{pmatrix}$ and two matrices of fundamental solutions of (9.1)-(9.2) as before

$$\check{\Phi} = \begin{pmatrix} \check{\phi}_1 & \check{\check{\phi}}_1 \\ \check{\phi}_2 & \check{\check{\phi}}_2 \end{pmatrix} \quad ; \quad \hat{\Phi} = \begin{pmatrix} \hat{\phi}_2 & \hat{\phi}_1 \\ \hat{\hat{\phi}}_2 & \hat{\hat{\phi}}_1 \end{pmatrix} \ , \tag{9.3}$$

the following equations are obtained

$$\frac{d}{dx}\check{\Phi} = V\cdot\check{\Phi} - i\zeta\sigma_3\check{\Phi} \ ;$$
$$\frac{d}{dx}\hat{\Phi} = \hat{\Phi}\cdot V - i\zeta\hat{\Phi}\sigma_3 \ . \tag{9.4}$$

By simply iterating equations (9.4) and two operator Schrödinger equations are obtained

$$\frac{d^2}{dx^2}\check{\Phi} = U\cdot\check{\Phi} - \zeta^2\check{\Phi}$$
$$\frac{d^2}{dx^2}\hat{\Phi} = \hat{\Phi}\cdot U - \zeta^2\hat{\Phi} \tag{9.5}$$

for U as above.

Corollary 9.2- There are fundamental solutions $\check{\Phi}_0$ and $\hat{\Phi}_0$ of (9.5) such that

$$U = \int \check{\Phi}_0(\zeta)\cdot\hat{\Phi}_0(\zeta)d\mu_\zeta \quad . \tag{9.6}$$

Here $\check{\Phi}_0$, $\hat{\Phi}_0$ are connected with $\check{\Phi}(\zeta)$, $\hat{\Phi}(\zeta)$ defined above in the theorem 9.1. E.g. we can put

$$\check{\Phi}_0(\zeta) = 2i\zeta\check{\Phi}(\zeta) \quad , \quad \hat{\Phi}_0(\zeta) = \hat{\Phi}(\zeta) \quad .$$

In this case there exists an infinite system of evolution equations. These equations are obtained from the one-dimensional operator Hamiltonian systems

$$\begin{aligned} \frac{d^2\check{\Phi}_0(\eta)}{dx^2} &= \int \check{\Phi}_0(\zeta)\hat{\Phi}_0(\zeta)d\mu_\zeta\cdot\check{\Phi}_0(\eta) - \eta^2\check{\Phi}_0(\eta) \\ \frac{d^2\hat{\Phi}_0(\eta)}{dx^2} &= \hat{\Phi}_0(\eta)\cdot\int \check{\Phi}_0(\zeta)\hat{\Phi}_0(\zeta)d\mu_\zeta - \eta^2\hat{\Phi}_0(\eta) \quad . \end{aligned} \tag{9.7}$$

Starting from the Hamiltonian systems (9.7) infinitely many commuting Hamiltonians $H_n: n\geq 1$, commuting with that of (9.7): H_0 can be obtained.

Now the evolution according to t_n for U has the form:

$$U_{t_n} = \frac{\partial}{\partial x}\left\{\int \zeta^n\check{\Phi}_0(\zeta)\hat{\Phi}_0(\zeta)d\mu_\zeta\right\} \quad : n=0,1,2,\ldots \quad .$$

E.g. for n=0, $U_{t_0}=U_x$.

The evolution of U_{t_n} (according to the Hamiltonian H_n) can be translated into the evolution equations for R, Q.

E.g.

$$\begin{aligned} &R_{t_0} = R_x \quad , \quad Q_{t_0} = Q_x \quad ; \\ &iR_{t_1} = R_{xx} - 2RQR \quad , \quad -iQ_{t_1} = Q_{xx} - 2\,QRQ \quad ; \\ &R_{t_2} = 3R_xQR + 3RQR_x - R_{xxx} \quad , \\ &Q_{t_2} = 3Q_xRQ + 3QRQ_x - Q_{xxx} \quad ; \quad \ldots \end{aligned}$$

It is easily seen that the evolution in t_0 determines U (or R, Q) as the functions on x ($t_0 \equiv x$), the combination of evolutions in (t_0,t_1) determine two-component non-linear Schrödinger and evolution in (t_0,t_2) produces a two-component operator modified KdV equation.

In order to get all flows t_n in terms of U, U_x, U_{xx},... only, an integro-differential operators Λ must be used. However, the flows t_{2n} (corresponding to H_{2m} with even 2m) can be represented as polynomials in U, U_x, U_{xx},... and produce exactly n-th operator KdV equations of evolution for U.

10. THREE-DIMENSIONAL COMPLETELY INTEGRABLE SYSTEMS

Now, for the first time some three-dimensional system can be referred to as completely integrable, because these systems can be: a) represented as Hamiltonian systems; b) can be indeed reduced to action-angle variables.

For 9 years already, there have existed examples of (x,y,t) systems that arise from conditions of commutativity of two linear differential operators in $\partial/\partial x$, $\partial/\partial y$, $\partial/\partial t$. These equations were called "completely integrable" because there was known a large class of solutions. However little is known about these systems in general. There is an open question about the existence of Hamiltonian structure.

We shall see how three dimensional completely integrable systems can be reduced to one dimensional operator system as a result of the motion via *three* Hamiltonian flows that commute. Of course, as one can guess, this one dimensional operator system is none other than an operator Russian chain

$$\frac{d^2\psi(\zeta)}{dx^2} = \int \psi(\eta)\Phi(\eta)d\mu_\eta\cdot\psi(\zeta) - \zeta^2\psi(\zeta)$$
$$\frac{d^2\Phi(\zeta)}{dx^2} = \Phi(\zeta)\int \psi(\eta)\Phi(\eta)d\mu_\eta - \zeta^2\Phi(\zeta) \quad . \qquad (OR)$$

Our way of thinking can be explained in a very simple way. It is known that all the higher KdV equations and, generally speaking, all two dimensional completely integrable systems of order two can be reduced to an infinite component Russian chain (infinite component non-linear stationary Schrödinger). In other words, the introduction of a new variable k (spectral variable) allows the elimination of variable t. Now it is quite natural, that if one variable is added in the (x,y,t) system variable t can be eliminated and if variable λ is introduced variable y can be eliminated. This way the two dimensional KdV equation is first change to multicomponent non-linear Schrödinger and then from multicomponent non-linear Schrödinger to operator stationary non-linear Schrödinger (operator Russian chain). Basically the rule here is the following: to construct the solution u(x,t) of two dimensional system of KdV type, it is to be written in the form

$$u(x,t) = \int f(x,t,k)f^*(x,t,k)d\mu_k$$

where f(k) are eigenfunctions of

$$f_{xx}(k) = u(x,t)f(k) - k^2 f(k) \quad ,$$

with the evolution in x determined by this Russian chain and the evolution in t is governed by any Hamiltonian flow commuting with Russian chain (or the Hamiltonian $\mathscr{H}$). Now if one wants to find the solution u(x,t,y) of (x,y,t)-dimensional system of KdV type, it is presented in the form

$$u(x,t,y) = \int tr[\Psi^{+}\Psi]d\mu_{\zeta}$$

where ψ^{+}, ψ arise from a stationary Russian chain:

$$\frac{d^2\psi(\zeta)}{dx^2} = \psi(\zeta)\int \psi^{+}(\eta)\psi(\eta)d\mu_{\eta} - \zeta^2\psi(\zeta) \quad ,$$

and two Hamiltonian flows in t and y commuting with the given one in x.

To start with an example take *three* commuting Hamiltonian flows arising from an operator Dirac equation :

$$\begin{aligned} &R_{t_0} = R_x \quad , \quad Q_{t_0} = Q \qquad (\text{i.e. } t_0 = x) \; , \\ &iR_y = R_{xx} - 2RQR \; ; \quad -iQ_y = Q_{xx} - 2QRQ \quad (\text{i.e. } t_1 = y) \\ &R_t = 3R_xQR + 3RQR_x - R_{xxx} \quad , \\ &Q_t = 3Q_xRQ + 3QRQ_x - Q_{xxx} \qquad (\text{i.e. } t_2 = t) \quad . \end{aligned} \tag{10.1}$$

According to the previous explanation, these three flows naturally act on an infinite-dimensional symplectic manifold, arising from the following Hamiltonian system: consider fundamental solutions $\check{\Phi}(\zeta)$, $\hat{\Phi}(\zeta)$ of the following system

$$\begin{aligned} &\frac{d}{dx}\check{\Phi}(\zeta) = V\cdot\check{\Phi}(\zeta) - i\zeta\sigma_3\check{\Phi}(\zeta) \\ &\frac{d}{dx}\hat{\Phi}(\zeta) = \hat{\Phi}(\zeta)\cdot V - i\zeta\hat{\Phi}(\zeta)\sigma_3 \end{aligned} \tag{10.2}$$

for

$$V = \begin{pmatrix} 0 & Q \\ R & 0 \end{pmatrix} \quad , \quad \sigma_3 = \begin{pmatrix} I & 0 \\ 0 & -I \end{pmatrix} \quad . \tag{10.3}$$

Then V and certain differential polynomials of V, V_x,... are represented in terms of momentae of product $\check{\Phi}(\eta)\hat{\Phi}(\eta)$, integrated over η by spectral measure $d\mu_{\eta}$.

More precisely this expression can be given for the lowest momentae of $\check{\Phi}(\eta)\hat{\Phi}(\eta)d\mu_\eta$:

Theorem 10.1 - For certain operator solutions $\check{\Phi}(\zeta)$, $\hat{\Phi}(\zeta)$ of (10.2) there are the following formulae:

$$V\sigma_3 = \int \check{\Phi}(\zeta)\hat{\Phi}(\zeta)d\mu_\zeta \quad , \tag{10.4}$$

$$V^2 + V_x = \int (2i)\zeta\check{\Phi}(\zeta)\hat{\Phi}(\zeta)d\mu_\zeta \quad , \tag{10.5}$$

$$\sigma_3(2V^3 - V_{xx} + VV_x - V_x V) = \int (2i)^2\zeta^2\check{\Phi}(\zeta)\hat{\Phi}(\zeta)d\mu_\zeta \ , \ \ldots \tag{10.6}$$

Here

$$V\sigma_3 = \begin{pmatrix} 0 & -Q \\ R & 0 \end{pmatrix} ,$$

$$V^2 + V_x = \begin{pmatrix} QR & Q_x \\ R_x & RQ \end{pmatrix} ,$$

$$\sigma_3(2V^3 - V_{xx} + VV_x - V_x V) = \begin{pmatrix} QR_x - Q_x R & 2QRQ - Q_{xx} \\ R_{xx} - 2RQR & R_x Q - RQ_x \end{pmatrix}$$

and for

$$\check{\Phi} = \begin{pmatrix} \check{\phi}_1 & \check{\bar{\phi}}_1 \\ \check{\phi}_2 & \check{\bar{\phi}}_2 \end{pmatrix} , \qquad \hat{\Phi} = \begin{pmatrix} \hat{\phi}_2 & \hat{\phi}_1 \\ \hat{\bar{\phi}}_2 & \hat{\bar{\phi}}_1 \end{pmatrix}$$

$$\check{\Phi}(\zeta)\hat{\Phi}(\zeta) = \begin{pmatrix} \check{\phi}_1\hat{\phi}_2 + \check{\bar{\phi}}_1\hat{\bar{\phi}}_2 & \check{\phi}_1\hat{\phi}_1 + \check{\bar{\phi}}_1\hat{\bar{\phi}}_1 \\ \check{\phi}_2\hat{\phi}_2 + \check{\bar{\phi}}_2\hat{\bar{\phi}}_2 & \check{\phi}_2\hat{\phi}_1 + \check{\bar{\phi}}_2\hat{\bar{\phi}}_1 \end{pmatrix} .$$

Now (10.4)-(10.6) mean that the evolution equations arise from the following Hamiltonians and systems commuting with this Hamiltonian:

$$\begin{aligned} \frac{d\check{\Phi}(\zeta)}{dx} &= \int \check{\Phi}(\eta)\hat{\Phi}(\eta)d\mu_\eta \cdot \sigma_3 \cdot \check{\Phi}(\zeta) - i\zeta\sigma_3\check{\Phi}(\zeta) \\ \frac{d\hat{\Phi}(\zeta)}{dx} &= \hat{\Phi}(\zeta) \cdot \int \check{\Phi}(\eta)\hat{\Phi}(\eta)d\mu_\eta \cdot \sigma_3 - i\zeta\hat{\Phi}(\zeta) \cdot \sigma_3 \quad . \end{aligned} \tag{10.7}$$

The following is a deduction of the two-dimensional KdV, or Kadomtsev-Petviashvili operator equations from the equations (10.1). Indeed, (10.1) can be viewed as the conditions of commutativity of equations in y and t. However, it is known that these equations *are* consistent as they correspond to commuting Hamiltonian flows.

Nevertheless we can write the conditions of consistency

$$R_{yt} = R_{ty} \ , \quad Q_{yt} = Q_{ty} \quad . \tag{10.8}$$

It can be immediately verified that the conditions in (10.8) can be reduced to the following ones: setting

$$U = QR, \quad W = QR_x, \quad Z = Q_xR, \tag{10.9}$$

so that $U_x = W+Z$.

Then (10.1) takes the form:

$$\begin{aligned} iR_y &= R_{xx} - 2RU \quad ; \quad -iQ_y = Q_{xx} - 2UQ \ ; \\ R_t &= 3R_xU + 3RW - R_{xxx} \quad ; \\ Q_t &= 3UQ_x + 3ZQ - Q_{xxx} \quad . \end{aligned} \tag{10.10}$$

Then the conditions in (10.8) take the form

$$R_x(3U_y - 3iU_{xx} + 6iW_x) + R(6iU_xU + 6i[U,W] + 3W_y - 2iU_{xxx} + 3iW_{xx} - 2iU_t) = 0 \tag{10.11}$$

and

$$(6iZ_x - 3U_y - 3iU_{xx})Q_x + (3iZ_{xx} - 2iU_t - 6i[U,Z] - 3Z_y + 6iUU_x - 2iU_{xxx})Q = 0 \ . \tag{10.11'}$$

and consequently:

Corollary 10.2 - If rank(R) and rank(Q) are larger than rank(U)+rank(W), then systems (10.11) or (10.11)' are equivalent to the following:

$$3U_y - 3iU_{xx} + 6iW_x = 0$$

$$6iU_xU + 6i[U,W] + 3W_y - 2iU_{xxx} + 3iW_{xx} - 2iU_t = 0 \ .$$

These equations can be written in a more convenient form:

$$U_x = W + Z, \quad U_y = (W - Z)_x$$

or

$$W = \tfrac{1}{2}U_x + \frac{i}{2}\int U_y dx, \quad Z = \tfrac{1}{2}U_x - \frac{i}{2}\int U_y dx \ . \tag{10.12}$$

The second equation in the corollary 10.2 can be written by taking into account (10.12) as a single (x,y,t)-operator equation:

$$3U_{yy} = \frac{\partial}{\partial x}(4U_t + U_{xxx} - 6(U_xU + UU_x) + 6i[\textstyle\int U_y dx, U]) \ . \tag{10.13}$$

It is natural to call this equation Operator Kadomtsev-Petviashvili equation.

In particular, if U is a scalar (or, e.g. $\int U_y dx$ commutes with U), then (10.13) becomes simply an ordinary Kadomtsev-Petviashvili equation

$$3u_{yy} = \frac{\partial}{\partial x}(4u_t + u_{xxx} - 12uu_x) \ . \tag{10.14}$$

It must be remembered that in order to get the equation (10.13) there was the restriction that rank(R)>rank(U). For this it is natural to take R and Q as operators with $\mathrm{Rank}(R)=\mathrm{Rank}(Q)=H_0$ with $\mathrm{codim}H_0$ in H to be infinite.

As a model consider H as an L_2-type space over H_0: $H=\oplus_{L^2(\Omega,d\sigma)} H_0$, for a certain space $L^2(\Omega,d\sigma)$.

Now choose operators R, Q in the following matrix form over H_0:

$$(R)_{ij} = R_i\alpha_j \quad ,$$
$$(Q)_{ij} = \beta_i Q_j \quad , \qquad i,j\in\Omega \tag{10.15}$$

for operators R_i, Q_j over H_0 and $\alpha_j=\alpha(j)$, $\beta_i=\beta(i)$ from $L^2(\Omega,d\sigma)$ such that

$$\int_\Omega \alpha(j)\beta(j)d\sigma = 1 \quad . \tag{10.16}$$

Here (10.15) means that R acts on element $\overline{\psi}=(\psi(j):j\in\Omega)$ of H as

$$R\overline{\psi} = \overline{\varphi} \;\; H \quad ,$$

$$\overline{\varphi}_j = R_j\int_\Omega \alpha(i)\psi(j)d\sigma \; : \; j\in\Omega \quad .$$

Analogous to this,

$$Q\overline{\psi}_1 = \overline{\varphi}_1$$

if

$$(\overline{\varphi}_1)_j = \beta_j\int_\Omega Q_i\psi(i)d\sigma \; : \; j\in\Omega \; .$$

Now

$$(Q\cdot R)_{ij} = \beta_i\alpha_j\int_\Omega Q_\lambda R_\lambda d\sigma \quad . \tag{10.17}$$

Defining a C-matrix (an operator on $L^2(\Omega,d\sigma)$) as

$$C = (\beta_i\alpha_j)$$

and (an operator on H_0) potential as

$$\mathcal{U} = \int_\Omega Q_\lambda R_\lambda d\sigma \; . \tag{10.18}$$

It follows

$$QR = C\cdot\mathcal{U} = \mathcal{U}\cdot C \quad .$$

Taking into account condition (10.16),

$$C^2 = C$$

is obtained and C is constant. Substituting

$$U = QR = C\cdot\mathcal{U}$$

into the equation (10.13) a single equation for $\mathcal{U}$ only is obtained:

$$3U_{yy} = \frac{d}{dx}(4U_t + U_{xxx} - 6(U_x U + U U_x) + 6i[\int U_y dx, U]) \ . \tag{10.19}$$

Now consider the solutions $\check{\Phi}(\zeta)$, $\hat{\Phi}(\zeta)$ (as operators on H) of the operator Dirac equations (10.2) fo a choice of R, Q in (10.15)-(10.16). In this case the spectral measure $d\mu$ depends on U *and* on $d\sigma$ as well.

According to the results above there exist equations

$$\begin{aligned} \frac{d^2}{dx^2}\check{\Phi}_0(\zeta) &= U\cdot\check{\Phi}_0(\zeta) - \zeta^2\check{\Phi}_0(\zeta) \ , \\ \frac{d^2}{dx^2}\hat{\Phi}_0(\zeta) &= \hat{\Phi}_0(\zeta)\cdot U - \zeta^2\hat{\Phi}_0(\zeta) \ , \end{aligned} \tag{10.20}$$

for

$$U = V^2 + V_x = \begin{pmatrix} QR & Q_x \\ R_x & RQ \end{pmatrix} \ . \tag{10.21}$$

By Theorem 10.1, after a proper norming of $\check{\Phi}(\zeta)$, $\hat{\Phi}(\zeta)$ we have

$$U = 2\int \check{\Phi}_0(\zeta)\hat{\Phi}_0(\zeta)d\mu_\zeta \ . \tag{10.22}$$

Having written an operator identity in H×H above it is now possible to take a trace with respect to $H^2 = H\times H_{/H_0}$ and get

$$tr_{H^2/H_0}(U) = 2\int tr_{H^2/H_0}(\check{\Phi}_0(\zeta)\hat{\Phi}_0(\zeta))d\mu_\zeta \ . \tag{10.23}$$

Now

$$tr_{H^2/H_0}(U) = tr_{H/H_0}(QR) + tr_{H/H_0}(RQ) = 2tr_{H/H_0}(QR) \ .$$

But by (10.16)

$$tr_{H/H_0}(QR) = U \ .$$

Thus from (10.23) it follows

$$\int tr_{H^2/H_0}(\check{\Phi}_0\hat{\Phi}_0(\zeta))d\mu_\zeta = U \ . \tag{10.24}$$

This formula gives the most general expression for the solution of the Operator (x,y,t)-Dimensional Completely Integrable System.

Corollary 10.3 - In order to get a solution $U(x,t,y)$ of a matrix (operator) three-dimensional system over the Hilbert space H_0 it's necessary to consider the Hilbert space H_2, infinite-dimensional over H (being L^2-ex= tension of H_0) and the corresponding operator Schrödinger equations over H_2:

$$\frac{d^2\check{\Phi}(\eta)}{dx^2} = U\check{\Phi}(\eta) - \eta^2\check{\Phi}(\eta) \quad ,$$

$$\frac{d^2\hat{\Phi}(\eta)}{dx^2} = \hat{\Phi}(\eta)U - \eta^2\hat{\Phi}(\eta) \qquad (\mathcal{OR})$$

being also "an operator Russian chain" with the potential

$$U = \int \check{\Phi}(\zeta)\hat{\Phi}(\zeta)d\mu_\zeta \quad .$$

Now we take the solution $\check{\Phi}(\zeta;x,t,y)$, $\hat{\Phi}(\zeta;x,t,y)$ of the system $(\mathcal{OR})$ together with two Hamiltonian systems in t and y commuting with $(\mathcal{OR})$. These systems can be taken in a standard form, such that

$$U_t = \frac{\partial}{\partial x}\int \zeta^n \check{\Phi}(\zeta)\hat{\Phi}(\zeta)d\mu_\zeta \quad \text{for } t=t_{2n} \quad \text{etc.}$$

Then one can write

$$\mathcal{U}(x,t,y) = \int tr_{H_{2/H_0}} \{\check{\Phi}(\zeta)\hat{\Phi}(\zeta)\}d\mu_\zeta$$

for these $\check{\Phi}(\zeta)=\check{\Phi}(\zeta;x,t,y)$, $\hat{\Phi}(\zeta)=\hat{\Phi}(\zeta;x,t,y)$. Moreover in the case of real $\mathcal{U}$, $\mathcal{U}^\dagger=\mathcal{U}$ and self-adjoint problem $(\mathcal{OR})$ we have

$$\hat{\Phi}(\zeta) = \check{\Phi}(\zeta)^\dagger \quad , \quad U^\dagger = U$$

and

$$\mathcal{U}(x,t,y) = \int \mathbf{tr}_{H_{2/H_0}} \{\Phi^\dagger(\zeta;x,t,y)\cdot\Phi(\zeta;x,t,y)\}d\mu_\zeta$$

for

$$\frac{d^2\Phi(\zeta)}{dx^2} = \Phi(\zeta)\int \Phi^\dagger(\eta)\Phi(\eta)d\mu_\eta - \zeta^2\Phi(\zeta) \quad .$$

REFERENCES

[1] Burchnall, J.L., Chaundy, T.W.: 1922, Proc.London Math.Soc. 21, pp.420-440; 1828, Proc.Soc.London Ser. A118, pp.557-573.

[2] Calogero, F., Degasperis, A.: 1976, Nuovo Cimento 32B, p.201; 1977, Nuovo Cimento 39B, p.1.

[3] Degasperis, A.: 1979, Lecture Notes in Physics (to appear).

[4] Y.M.Berezansky, Eigenfunction expansions , AMS, Providence, Rhode Island, 1970.

[5] Chudnovsky, D.V.: 1979, Phys.Rev.D (to appear). Service de Physique Théorique, DPh.T. 79/80.

[6] Chudnovsky, D.V.: 1979, C.R.Acad.Sci.Paris, (to appear).

[7] Chudnovsky, D.V.: 1979, C.R. Acad.Sci.Paris, (to appear).

[8] Chudnovsky, D.V. and Chudnovsky, G.V.: (1977-1978) Seminaire sur les équations non-linéaires I, Centre de Mathématiques, Ecole Polytechnique, p.300.

[9] Chudnovsky, D.V. and Chudnovsky, G.V.: 1978, Lett.Nuovo Cimento 22, pp.31-36.

[10] Chudnovsky, D.V. and Chudnovsky, G.V.: 1978, Lett.Nuovo Cimento 22, pp.47-51.

[11] Chudnovsky, D.V. and Chudnovsky, G.V.: 1978, C.R.Acad.Sci.Paris 286A, pp.A1075-A1078.

[12] Chudnovsky, D.V. and Chudnovsky, G.V.: 1979, Phys.Lett.72A, pp.291-293 (Utrecht preprint).

[13] Chudnovsky, D.V. and Chudnovsky, G.V.: 1979, Phys.Lett.73A, pp.292-294.

[14] Chudnovsky, D.V. and Chudnovsky, G.V.: 1979, Seminar on Spectral Theory, Riemann Problem and Complete Integrability IHES, Paris.

[15] Drach, J.: 1919, C.R.Acad.Sci.Paris 168, pp.47-50; 1919, C.R.Acad.Sci.Paris 168, pp.337-340.

[16] Fuchs, R.: 1905, C.R.Acad.Sci.Paris 141, pp.555-558; 1907, Math.Ann. 63,pp.301-321.

[17] Gardner, G., Greene, J.M., Kruskal, M.D., Miura, R.V.: 1974, Comm.Pure Appl.Math. 27, pp.97-120.

[18] Garnier, R.: 1919, Circolo Mat. Palermo, 43, pp.155-191.

[19] Gelfand, J.M., Dikij, L.A.: 1975, Russian Math.Survey 30, pp.37-100.

[20] Hilbert, D.: Grundzüge der Integralgleichungen, Leipzig-Berlin, 2-te Aufl., 1924.

[21] Ince, E.L.: Ordinary differential equations, Dover, 1956.

[22] Lappo-Danilevsky, J.A.: Mémoires sur la Theéorie des systèmes des Equations Différentielles Linéaires, Chelsea Publishing Company, 1953.

[23] Lax, P.: 1968, Comm.Pure Appl.Math. 21, pp.467-490.

[24] Painlevé, P.: Oeuvres Mathématiques, V 1-3, Paris, 1972.

[25] Sato, M., Miwa, T., and Jimbo, M.: 1977, Proc.Japan Acad.53A, pp.6-10, 147-152, 153-158, 183-185, 219-224; 1978 Proc.Japan Acad.54A, pp.1-5, 36-41, 309-312.

[26] Schlesinger, L.: 1912, J.für Reine Angew.M. 141, pp.96-145.

[27] Kaup, D.: 1976, J.Math.Analy.Appl. 54, p.1.

[28] Zakharov, V.E., Mikhailov, A.V.: 1978, JETP 74, pp.1953-1970.

[29] Zakharov, V.Z., Shabat, A.B.: 1974, Funct.Anal.Appl. 8, pp.43-57.

[30] Riemann, B.: Oeuvres Mathémathiques, Albert Blanchard, Paris, 1968, pp.353-363.

[31] Krein, M.G., Daleekij, Y.L.: Stability of solutions linear differential equations in the Hilbert space, AMS, Providence, 1972.

[32] Krein, S.G.: Differential equations in Banach spaces, AMS, Providence,1971.

[33] Thacker, H.B., Wilkinson, D.: 1979, Fermilab-Pub-79, 19-THY.

[34] Honerkamp : 1979, Freiburg Preprint.

[35] Ablowitz, M.J., Kaup, D.J., Newell, A.C. and Segur, H.: 1974, J.Math.Phys. 15, p.1852.

THE INVERSE SCATTERING PROBLEM AND APPLICATIONS TO ARITHMETICS, APPROXIMATION THEORY AND TRANSCENDENTAL NUMBERS

G.V. Chudnovsky*

CNRS, IHES, 35 route de Chartres,

91440, Bures-sur-Yvette, France

0. INTRODUCTION

The purpose of this lecture is to demonstrate how the methods of Mathematical Physics are applied in Arithmetics and, especially, to Transcendental Numbers Theory. Of course, Transcendental Number Theory itself needs analytic methods such as the methods of the Approximation Theory. Here we show how the problems of the Approximation Theory appear within classical problems of modern Mathematical Physics. These problems are, first of all, different forms of the Inverse Scattering Method and Padé approximations.

As is clear in the discussion below, one of the main problems in Transcendental Number Theory is the construction of the so called auxiliary functions $F(z)=P(z,f_1(z),\ldots,f_n(z))$ for $P(\bar{x})\in\mathbb{C}[\bar{x}]$ or for a polynomial approximation to analytic functions $f_1(z),\ldots,f_n(z)$. This auxiliary function $F(z)$ is choosen by conditions of $F(z)$ having zeroes of high multiplicities at given points $z=z_1,\ldots,z=z_N$. In the case of $F(z) = \sum_{i=1}^{n} P_i(z) f_i(z)$ being a linear form in $f_1(z),\ldots,f_n(z)$ with polynomial coefficients, we come to the classical problem of the N-point Padé approximation to $f_1(z),\ldots,f_n(z)$.[1],[23]

---

* On leave from Dept.of Math., Columbia University, New York, N.Y.10027, USA. The work was supported in part by the Office of the Naval Research and NSF.

For some choice of functions $f_i(z)$ (usually connected with the hypergeometric function) the explicit construction of the Padé approximants is known in terms of classical special functions (generalized hypergeometric polynomials). The most important feature of this explicit construction is the existence of explicit simple expressions of recurrent formulae which connect the remainder function and polynomial approximants with different indices. For an arbitrary system of functions the corresponding recurrent formulae are not explicit, in the sense that they are expressed in terms of complicated ratios of Hankel-type determinants.

The existence of the recurrent formulae connecting Padé approximants for different integer weights N is related to the Darboux-Bäcklund transformation and to contiguos systems of functions in Riemann's sense. Darboux and Bäcklund's idea of getting from one solution of a linear differential equation some other ones is materialized as a recurrence formula for the Padé approximants. In particular cases it has the form of "addition of indices", for example when recurrences relate classical orthogonal polynomials $P_n^{(\alpha,\beta)}(x)$ with different integer indices n [(+)].

Together with an exposition of the Darboux-Bäcklund transformation, we explain for the example of the Bessel function this procedure of "adding of indices": $J_\nu \longmapsto J_{\nu+n}$ that furnish us with Padé approximants to $\{J_\nu(x), J_{\nu+1}(x)\}$ (so-called Bessel polynomials [32]). Simultaneously we give arithmetic corollaries of the construction of these polynomials in the form of the measure of irrationality of $J'_\nu(x)/J_\nu(x)$ for $\nu \in \mathbb{Q}(\sqrt{-D})$ and imaginary quadratic x, $x \neq 0$ [10] (appendix to §3, theorem 3.1).

Riemann's notion of contiguous functions [28] (coinciding in many interesting cases with the special Darboux-Bäcklund transformation [18] or [14], [17]) is used on the other hand, to provide us with the explicit construction of the Padé approximation to the solutions of Fuchsian linear differential equations. This explicit formula for remainder function of Padé approximation problem is given in §9, theorem 9.1. In §9 (theorem 9.2) we present also the remainder function of Padé approximation problem for the system of ${}_qF_p$ hypergeometric-functions.

(+) For classical orthogonal polynomials these recurrences are three term linear relations between P_{n-1}, P_n, P_{n+1} or P_n, P'_n, P_{n+1} or P_n, P'_n, P_{n-1}. See [1], [2] for a detailed discussion.

In §1 we give a general definition of Padé approximations and consider the exponential case using the Newton interpolation series. In §§ 2,4 we apply Padé approximation methods to the study of arithmetic nature of values of functions $y(x)=\sum_{m=1}^{\infty} x^m/\prod_{j=1}^{m}\theta(j)$: $\theta(x)\in\bar{\mathbb{Q}}[x]$.

In §§3,5,6 we consider different applications of Darboux-Bäcklund transformations and the study of Fuchsian linear differential equations to diophantine approximations. §7 contains multidimensional interpolation formulae and different forms of the Global residue formula that is used in §9 for the construction of Padé approximants. In §8 we formulate major problems of transcendental numbers and relate them with solutions of linear differential equations. At last, in §9 the effective construction of Padé approximations to solutions of linear differential equations is presented. We would like to note that theorem 6.3 and corollary 9.3 contain very good new bounds for the measure of irrationality of $\pi/\sqrt{3}$, π.

I would like to express my sincere gratitude to Prof.Boiti and Prof. Soliani for their warm hospitality in Lecce and for their patience in dealing with this paper.

1. THE DEFINITION OF THE N-POINT PADÉ APPROXIMATION. THE HERMITE INTERPOLATION FORMULAE AND THE NEWTON INTERPOLATION SERIES.

We present the most general definition of the N-point Padé approximation in C^k (the so-called Padé approximation of type I [39]).

<u>*Definition 1.1*</u> - *Let* $z_1,\ldots,z_N$ *be* N *distinct points in* $\mathbb{C}^1$ *(or* $\mathbb{C}^k$*) and* $f_1(z),\ldots,f_n(z)$ *be functions analytic in the neighbourhood of* $z=z_1,\ldots,z=z_n$ *. For* n *non-negative integers* $m_1,\ldots,m_n$*, we consider polynomials* $A_i(x)=A_i(x|f_1,\ldots,f_n;m_1,\ldots,m_n)$ *of degree* $\le m_i$: $i=1,\ldots,n$ *such that the remainder function:*

$$R(z) = R(z|f_1,\ldots,f_n;m_1,\ldots,m_n) \overset{\text{def}}{=} \sum_{i=1}^{n} A_i(z)f_i(z)$$

has zeroes at z_i: $i=1,\ldots,N$:

$$\sum_{i=1}^{N} \operatorname{ord}_{z_i}(R) \ge M,$$

where

$$M = \sum_{i=1}^{n}(m_i+1)-1$$

in the $\mathbb{C}^1$*-case and*

$$M = \sum_{i=1}^{n} \binom{m_i+k}{k} - 1$$

in the $\mathbf{C}^k$*-case.*

Then $A_i(z|f_1,\ldots,f_n;m_1,\ldots,m_n)$ *are called* $(m_1,\ldots,m_n)$ *Padé approximants to* $(f_1(z),\ldots,f_n(z))$ *at the points* $\{z_1,\ldots,z_N\}$. *A Padé approximation is perfect if always* $\sum_{i=1}^{N} \mathrm{ord}_{z_i}(R)=M$, *and almost perfect if* $|\sum_{i=1}^{N} \mathrm{ord}_{z_i}(R)-M| \le C(\bar{f})$ *for some absolute constant* $C(\bar{f}) \ge 0$.

Though the name Padé is closely connected with rational approximations to analytic functions [27], Hermite's contribution to this field can be considered the most important. What is the most interesting from the point of view of this paper, is the Hermite's creation of the method of Padé approximations for the purposes of transcendental number theory [22]. The first general example of a Padé approximation was constructed by Hermite around 1873 [22], and this was an example of $(m_1,\ldots,m_n)$ Padé approximants to $(e^{\omega_1 z},\ldots,e^{\omega_n z})$ at $z=0$. Hermite himself used these Padé approximants to prove that e is a transcendental number [22], and Hermite's system of Padé approximants was (and *is still*) the only method of proof of the Lindemann-Weierstrass theorem (see [30]) that $e^{\alpha_1},\ldots,e^{\alpha_n}$ are algebraically independent for algebraic numbers $\alpha_1,\ldots,\alpha_n$ linearly independent over $\mathbf{Q}$.

We now present the Hermite system of Padé approximants in terms of a contour integral in $\mathbf{C}^1$ (cf.[10], [25], [30]). These particular formulae will be generalized in $\mathbf{C}^1$ using the Hermite interpolation formula and the Newton interpolation series, and in $\mathbf{C}^k$ using the Grothendick residue formula (see later in §7).

Let $\omega_1,\ldots,\omega_m$ be m different complex numbers, $n_1,\ldots,n_m$ be m non-negative integers, and

$$N = \{\sum_{k=1}^{m} (n_k + 1)\} - 1 . \tag{1.1}$$

We define the following function:

$$R(x) = R(x|e^{\omega_1 x},\ldots,e^{\omega_m x};n_1,\ldots,n_m) \overset{\text{def}}{=} \frac{1}{2\pi i}\int_C \frac{e^{xz}\,dz}{Q(z)} \tag{1.2}$$

$$\text{for} \quad Q(z) = \prod_{k=1}^{m}(z-\omega_k)^{n_k+1},$$

where C is a simple closed curve with positive orientation which contains the whole circles $|z|\le|\omega_k|:k=1,\ldots,m$ in its interior. Then by the Cauchy residue theorem we have

$$R(x) = Q_1(x)e^{\omega_1 x} + \ldots + Q_m(x)e^{\omega_m x} \qquad (1.3)$$

where $Q_1(x),\ldots,Q_m(x)$ are polynomials of degree $n_1,\ldots,n_m$, respectively having the form

$$Q_k(x) = \prod_{\substack{i=1\\ i\neq k}}^{m} (\omega_k-\omega_i+\frac{d}{dx})^{-n_i-1} \frac{x^{n_i}}{n_i!} \quad , \; k=1,\ldots,m. \qquad (1.4)$$

Moreover at $x=0$ we have the following Taylor expansion of $R(x)$:

$$R(x) = \frac{x^N}{N!} + O(x^{N+1}) \quad ,$$

i.e., $R(x)$ is the remainder function and $(Q_1(x),\ldots,Q_m(x)$ are $(n_1,\ldots,n_m)$ Padé approximants to $(e^{\omega_1 x},\ldots,e^{\omega_m x})$ at $x=0$.

We explain now how to deduce the Hermite formulae for Padé approximations from the contour integral representation of the Hermite interpolation formula. Different forms of the Hermite interpolation formula give us the possibility of certain generalizations of Hermite's method to other functions.

We use in this situation a rather unexpected relation of the Padé approximation to $(e^{\omega_1 x},\ldots,e^{\omega_m x})$ at $x=0$ with an m-point polynomial approximation to the function e^{xZ} at points $\omega_1,\ldots\omega_m$ and, thus, with the Hermite interpolation formula.

This is not obvious at all! Perhaps, the first to realize this was A.O.Gelfond, who used this formalism in 1929 to prove the transcendence of e^{π}. Then, later on, in 1931, K.Mahler used this connection [25].

We explain the Hermite interpolation formula and its relations with the Newton interpolation series.

We consider the sequence of divided differences, corresponding a given function $f(x)$. Let $\lambda_1,\lambda_2,\lambda_3,\ldots$ be the sequence of distinct points; we set

$$[\lambda_1]_f = f(\lambda_1) \quad ;$$

$$[\lambda_1,\lambda_2]_f = \frac{f(\lambda_1)-f(\lambda_2)}{\lambda_1-\lambda_2} \quad ;$$

and inductively

$$[\lambda_1,\ldots,\lambda_{k+1}]_f = \frac{[\lambda_1,\ldots,\lambda_k]_f - [\lambda_2,\ldots,\lambda_{k+1}]_f}{\lambda_1-\lambda_{k+1}}$$

Of course, if f(x) is smooth enough (e.g.C^∞), then this formula becomes a continuous functions of the variables λ_i. In this case we can write an expression for divided differences when some of λ_i are equal. Suppose that in sequence $(x_1,\ldots,x_n)$ we have only m distinct numbers $\xi_1,\ldots,\xi_m$, where ξ_j is repeated ν_j times: $j=1,\ldots,m$;

$$\nu_1 + \ldots + \nu_m = n \qquad \text{and}$$

$$P(x) = \prod_{i=1}^{n}(x-x_i) = \prod_{j=1}^{m}(x-\xi_j)^{\nu_j} .$$

Then

$$[x_1,\ldots,x_n]_f = \sum_{j=1}^{m}\frac{1}{(\nu_j-1)!}\left(\frac{\partial}{\partial x}\right)^{\nu_j-1}\left\{\frac{f(x)(x-\xi_j)^{\nu_j}}{P(x)}\right\}\Bigg|_{x=\xi_j} .$$

In this case we have the following useful Hermite interpolation formulae:

Lemma 1.2 - For sufficiently smooth f(x) *and arbitrary* $\lambda_1,\ldots,\lambda_n$ *we have the following representation*

$$f(x) = \sum_{i=0}^{n-1}[\lambda_1,\ldots,\lambda_{i+1}]_f P_i(x) + R_n(x) ,$$

where

$$R_n(x) = [\lambda_1,\ldots,\lambda_n,x]_f\, P_n(x)$$

$$P_0(x)=1 , \qquad P_1(x)=x-\lambda_1, \;\ldots,\; P_k(x)=\prod_{j=1}^{k}(x-\lambda_j), \;\ldots$$

In particular, $R_n(x)$ *vanishes at any point* $\lambda_i:i=1,\ldots,n$ *with the multiplicity equal to the number of occurence of* λ_i *in* $\{\lambda_1,\ldots,\lambda_n\}$.

In other words we obtain closed expression for polynomial Padé approximation to f(x) at any set of points; this Padé approximation is, of course, unique.

Under some conditions on f(x) and $\lambda_1,\ldots,\lambda_n,\ldots$ we can prove that for an infinite system of points $\lambda_1,\ldots,\lambda_n,\ldots$ the series

$$\sum_{i=0}^{\infty}[\lambda_1,\ldots,\lambda_{i+1}]_f\, P_i(x)$$

converges to f(x). For an entire function f(x) this can be uniform convergence; for meromorphic ones only convergence is measure.

Let us present the Hermite interpolation formula in terms of the residue theorem (cf.[1], [30]). Let our function f(x) be regular in a domain D of the complex z-plane and n points $z_1,\ldots,z_n$ are given. We want (as in all approximation problems) to determine a polynomial $H_{n-1}(z)$ of degree $\le n-1$ such that the fraction

$$T_n(z) = \frac{f(z)-H_{n-1}(z)}{\prod_{i=1}^{n}(z-z_i)}$$

is regular in D. It's clear that the solution is unique. We put

$$P_k(z) = (z-z_1) \ \dots \ (z-z_k) \ : \ k=0,1,\dots,n.$$

Then

$$P_k(z) = (z-z_k)\, P_{k-1}(z)$$

and

$$(z-z_k)P_{k-1}(\zeta) - P_k(\zeta) = (z-\zeta)P_{k-1}(\zeta) ,$$

so

$$\frac{1}{z-\zeta}\left(\frac{P_{k-1}(\zeta)}{P_{k-1}(z)} - \frac{P_k(\zeta)}{P_k(z)}\right) = \frac{P_{k-1}(\zeta)}{P_k(z)} . \tag{1.5}$$

Suppose that ζ lies in D and consider a simply closed curve C in D whose interior lies in D and contains the n+1 points $z_1,\dots,z_n,\zeta$. We define then

$$a_{k-1} = \frac{1}{2\pi i}\int_C \frac{f(z)}{P_k(z)}\,dz \quad : k=1,\dots,n \tag{1.6}$$

and

$$R_k(\zeta) = \frac{1}{2\pi i}\int_C \frac{P_k(\zeta)}{P_k(z)}\,\frac{f(z)}{z-\zeta}\,dz . \tag{1.7}$$

Then, by the Cauchy theorem,

$$R_0(\zeta) = f(\zeta)$$

and by (1.5)-(1.7) we have

$$R_{k-1}(\zeta) - R_k(\zeta) = a_{k-1}\, P_{k-1}(\zeta) \ : \ k=1,\dots,n .$$

Adding these expressions we obtain, at last,

$$H_{n-1}(\zeta) = a_0 P(\zeta) + \dots + a_{n-1}\, P_{n-1}(\zeta)$$

and

$$f(\zeta) = H_{n-1}(\zeta) + R_n(\zeta) ,$$

where

$$T_n(\zeta) = \frac{R_n(\zeta)}{P_n(\zeta)} = \frac{1}{2\pi i}\int_C \frac{f(z)}{P_n(z)}\,\frac{dz}{z-\zeta} .$$

Here $H_{n-1}(z)$ is an approximation polynomial and $R_n(z)$ is the remainder function. If the sequence $z_1,\dots,z_n,\dots$ is an infinite one and $\lim_{n\to\infty} R_n(z)=0$ for all $z\in D_0\subset D$, then

$$f(z) = a_0P_0(z) + a_1P_1(z) + \ldots \qquad (z\in D_0)$$

In particular, if f(z) is not a polynomial, then $a_n \neq 0$ for infinitely many n.

Now we realize that in the special case

$$f(z) = e^{xz}$$

the expression $a_{k-1}(x)$,

$$a_{k-1}(x) = \frac{1}{2\pi i}\int_C \frac{e^{xz}dz}{P_k(z)}$$

is exactly the function R(x) of Hermite constructed for the Padé approximation to $(e^{\omega_1 x},\ldots,e^{\omega_m x})$. This corresponds to the case

$$K = \sum_{i=1}^{m}(n_i+1) = N+1$$

when we take for $\{z_1,\ldots,z_{N+1}\}$ the set containing n_i+1 times ω_i:i=1,...,m:

$$(\underbrace{\omega_1,\ldots,\omega_1}_{n_1+1};\ldots;\underbrace{\omega_m,\ldots,\omega_m}_{n_m+1})$$

Why should the coefficient $a_{k-1}(x)$ be the remainder function for one-point Padé approximation (see [30])?

2. THE EXPRESSION FOR ONE-POINT PADE APPROXIMATION FOR SOME SPECIAL CASES IN TERMS OF "GLOBAL RESIDUE FORMULA"

First of all, we want to answer the question from the end of the previous chapter and present you with one of the possible expressions for the remainder function in Padé approximation problem.

The connection between the N-point Padé for one function and the one-point Padé for many functions lies in the following trivial one-dimensional (in $\mathbb{C}^1$).

Corollary 2.1 - Let f(z) be regular in the interior of the closed curve C*, where the interior of* C *contains the origin* z=0 *and* n *points* $z_1,\ldots,z_n$. *Then for the function*

$$R(x) = \frac{1}{2\pi i}\int_C \frac{f(xz)}{P_n(z)}\,dz \quad , \tag{2.1}$$

$$P_n(z) = \prod_{i=1}^{n} (z-z_i)$$

R(x) *has a zero at* x=0 *of multiplicity* $\geq n-1$.

Also we have the finite expression for R(z) in terms of $f(xz_i)$ and their derivatives. Let $\{z_1,\ldots,z_n\}$, as before, consist of m different points $\omega_1,\ldots,\omega_m$, where ω_j is repeated n_j+1 times: $j=1,\ldots,m$:

$$P_n(z) = \prod_{i=1}^{n} (z-z_i) = \prod_{j=1}^{m} (z-\omega_j)^{n_j+1} .$$

Then we have

$$R(x) = \sum_{j=1}^{m} \frac{1}{n_j!} \left(\frac{\partial}{\partial z}\right)^{n_j} \left\{ \frac{f(xz)\cdot(z-\omega_j)^{n_j+1}}{P_n(z)} \right\} \Bigg|_{z=\omega_j} .$$

Now R(x) is a linear combination of the expressions

$$x^i f^{(i)}(\omega_j x) \quad : \quad i=0,\ldots,n_j,\ j=1,\ldots,m$$

with the coefficients depending on ω_j and n_j: $j=1,\ldots,m$ only.

In other words, for special functions f(x), satisfying certain types of linear differential equations with the coefficients being rational functions we can obtain by the "Global residue formula" (corollary supra) a nice Padé approximation.

In the case of f(x) satisfying an inhomogenuous linear differential equation of the first order there are now *two* ways of the representation of Padé approximants. One, algebraic, belongs to Laguerre [24] and, another, based on monodromy group, is explained in [9] and below. We propose to use "Global residue formula" to get another representation of Padé approximants.

Proposition for Research. It would be nice to study Padé approximants via contour integral formulas for f(x) which satisfy inhomogenuous linear equations of the first order, when we do know to construct the Padé approximation algebraically.

Although we don't know exactly how to generalize the "Global Residue Formula" to get auxiliary functions in *every* transcendence problem, we wouldn't be surprised if one day such universal representations (even if not very convenient) appeared.

We'll present below several examples showing the importance of the

"General Residue Formula" in transcendence applications. Now we'll speak briefly about special hypergeometric functions, which are natural generalizations of Bessel functions.

These are the following functions:

$$y(x) = 1 + \sum_{n=1}^{\infty} \frac{x^n}{\prod_{i=1}^{n} \theta(i)}$$

for polynomial $\theta(x)=x^m+\ldots+\theta_0$ of degree $m\geq 1$ with algebraic coefficients. This function satisfies a linear differential equation of the order m:

$$\theta(x\frac{d}{dx})\, y = xy + \theta_0 \ .$$

The first results on the irrationality of $y(z),\ldots,\ldots,y^{(m-1)}(z)$ at rational $z\neq 0$ belong to A.Hurwitz (1883 and 1888). He proved, e.g. that for $\theta(x)\in \mathbb{Q}[x]$ and $z\neq 0$ in the imaginary quadratic field $\mathbb{Q}(\sqrt{-D})$, two of the numbers $y(z),\ldots,y^{(m)}(z)$ have a ratio which is not in $\mathbb{Q}(\sqrt{-D})$. Then Ch.Osgood (1966, [26]) treated this problem for the case of $\theta(x)$ having only rational roots, using the "Global Residue Formula". However in that case $y(x)$ was an E-function. Then Osgood and Galochkin (1970-1978), see [21], estimated linear forms in $y^{(i)}(\omega_j)$: $i=0,\ldots,m-1$, $j=1,\ldots,N$ for different ω_j from the imaginary quadratic field Π.

It is very easy to find that the "General Residue Formula" can be indeed applied to the construction of the auxiliary function $R(x)$ in the problem of the linear independence of $y^{(i)}(\omega_j)$: $i=0,\ldots,m-1$, $j=1,\ldots,N$ *even when* $\theta(x)$ *has nonrational roots.*

Theorem 2.2 - *(Hurwitz-Osgood-Galochkin-...) Let* $\theta(x)$ *be the polynomial from* $\Pi[x]$, *where* Π *is the imaginary quadratic field with roots different from* $-1,-2,\ldots$ *Then for any* N *distinct non-zero numbers* $\omega_1,\ldots,\omega_N$ *from* Π, *the numbers*

$$1,\ y^{(i)}(\omega_j):\quad i=0,\ldots,m-1,\quad j=1,\ldots,N$$

are linearly independent over Π.

For the proof we use simply the "Global Residue Formula". Namely the auxiliary function $R(x)$ is of the form (2.1):

$$R(x) = \frac{1}{2\pi i}\int_C \frac{y(xz)}{z^{N_0}\prod_{j=1}^{N}(z-\omega_j)^{N_j}}\, dz \quad .$$

The *crucial* point here is the following observation, based completely on the properties of the differential equation, satisfied by $y(x)$.

Lemma 2.3 - *Let* $\theta(x)\in\mathbb{K}[x]$ *and, for an integer* $b\geq 1$, $b\cdot\theta(x)$ *have algebraic integer coefficients.*

Then for $k=nm+r$, $0\leq r\leq m-1$,

$$x^k y^{(k)} = \sum_{j=0}^{m} B_{k,j}(x)\cdot\left(x\frac{d}{dx}\right)^j y$$

for polynomials $B_{kj}(x)\in\mathbb{K}[x]$, *where* $b^k B_{k,j}(x)$ *have integer coefficients and*

1) $\deg(B_{k,0})\leq n-1$, $\deg(B_{k,j})\leq n$:

2) $|\overline{B_{k,j}}|\leq\prod_{i=0}^{k}(i+1+|\bar{g}|)$:

$k=nm+r$, $j=0,\ldots,m$.

According to this lemma we can change in the Hermite expression for $R(x)$ all non-constant terms

$$x^i y^{(i)}(\omega_j x) \quad : \quad 0\leq i\leq N_j-1 \quad : \quad j=0,\ldots,N$$

into terms

$$Q_{i,j}(x)\left(x\frac{d}{dx}\right)^i y(\omega_j x) \quad : \; i=0,\ldots,m, \; j=0,\ldots,N \text{ and } \deg(Q_{ij})\leq\left[\frac{N_j-1}{m}\right].$$

In other words, the function $R(x)$ represents a perfect Padé approximation for the system of functions $\left(x\frac{d}{dx}\right)^i y(\omega_j x)$: $i=0,\ldots,m$, $j=0,\ldots,N$ or for the system $y^{(i)}(\omega_j x)$: $i=0,\ldots,m$, $j=0,\ldots,N$.

The rest of the proof is already the same as for $e^{\omega_j x}$, only the second algebraic part is changed a little in view of the previous lemma.

This theorem *cannot* be immediately generalized to the algebraic independence of $y^{(i)}(\omega_j)$ rather than to linear independence. The reason for this is simple: the product of the functions $y(\omega_j x)$ is *not* a function of the same type. However the measure of the linear independence of $y^{(i)}(\omega_j)$ can be proved to be rather good.

Despite this caution, this theorem is indeed nontrivial.

Indeed, if one of the roots of $\theta(x)$ is an irrational number then $y(x)$ is not an E-function in Siegel's sense [29], [30]. In other words, none of the general statements of Siegel (see [19], [29], [30]) can be applied for such $y(x)$.

The studies of arithmetic properties of the values of $y(x)$ will be continued in this paper.

3. PADÉ APPROXIMATION AND RELATIONS WITH DARBOUX-BÄCKLUND TRANSFORMATION

As we explained before, even for the exponential function the general N-point Padé approximation problem is still unresolved. Hermite formulae give us only an N-point polynomial approximation to $e^{\omega x}$. However this approximation is not "almost perfect" at all: in the Hermite formula

$$f(z) - H_{n-1}(z) = T_n(z)\prod_{i=1}^{n}(z-z_i)$$

the degree of $H_{n-1}(z)$ can be much smaller than n-1, because $f(z)-H_{n-1}(z)$ is a transcendental function with infinite number of zeroes. The distribution of these zeroes in a very complicated problem and is crucial for the closed expression of the auxiliary functions $F(z)=P(z,e^{\omega_1 z},\dots,e^{\omega_n z})$, having zeroes at points $z=z_0,z_1,\dots,z_n$.

<u>Problem</u>. Let $\omega_1,\dots,\omega_n$ be distinct complex numbers. Find the distribution of the zeroes of the entire functions $P(z,e^{\omega_1 z},\dots,e^{\omega_n z})$ or $P(e^{\omega_1 z},\dots,e^{\omega_n z})$ in terms of deg(P) for a polynomial $P(\bar{X})\in\mathbb{C}[\bar{X}]$. This distribution should depend heavily on the measure of the linear approximation of $\omega_1,\dots,\omega_n$.

Already for the sum of three exponents the situation becomes formidable. We can put

$$\Delta(z) = a_1 e^{\omega_1 z} + b_1 e^{\omega_2 z} + e^{\omega_3 z}$$

and ask about the position of zeroes of $\Delta(z)$ according to the minima of the linear forms $|n\omega_1+m\omega_2+k\omega_3|$. We know only some numerical experiments, showing close relations between the distribution of zeroes and continued fraction expansions of numbers like $(\omega_1-\omega_3)/(\omega_2-\omega_3)$.

There exists a very powerful method to treat this kind of problem, which was suggested by the example of $\Delta(z)$. We can put

$$u(z) = -\frac{d^2}{dz^2}\log\Delta(z) .$$

Then we see that zeroes of $\Delta(z)$ are poles of $u(z)$. For a special choice of the coefficients of $\Delta(z)$ we find that $u(z)$ is a so-called reflectionless potential of Schrödinger operator with finitely many bound states.

We remember, however, that reflectionless potentials with N bound states (negative eigenvalues) are N-soliton solutions for the Korteweg-de Vries equation

$$u_t = 12uu_x - u_{xxx} .$$

This one particular example suggests a good analogy between Padé approximation and the inverse scattering technique in completely integrable systems.

Although we have now very general approach to the construction of Padé approximation and auxiliary functions using inverse scattering for an arbitrary (matrix) linear differential equation, we'll present an exposition only for the case of the equation of the second degree: Schrödinger equation.

We are searching for a function $\psi(z)$ (the remainder function in Padé approximation problem) satisfying an equation

$$-\psi_{zz} + u(z)\psi = \lambda\psi .$$

Then the poles (or zeroes) of ψ are determined by the poles of u(z). In the particular case of u(z) being a rational function in z or sinz, sinhz, we come to a very interesting many-particle problem. E.g. if

$$u(z) = 2 \sum_{i\in I} (z-a_i)^{-2} ,$$

then $\psi(z)$ is meromorphic function of z iff the system of particles a_i is the stationary configuration for the Hamiltonian with the potential $-x^2$, i.e.

$$\mathcal{U}: \sum_{j\neq i} (a_i-a_j)^{-3} = 0 \quad : \quad i\in I .^{*}$$

For the finite I the set of such potentials is non-zero if and only if

$$|I| = \frac{n(n+1)}{2}$$

for some integer $n\geq 1$ and is n-dimensional in this case [14]. Now $\psi(z)$ has a zero (or pole) of high order at $z=a_i$ if some of the particles a_j collapse at the point a_i. We can investigate such situations, because we have operating on the manifold M n-commuting Hamiltonian flows, generated by higher KdV equations [12], [13], [16]. The simplest case is the case

* See papers [12], [13], [17] for a discussion of these many-particle problems.

$$a_i = a \quad : \quad i \in I$$

and, of course,

$$|I| = \frac{n(n+1)}{2} \quad .$$

In other words we are considering an equation

$$-\psi_{zz} + \frac{n(n+1)}{z^2}\psi = \lambda\psi \ . \tag{3.1}$$

Similarly in the case of the stationary system of particles corresponding to the potential $\sin^{-2}x$ we obtain the following system

$$-\psi_{zz} + \frac{n(n+1)}{\sin^2 z}\psi = \lambda\psi \ . \tag{3.2}$$

Both these systems are degenerate versions of Lamé equation

$$-\psi_{zz} + n(n+1)\mathscr{P}(z)\psi = \lambda\psi$$

for an elliptic function $\mathscr{P}(z)$.

For n=0 both of the equations (3.1) and (3.2) have trivial solutions $\psi = e^{kz}$, $k^2 = \lambda$. However for $n \geq 1$ we found that (3.1) and (3.2) give us already the solution of the Padé approximation problem

1) (3.1) for e^{kx} or $\frac{tg\ kx}{k}$

2) (3.2) for $(1-x)^{\omega}$, $\omega \notin \mathbb{Z}$.

Indeed, according to the very simple consequence of the study of the stationary solutions of the Korteweg-de Vries equations we found that the equation (3.1) admits the solution

$$\psi(x) = A(x)e^{x} + B(x)e^{-x} \qquad (\lambda = 1)$$

having zero at x=0 of order exactly 2n+1 and with A(x), B(x) being polynomials of degree n. Then, of course, $\psi(x)e^{x} = R(x)$ is the remainder function of Padé approximation to e^{2x} of the type we constructed before (Bessel polynomials) in §1: (1.2)-(1.4) for $\omega_1 = -1$, $\omega_2 = 1$ (cf.[10], [32]).

Analogically for (3.2) we can construct the function

$$\psi(x) = A'(e^{x})e^{kx} + B'(e^{x})e^{-kx} \quad , \quad k^2 = \lambda$$

having zero at x=0 of order $\geq 2n+1$ for polynomials A'(z), B'(z) of degree $\leq n$.

Of course, equations of the type (3.2) are much more perspective ones than those of the type (3.1). In principle, equations $L\psi = \lambda\psi$ with linear operator $L=L(\sin^{-1}z;\partial/\partial z)$ of order $n\geq 2$ with typical coefficients being polynomials of $\sin^{-1}z$, should give us the construction of the auxiliary function $P(e^{\omega_1 z},\dots,e^{\omega_n z})$ for N-point approximation problem.

However, we can ask: are these examples nothing more than examples?

No, explicite construction of eigenfunctions $\psi(z)$ is possible using inverse scattering method and/or Gelfand-Levitan integral equation. For our problem the situation is even increadibly simplified.

First of all, for potentials $u(z)$ connected *with* many-particle problem (like $n(n+1)\mathcal{P}(x)$ etc.) the eigenfunctions $\psi(z)$ which we try to find is the *common* eigenfunction for two *commuting* linear operators

$$L\psi = \lambda\psi \ ; \qquad L = -\frac{d^2}{dx^2} + u(x) \qquad (3.3)$$
$$A\psi = \mu\psi \ .$$

(Here $A=A(u,u',\dots,d/dx)$ is of order $2n+1$ where $n(n+1)/2$ is "the number of particles" in $u(x)$).

According to the Burchuall-Chaundy-...-Novikov-Lax-... theory, both $u(x)$ and $\psi(x)$ are easily expressed in terms of θ-function of certain hyperelliptic curve $y^2=P_{2n+1}(x)$. [This curve is defined by very simple rule: L, A satisfy an equation $A^2=P_{2n+1}(L)$].

If this curve is rational, then $\psi(z)$ is expressed in terms of $e^{\omega_i z}$ only: multisoliton solutions like for (3.2) above. In the degenerate case $y^2=x^{2n+1}$ we obtain $u(x)$ as a rational function with the poles being particles: case (3.1) above.

In all these cases $\psi(z)$ is the solution of some problem of approximation for Abelian (or θ-) functions, corresponding to an elgebraic curve. E.g. in the degenerate cases we arrive to Padé approximation for exponential and linear function.

Problem - To write down the expression for $\psi(z)$ as solution of Padé approximation problem for the system of Abelian functions, corresponding to a given hyperelliptic case.

This problem is non-trivial even though we know already the expression for $\psi(z)$. In order to get something useful for applications we need first of all to prove that for a given hyperelliptic curve Γ of genus $g>0$ the product $J(\Gamma)\times\dots\times J(\Gamma)$ of its Jacobians $J(F)$ n times can be represented (up to the isogeny) as the Jacobians $J(F)$ of the hyperelliptic curve F of genus n·g and find this F! I don't know even the possibility of such construction (besides the case g=1, n=2 worked out by P.Deligne, A.Brumer).

We see, however, that the possibilities of construction of "auxiliary functions" $\psi(z)$ using commuting differential operators are limitated to the very narrow class of functions and the corresponding potentials u(x) have in this case very special form. We would like to consider e.g. potential

$$u(x) = \frac{w(w+1)}{x^2}$$

with non-integer value of w. This corresponds to the classical case of the Bessel function

$$J_{w+\frac{1}{2}}(x) .$$

The question now is different: not to construct simultaneously u(z), $\psi(z)$ and Padé approximation to $\psi(z)$, but, instead, knowing (z) to construct its Padé approximants.

The idea now is to use Darboux-Bäcklund transformation from the theory of linear differential operators to "generate" starting from the given solution $\psi(z)$ of

$$L_n\psi = 0$$

another solution ψ^{+1} of $L_n^{+1}(\psi^{+1})=0$, etc. such that for m-th Darboux transform ψ^{+m} of ψ, satisfying

$$L_n^{+m}(\psi^{+m}) = 0$$

we have ψ^{+m} as the remainder function of the Padé approximation problem for $\psi, \psi', \ldots, \psi^{(n-1)}$. We want also to get recurrent formulae between ψ^{+m}: m=0,1,2,... This approach is especially clear on the example of the Bessel function $J_\nu(x)$, where it's simply reduced to *the adding of integers* to the index ν

$$J_\nu^{+m} \longrightarrow J_{\nu+m} .$$

<u>Darboux transformation</u>. Let we have an equation

$$L\varphi = \varphi_t, \quad L = (d^n/dx^n) + \sum_{j=0}^{n-2} u_j(d^j/dx^j) . \tag{3.4}$$

Then for any solution ψ of (3.4) the function f defined by

$$f = \frac{\psi\varphi' - \varphi\psi'}{\psi} \tag{3.5}$$

satisfies another equation of the type (3.4)

$$L_n^* f = f_t ,$$

where coefficients of L_n^* are defined through $\psi, \psi', \ldots$.

In the stationary case $\varphi = e^{\lambda t}\varphi_0$ this transformation means that we add to the spectrum of L_n one eigenvalue λ_0, if $\psi = e^{\lambda_0 t}\psi_0$ and we obtain new operator L_n^*, whose coefficients and spectral data are easily recovered from L_n and (λ_0, ψ_0). The procedure of "adding of the indexes" means that we apply Darboux transformation to the *same* eigenvalue λ_0.

If we apply Darboux transformation first to λ_0, then to $\lambda_1, \lambda_2, \ldots$ we can reconstruct nice Wronskian formulae of D.Chudnovsky [14], [17], for the solutions of completely integrable systems. These formulae give us, e.g. Padé approximation forms to

$$J_{\nu_i}(\omega_j x) \; : \quad i=1,\ldots,n; \quad j=0,\ldots,N$$

(cf.supra).

Let's present formulae for Darboux transformation for the operator of the second order.

We start with the Sturm-Liouville problem

$$-f^{\alpha}{}'' + uf^{\alpha} = \alpha f^{\alpha} \tag{3.6}$$

for $u=u(x)$, $f^{\alpha}=f^{\alpha}(x)$. Then the Darboux transformation, applied *only to the point* $\alpha=\alpha_0$, gives us such a recurrent formula.

For the n-th potential

$$u_{n+1}^{\alpha} = -u_{n+2}^{\alpha}\left(\frac{f_n^{\alpha}{}'}{f_n^{\alpha}}\right)^2 + 2\alpha \quad ; \tag{3.7}$$

for the $n+1^{th}$ eigenfunction, corresponding to the eigenvalue α we have

$$f_{n+1}^{\alpha} = \frac{\int (f_n^{\alpha})^2 dx}{f_n^{\alpha}} \tag{3.8}$$

and for the general eigenfunction corresponding to the eigenvalue β:

$$f_{n+1}^{\beta} = f_n^{\beta}{}' - \frac{f_n^{\alpha}{}'}{f_n^{\alpha}} \cdot f_n^{\beta} \quad . \tag{3.9}$$

Then we have again

$$-f_n^{\beta\,''} + u_n^{\alpha}\cdot f_n^{\beta} = \beta f_n^{\beta} \quad , \tag{3.10}$$

where

$$u_0^{\alpha} = u \,, \qquad f_0^{\beta} = f^{\beta} \quad .$$

Example - For the Bessel function $J_\nu(x)$ we can always consider $J_\nu(x)$ as the solution of the equation of the type (3.6). The general solution of

$$-y'' + \frac{n(n-1)}{x^2}\,y = \alpha y \tag{3.11}$$

has the form

$$y = J_\nu(\sqrt{\alpha x})\cdot\sqrt{x} \tag{3.12}$$

where $n=\nu+\frac{1}{2}$.

We now can apply transformation (3.6)-(3.9) to $\alpha=0$, where $f^0=x^n$. Then

$$f_m^0 = x^{n+m} \quad ;$$

$$u_m^0 = \frac{(n+m)\ (n+m-1)}{x^2} \quad ;$$

$$f_0^{\beta} = J_{\nu+m}(\sqrt{\beta x})\cdot\sqrt{x} \quad .$$

In particular, for the half-integer ν we obtain expressions in terms of elementary functions.

In general the transformations (3.6)-(3.9) has a very nice spectral interpretation for rapidly decreasing on infinity $u(x)$ (cf.[14]).

We want to show how this method of "adding of indexes" is working for the diophantine approximations to $J'_\nu(x)/J_\nu(x)$ for imaginary quadratic $x\neq 0$. We follow our paper [10].

Appendix to §3. - *The measures of irrationality of $e^{2/n}$ and of $J'_\nu(x)/J_\nu(x)$ based on Darboux-Bäcklund transformation.*

We start with an arbitrary Bessel function $J_\nu(x)$. Then using the recurrent formula

$$J_{\nu-1}(x) + J_{\nu+1}(x) = 2\nu x^{-1}J_\nu(x)$$

(the addition of the index), we obtain in general the formula

$$J_{\nu+m}(x) = J_\nu(x)R_{m,\nu}(x) - J_{\nu-1}(x)R_{m-1,\nu+1}(x) \tag{*}$$

where $R_{m,\nu}(x)$ is the polynomial of degree m of x^{-1} (Lommel polynomial, [2]).

Then we have the classical representation [2]:

$$1)\quad R_{m,\nu}(z) = \sum_{n=0}^{[m/2]} \frac{(-1)^m (m-n)!\Gamma(\nu+m-n)}{n!(m-2n)!\Gamma(\nu+n)} \left(\frac{z}{2}\right)^{-m+2n} ;$$

$$2)\quad R_{m,\nu}(z) = \frac{\pi z}{2\sin(\nu\pi)} \left[J_{\nu+m}(z)J_{-\nu+1}(z)+(-1)^m J_{-\nu-m}(z)J_{\nu-1}(z)\right].$$

From the formula 2) we get an asymptotic formula

$$3)\quad \lim_{m\to\infty} \frac{(z/2)^{m+\nu-1} R_{m,\nu}(z)}{\Gamma(\nu+m)} = J_{\nu-1}(z)$$

$$4)\quad |R_{m,\nu}(x)| \sim |J_{\nu-1}(x)|\cdot\left(\frac{2}{x}\right)^{m+\nu-1}\Gamma(\nu+m) \quad .$$

We have classical estimates of $J_\nu(z)$ [2], e.g.

$$5)\quad J_w(z) \sim \left(\frac{z}{2}\right)^w \frac{1}{\Gamma(w+1)}$$

or

$$|J_w(z)| \le \left|\frac{z}{2}\right|^w \frac{e^{|y|}}{\Gamma(w+1)} \quad \text{for } w>-\tfrac{1}{2},\ z=x+iy .$$

Let $\mathbb{K} = Q(\sqrt{-D})$ be a quadratic imaginary field and $\nu\in\mathbb{K}$. Let $\mathbb{Z}_{\mathbb{K}}$ be the ring of integers of $\mathbb{K}$ and $d\in\mathbb{Z}_{\mathbb{K}}$ be such that $d\nu\in\mathbb{Z}_{\mathbb{K}}$.

6) $R_{m,\nu}(x)$ is an integer from $\mathbb{Z}_{\mathbb{K}}$ if $x\in\mathbb{K}$ and $2/_{x\cdot d}\in\mathbb{Z}_{\mathbb{K}}$;

7) for $4/_{x^2\cdot d^2}\in\mathbb{Z}_{\mathbb{K}}$ the number $(2/_{xd})^m R_{m,\nu}(x)$ is an algebraic integer from $\mathbb{Z}_{\mathbb{K}}$.

Of course, 6) and 7) follows trivially from the expressions 1), 2). From 4) and 5) we get

$$8)\quad \left|\frac{J_\nu(x)}{J_{\nu-1}(x)} R_{m,\nu}(x) - R_{m-1,\nu+1}(x)\right| < \left|R_{m,\nu}(x)\right|^{-1} \frac{\Gamma(\nu+m)}{\Gamma(\nu+m+1)}\cdot c$$

for $c=c(\nu)>0$.

Now 4), 6), 7), 8) implies immediately

Theorem 3.1 (see [10]) - Let $\mathbb{K}=Q(\sqrt{-D})$, $d\in\mathbb{Z}_{\mathbb{K}}$ *and* $d\nu\in\mathbb{Z}_{\mathbb{K}}$. *Let* $2/_{xd}\in\mathbb{Z}_{\mathbb{K}}$, *then for integers* $p, q\in\mathbb{Z}_{\mathbb{K}}$, $q\neq 0$,

$$\left| J'_\nu(x)/J_\nu(x) - p/q \right| > c' \cdot |q|^{-2} \frac{\log\log(|q|+2)}{\log(|q|+2)} \quad .$$

Let $4/x^2 d^2 \in \mathbb{Z}_{\mathbb{K}}$. *Then for integers* $p,q \in \mathbb{Z}_{\mathbb{K}}$, $q \neq 0$,

$$\left| x \cdot J'_\nu(x)/J_\nu(x) - p/q \right| > c' \cdot |q|^{-2} \frac{\log\log(|q|+2)}{\log(|q|+2)}$$

for $c' = c'(\nu) > 0$.

For example, for $\nu = \frac{1}{2}$ we obtain the same bounds for the rational approximations to the following numbers (considered by W.Adams and P.Bundschuh):

$$tg(1/n), \quad tg(1/n)/\sqrt{n}, \quad e^{2/n} \qquad \text{for} \quad n \in \mathbb{Z},\ n \neq 0 \ .$$

For some numbers of the nature considered in 3.1 above the same bound was established independently by P.Bundshuh.

All these statements follow immediately from the procedure of the "adding of indexes" by Darboux-Bäcklund transformation and are particular cases of the applications of the inverse scattering method to arithmetics.

4. BOUNDS FOR THE MEASURE OF LINEAR INDEPENDENCE OF VALUES OF $y(x),\ldots,y^{(m-1)}(x)$.

We return to the function $y(x)$ from §2. We obtain a bound for the measure of linear independence of $y(x),\ldots,y^{(m-1)}(x)$ at rational $x \neq 0$ using our knowledge of explicit Padé approximants and Darboux-Bäcklund transformation of $(d^m/dx^m)+(-1)^m\lambda$.

We'll apply now formulae of Padé approximations from §§1-3 to some fine estimates in the linear forms. We consider two classes of functions:

1) linear forms in

$$y(1/q),\ y'(1/q),\ \ldots\ ,\ y^{(m-1)}(1/q)$$

for $y(x) = \sum_{n=0}^{\infty} \{\prod_{j=1}^{n} \theta(j)\}^{-1} x^n$ and $\theta(x) \in \mathbb{I}[x]$, $d(\theta) = m$;

2) linear forms in

$$\zeta_m^\kappa e^{\zeta_m^\kappa x} : \kappa = 1,\ldots,m \qquad \text{for } \zeta_m = e^{2\pi i/m} \quad \text{for } m \geq 2.$$

For linear forms in both sequences of numbers we show results that are better than those for "almost all" sequences of numbers, predicted by Khirtchin theorem [34].

For both cases we can apply the Bäcklund-Darboux transformation to the linear operator of m-th order.

Especially interesting is the case 2). We start from the linear operator of the m-th order

$$\mathcal{L} = \frac{d^m}{dx^m} + (-1)^m \lambda$$

and we apply Bäcklund transformation to the point $\lambda_0=0$. In this way we obtain systems of linear forms

$$\sum_{\kappa=1}^{m} P_{n,\kappa} e^{\zeta_m^\kappa x} = R_n(x)$$

for $P_{n,\kappa}(x) \in \mathbb{Z}[\zeta_m, x]$: $\kappa=1,\dots,m$ that give Padé approximation to

$$\{e^{\zeta_m x},\dots,e^{\zeta_m^m x}\} \text{ at } x=0 .$$

Here the function $e^{\zeta_m^\kappa x}$ naturally arises from the operator $\mathcal{L}$ for $\lambda=1$ (or any $\lambda \neq 0$).

<u>Remark</u>. Because Padé approximation to the exponential function is perfect, then the expressions for this Padé approximations are *the same* as in the §1: (1.2)-(1.4). However the general estimates based on (1.2)-(1.4) are unsatisfactory [25], [29], [30]. We present the estimates that are already best possible.

First of all we present the measure of the linear independence of the numbers

$$y(1/q),\dots,y^{(m-1)}(1/q)$$

from the theorem 2.2 in the case of the integer $q \geq 1$ following Galochkin.

<u>*Theorem 4.1*</u> - *Let* $\theta(x)$ *be the polynomial from* $\mathbb{I}[x]$ *of the degree* m *with roots different from* -1,-2,... ,*where* $\mathbb{I}$ *is the imaginary quadratic field and*

$$y(x) = \sum_{n=1}^{\infty} \{\prod_{j=1}^{n} \theta(j)\}^{-1} x^n \quad .$$

Then for an integer $q \geq 1$ *and integers* $n_1,\dots,n_m$ *with* $0<\max_{i=1,\dots,m}|n_i|=H$ *we have*

$$\left|\sum_{i=1}^{m} n_i\, y^{(i-1)}(1/q)\right| > \gamma_1\cdot|q|^{1-m}\cdot H^{1-m}\cdot\left\{\frac{\log\log(H+2)}{\log(H+2)}\right\}^{x}$$

for $\gamma_1=\gamma_1(y)>0$ *and* $x=(m-1)^2H(\theta)$.

Moreover for integers $n_0,\dots,n_m$ *with* $0<\max_{i=0,\dots,m}|n_i|=H$ *we have the same type of bound*

$$|n_0+n_1y(1/q)+\dots+n_my^{(m-1)}(1/q)| > \gamma_1'\cdot|q|^{-m}\cdot H^{-m}\cdot\left\{\frac{\log\log(H+2)}{\log(H+2)}\right\}^{x}$$

for $\gamma_1'=\gamma_1'(y)>0$, $x=(m-1)^2\cdot H(\theta)$.

Remark 1. Again we remark that if one of the roots of $\theta(x)$ is not a rational number then $y(x)$ is not a E-function in Siegel's [30] sense.

Remark 2. If q is sufficiently large, $q\geq q_0(H(\theta),m)$, then the exponent x can be changed to $x'=(m-1)^2$.

For example, in the case of the Bessel function $J_\nu(x)$ (see §3) this is much better than metric assertion like the Khintchin theorem.

The proof is the same as for the theorem 2.2. We consider the same auxiliary function

$$R(x) = \frac{1}{2\pi i}\int_C \frac{y(xz)\,dz}{z^{N_0}(z-1)^{N_1}} .$$

However, in this very simple case, we can always present $R(x)$ in the differential form (cf. with Hermite formalism). We have a nice differential representation for the auxiliary functions $R(x)$ (belonging to Osgood):

$$R(x) = \prod_{i=0}^{N_2}(\delta-i)\,y(x)$$

for $\delta=x\frac{d}{dx}$, where $N_0=N_1$ and $N_2=N_0+N_1$. Now we can rewrite $R(x)=R_{N_2}(x)$ as

$$R_{N_2}(x) = B_{N_2,1}(x)y(x) + \dots + B_{N_2,m}(x)y^{(m-1)}(x)$$

for polynomials $B_{N_2,j}(x)\in\mathbb{Z}[x]$: $j=1,\dots,m$ on the basis of lemma 2.3. Again we have

$$\deg(B_{N_2,j}) \leq \left[\frac{N_2+m-1}{m}\right] : \quad j=1,\dots,m$$

and the upper bounds for $R(1/q)$ can be deduced e.g. from the representation of $R(x)$ as a contour integral.

Question. Is it possible to have $x=1$ always (i.e. for $m\geq 3$)?

We'll present now a result about the linear form in $e^{\zeta_m^\kappa x}$: $\kappa=1,\dots,m$, $\zeta_m^m=1$. We have

Theorem 4.2 - *Let* $m\geq 2$ *and* $\zeta_m=e^{2\pi i/m}$. *We define* m *functions* $\psi_k(x):k=1,\dots,m$ *in the following way*

$$\psi_k(x) = \sum_{j=1}^{m} \zeta_m^{j\cdot k} e^{\zeta_m^j x} \quad : \quad k=1,\dots,m.$$

Then for any integer $q\geq 1$ *and the set* $\{n_1,\dots,n_m\}$ *of integers,* $0<\max_{i=1,\dots,m}|n_i|=H$ *we have*

$$\left|\sum_{\kappa=1}^{m} n_\kappa \psi_\kappa(1/q)\right| > \gamma_3(q)^{1-m} H^{1-m} \left\{\frac{\log\log(H+2)}{\log(H+2)}\right\}^{(m-1)^2}$$

for $\gamma_3=\gamma_3(m)>0$.

For the proof we simply use the Hermite formulae from the §1 in the case

$$\omega_\kappa = \zeta_m^\kappa \quad : \quad \kappa=1,\dots,m;$$

$$n_1 = n_2 = \dots = n_m = n \quad .$$

If we use for the polynomials $P_k(x)$ in

$$R(x;n) = \sum_{\kappa=1}^{m} P_\kappa(x) e^{\zeta_m^\kappa x}$$

the expression (1.4), then we find, after natural simplifications, that

$$P_k(x) = m^{-n-1} \zeta_m^{\kappa(-n-1)} \sum_{\lambda=0}^{n} \sum_{\lambda_1+\dots+\lambda_{m-1}=\lambda} \prod_{i=1}^{m-1} \binom{-n-1}{\lambda_i} (1-\zeta_m^i)^{-\lambda_i} \cdot$$

$$\zeta_m^{-\lambda\kappa} \partial_x^\lambda \{x^n/n!\} =$$

$$= m^{-n-1} \zeta_m^{\kappa(-n-1)} \sum_{\lambda=0}^{n} p_{n,\lambda} \cdot \zeta_m^{-\lambda\kappa} x^{n-\lambda} \quad .$$

Now, if we take $n+1\equiv 0 \pmod m$ (*) we can write $R(x;n)$ as

$$R(x;n) = m^{-n-1} \cdot \sum_{\lambda=0}^{n} x^{n-\lambda} p_{\lambda,n} \sum_{\kappa=1}^{m} \zeta_m^{-\kappa\lambda} e^{\zeta_m^\kappa x}$$

for rational numbers $p_{\lambda,n}$. Again $R(x;n)$ can be written as

$$R(x;n) = \sum_{\kappa=1}^{m} R_\kappa(x) \psi_\kappa(x)$$

for $R_\kappa(x)\in Q[x]$: $\kappa=1,\dots,m$. For the Arithmetic and Analytic End of the Proof we can use the standard method and bounds of Hermite from [30].

Remark. There are, of course, relations between $\psi_k(x)$ and $y(x)$. E.g. for

---

(*) This point explains why for $m>2$ we obtain $x=(m-1)^2$!

m=2 we have $\psi_1(x)=2\mathrm{sh}x$, $\psi_2(x)=2\mathrm{ch}x$ and $y(x)$ arises from $J_{\frac{1}{2}}(x)$.

Analogously, $\psi_k(x)$: $k=1,\dots,m$ can be expressed in the terms of $y(x),\dots,y^{(m-1)}(x)$ for

$$\theta(x) = m^{m-1}x(x+\frac{1}{m})\dots(x+\frac{m-1}{m}) \quad .$$

5. LINEAR DIFFERENTIAL EQUATIONS ASSOCIATED WITH THE CONSTRUCTION OF THE PADÉ APPROXIMATION. DIFFERENT APPROACHES TO THE PADÉ APPROXIMATION PROBLEM.

We start with the remainder function

$$R_N(x) = f_1(x)P_{1,N}(x) + \dots + f_n(x)P_{n,N} \quad : \quad N\geq 1$$

in the Padé approximation problem for $(f_1(x),\dots,f_n(x))$. Then, in general, all $P_{i,N}(x)$ and $R_N(x)$ satisfy certain linear recurrent formulae. In other words all generating functions

$$y_i(z) = \sum_{N=1}^{\infty} P_{i,N}(x)z^N \quad , \qquad y(z) = \sum_{N=1}^{\infty} R_N(x)z^N$$

satisfy some linear differential in $\partial/\partial x$ equation

$$\mathcal{D}(\partial_z,x)\,y = 0 \ .$$

Then, say by Szegö's theorem, the asymptotics of $|P_{i,N}(x)|$, $|R_N(x)|$ with $N\to\infty$ -the most interesting from the point of view of transcendence theory- are determined by the singularities of $y_i(z)$, $y(z)$, i.e. by the singularities of $\mathcal{D}(\partial_z,x)$. The most interesting case is that of $\mathcal{D}(\partial_z,\cdot)$ having only regular singularities -a Fuchsian system.

<u>Problem.</u> If $f_1(z),\dots,f_n(z)$ is a solution of the system of linear differential equations, is it possible then to find a Fuchsian equation $\mathcal{D}(\partial_z,x)y=0$ such that the generating functions for the Padé approximations to $f_1(z),\dots,f_n(z)$ are the solutions to $\mathcal{D}(\partial_z,x)y=0$?

It is possible to reverse this problem and to start with the Fuchsian linear differential equation $\mathcal{D}(\partial_z,z)y=0$ with rational exponents and $\mathcal{D}(\partial_z,z)\in Q[\partial_z,z]$. We want to construct solutions $y_1,\dots,y_n$ of $\mathcal{D}y=0$ regular at $z=0$ and numbers $\omega_1,\dots,\omega_n$ such that

1) $y_i \in Q[[z]]$ and y_i are G-functions

2) the radius of the convergence of $y=\omega_1 y_1+\ldots+\omega_n y_n$ is larger than those of y_i: $i=1,\ldots,n$.

Not for any equation $\mathcal{D}y=0$ can this be done. But this is, in particular, the case that Apery treated for $\zeta(2)$, $\zeta(3)$. Such a formulation in terms of differential equations belongs to Van-der-Poorten [33].

Problem. For the hypergeometric functions

$$
{}_{p+1}F_p\left(\begin{matrix}\alpha_1,\ldots,\alpha_{p+1} \\ \beta_1,\ldots,\beta_p\end{matrix} ; z\right)
$$

with rational α_i, β_j what are the values of these functions at ± 1, $\pm\frac{1}{2}$?

We know that periods of Ferma varieties, products of Γ-factors, values of polylogarithms and logarithms of certain numbers like log2,... -numbers arising as the periods of special varieties- appear in such a way. How about $L(k,x)$ for different L-functions? How about periods of certain modular forms?

Now we want to treat some linear differential equations in order to get new measures of the irrationality of certain numbers. We know two possible (maybe connected) approaches towards this problem, arising from our discussion in the beginning of this chapter.

I. Starting from the function $f(x)$ satisfying the linear differential equation $\mathcal{D}_n f=0$ we construct Padé approximations $R_m(x)=A_0(x)f(x)+\ldots+A_{n-1}(x)f^{(n-1)}(x)$ satisfying another differential equation $\mathcal{D}_m R_m=0$. Then we find the relations between R_m, R_{m+1}, ... and investigate the arithmetic properties of the coefficients of $A_0(x),\ldots,A_{n-1}(x)$ of $R_m(x)$ (their size, denominator,...). On the basis of this, it is possible to establish a bound for the measure of approximation of linear forms in $f(\alpha),\ldots,f^{(n-1)}(\alpha)$ for $\alpha\neq 0$ inside the radius of the convergence of $f(z)$.

We can realize the analytic part of this program in the sense that we can construct an explicit Padé approximation to any system of functions satisfying linear differential equations [9], [7]. In this construction we use a certain method of "adding indeces" in the form of Riemann's notion of contiguous functions [9], [28]. This means that we consider systems of solutions of, say, Fuchsian linear differential equations of order m with fixed regular singularities and a fixed monodromy group, whose local exponents differ by rational integers. Precise expressions for the remainder function of the Padé approximation problem for the solutions of Fuchsian linear differential equations are given below in terms of the Riemann monodromy problem. In particular we can present an effective form of recurrences and linear differential equations satisfied by the remainder function and Padé approximants. Unfortunately these expressions will enter the notorious accessory parameters [23] characterizing the global monodromy in addition to the local monodromy (easily observable from the form of the equation). There are however two famous cases when the equation doesn't contain any accessory parameters and when our formulae for Padé approximants are especially efficient (e.g. as simple n-fold integrals with algebraic kernels).

The best examples are given by Jordan-Pochhammer equations

$$Q(z)y^{(n)}+\mu Q'(z)y^{(n-1)}-\frac{\mu(\mu+1)}{2}Q''y^{(n-2)}+\ldots+pR(z)y^{(n-1)}+(\mu+1)R'(z)y^{(n-2)}-\ldots=0,$$

where $Q(z)$ and $zR(z)$ are polynomials of degrees $\leq n$. In this case the monodromy group can be computed [23] and the solution is, in general, of the form

$$y=\int_C(\zeta-a_1)^{\alpha_1-1}\ldots(\zeta-a_n)^{\alpha_n-1}(\zeta-z)^{\mu+n-1}\,d\zeta\quad.$$

In this case rather nice results can be obtained for the measure of irrationality of periods of special Abelian Varieties.

Another, very important case, is the case of generalized hypergeometric ${}_qF_p$-functions, already considered in [6], [7], [9]. We also present below the simplest expression for the remainder function in the diagonal or sub-diagonal case and we present also applications of our analytic Padé approximation methods to the evaluation of the measure of irrationality.

There exists, however, another approach, that uses the Fuchsian equation as well.

II. Starting from the number $\alpha=f(a)$ for some function $f(x)$ we can try to present this number α as a limit

$$\lim_{n\to\infty} a_n/b_n = \alpha$$

for two solutions $\varphi(z)=\Sigma a_n z^n$, $\psi(z)=\Sigma b_n z^n \in Q[[z]]$ of some Fuchsian linear differential equation $\mathcal{D}y=0$. If a is a parameter, we consider the Fuchsian equation $\mathcal{D}(a)y=0$ and two its solutions $\varphi(z,a)=\Sigma A_n(a)z^n$, $\psi(z,a)=\Sigma B_n(a)z^n$ with radius of convergence $d_1(a)$ such that the function

$$y(z,a) = \varphi(z,a) - f(a)\psi(z,a)$$

has the larger radius of convergence $d_2(a)>d_1(a)$. Then

$$R_n(a) = A_n(a) - f(a)\cdot B_n(a)$$

gives us the sequence of "Padé-like" approximations for f(a). Then this system of Padé approximations can be used for the measure of irrationality of $\alpha=f(a)$.

Of course we suspect that two of these approaches should be equivalent.

As an example we treat in detail ip next chapter one of the cases, when all the different approaches are equivalent. This is the case of rational approximations to a logarithm $\log\alpha$ for a rational number α close to 1.

6. RATIONAL APPROXIMATIONS TO LOGARITHMIC FUNCTION. DIOPHANTINE APPROXIMATIONS TO LOGARITHMS OF SPECIAL ALGEBRAIC NUMBERS.

The diophantine approximations to logarithms $\log\alpha$ of algebraic α ($\alpha\neq 0,1$ and α close to 1) is immediately reduced to the Padé approximation to, say, $\log(1-x)$. This Padé approximation was, in principle, constructed by the Hermite formulae: take ω_k to be integers ≥ 0 and make the subsitution $e^x=1-z$. Also we know that

$$\log\frac{1+z}{1-z} = 2xF(\tfrac{1}{2},1;\tfrac{3}{2};x^2),$$

so Padé approximations can be effectively written in terms of hypergeometric polynomials: namely, Legendre polynomials $P_n(x)$. This possibility to use Legendre polynomials has great advantages, e.g. recurrent formulae for Legendre polynomials that immediately give us bounds for the sizes of coefficients.

The Algebraic and Analytic Parts of the Proofs are greatly amplified and the estimates are sharpened if you know the recurrent formulae for

Padé approximations. On the other hand, the existence of these formulae also shows that the new function

$$f(x,z) = \sum_{n=0}^{\infty} P_n(x) z^n$$

satisfies a linear differential equation of the second degree in $\partial/\partial z$:

$$\left[(z^2 - 2xz + 1)\partial_z^2 + 3(z - x)\partial_z + 1\right] f = 0 \quad .$$

We can now realize, that this equation is Fuchsian, has one algebraic solution,

$$f(x,z) = (z^2 - 2xz + 1)^{-\frac{1}{2}} \quad ,$$

and so the second solution f_1 can be found, and the function

$$g_0(x) = \tfrac{1}{2} \log \frac{x+1}{x-1}$$

can be interpreted in a way such that the coefficient

$$f_1 - g_0 f$$

has radius of convergence $|z|=r_2$ for r_2 being the larger of two roots of

$$z^2 - 2xz + 1, \qquad r_{1,2} = x \pm \sqrt{x^2-1} \quad .$$

Now it is clear that for the evaluation of the measure of irrationality of

$$\tfrac{1}{2}\log\frac{x+1}{x-1}$$

for, say rational, x, $x\to\infty$, it's better to use *only* Legendre functions (see [2] for a complete discussion).

Legendre polynomials. Legendre polynomials $P_n(x)$ arise from the recurrent formula

$$(n+1)P_{n+1}(x) = (2n+1)xP_n(x) - nP_{n-1}(x): \ P_0(x)=1,\ P_1(x)=x,\ \dots \tag{6.1}$$

and the differential equation

$$(1-x^2)y'' - 2xy' + n(n+1)y = 0 \ . \tag{6.2}$$

These polynomials have nice expressions in terms of hypergeometrix functions, e.g.

$$P_n(x) = F(-n,n+1,1;\frac{1-x}{2}) = \sum_{m=o}^{n} \binom{n}{m}\binom{n+m}{m}\left(\frac{x-1}{2}\right)^m$$

or

$$P_n(x) = 2^{-n} \sum_{m=0}^{n/2} (-1)^m \binom{n}{m}\binom{2n-2m}{n} x^{n-2m} \quad .$$

E.g. $P_n(x)$ as a polynomial in $\frac{1-x}{2}$ has integer coefficients!

The second (independent) solution of (6.1) and (6.2) is the Legendre function of the second kind: $Q_n(x)$. This function has zero at ∞ of multiplicity n+1 and can be used for Padé approximation.

For $Q_n(x)$ there exists a natural (for the theory of orthogonal polynomials) representation in terms of the Hilbert (or Heine) transform:

$$Q_n(x) = \tfrac{1}{2}\int_{-1}^{1} \frac{P_n(t)}{x-t}\,dt \quad .$$

Then

$$Q_0(x) = \tfrac{1}{2}\log\frac{x+1}{x-1} \tag{6.3}$$

and

$$Q_n(x) = Q_0(x)P_n(x) + \tfrac{1}{2}\int_{-1}^{1} \frac{P_n(t)-P_n(x)}{x-t}\,dt \quad . \tag{6.4}$$

Then the polynomial at the right side of (6.4) is a polynomial with rational coefficients; it's easy to see that this polynomials, as a polynomial in $\frac{1-x}{2}$, after multiplying by

$$[1,\dots,n] = l.c.m.\{1,\dots,n\} \quad ,$$

has already integer coefficients. For this we can use even the explicit Christoffel-Hobbson formula

$$Q_n(x) = Q_0(x)P_n(x) - \sum_{k+1}^{n+\frac{1}{2}} \left(\frac{2}{2k-1} - \frac{1}{n-k+1}\right) P_{n-2k+1}(x) \quad . \tag{6.5}$$

Now the recurrent formula (6.1) gives us asymptotics for $|P_n(x)|$, $|Q_n(x)|$ for fixed x and $n\to\infty$.

If $|x\pm\sqrt{x^2-1}|\neq 1$ and, say $|x+\sqrt{x^2-1}|>1$, then

$$\begin{aligned} &\frac{1}{n}\log|P_n(x)| \to \log|x+\sqrt{x^2-1}| \quad , \\ &\frac{1}{n}\log|Q_n(x)| \to \log|x-\sqrt{x^2-1}| \quad \text{for } n\to\infty . \end{aligned} \tag{6.6}$$

It is possible to get even more precise asymptotics of $|P_n(x)|$, $|Q_n(x)|$ based on the Szego theorem [31]. From the point of view of the measure of irrationality, (6.6) is enough.

Now if $\frac{x-1}{2}=\frac{\ell}{q}$ for integers ℓ, q, $q\geq 1$, then according to (6.4),

$$q^n P_n(x) \quad \text{and} \quad q^n[1,\dots,n](Q_n(x)-Q_0P_n(x))$$

are rational integers. These results enable us to obtain the measure of irrationality of $\log(1+q/\ell)$, for $\ell, q\in\mathbb{Z}[i]$.

Let us present a general result on the rational approximation to the number log(1+x) for rational x, close to zero.

Theorem 6.1 - [10]. *Let*

$$x = \frac{q+2\ell}{q} \text{ for integers } \ell, q \geq 1 .$$

We obtain the measure of irrationality of the number

$$\log(1+\frac{q}{\ell}) ,$$

assuming

$$(*) \qquad 2\ell + q + 2\sqrt{\ell(\ell+q)} > q^2 \cdot e \quad .$$

We have under the assumption (*),

$$|\log(1+\frac{q}{\ell}) - \frac{P}{Q}| \geq c \cdot |Q|^{-\chi}$$

for $c = c(\ell, q) \geq 0$ *and any integers* P, Q, *provided that*

$$\chi = \frac{2 \log x}{\log(x/q\ell)} < \infty \quad \text{for} \quad x = x + \sqrt{x^2-1}$$
$$= q^{-1}(2\ell + q + 2\sqrt{\ell(\ell+q)} \quad .$$

We would like to mention another interesting application of Legendre polynomials.

Legendre polynomials can be used for the proof of irrationality of special values of logarithms at points from $\mathbf{Q}[i]$, see [8]. In particular, for the number $\pi/\sqrt{3}$ we obtain a very good measure of irrationality:

Proposition 6.2 - [8]. *Let* p, q *be rational integers*, $|q| \geq 2$. *Then*

$$|\frac{\pi}{\sqrt{3}} - P/q| > |q|^{-8.31}$$

and

$$|\frac{\pi}{\sqrt{3}} - P/q| > q^{-8.309986341\ldots} \qquad \text{for } |q| \geq q_0 \quad .$$

The bound $|\pi/\sqrt{3} - P/q| > |q|^{-8.31}$ for all $|q| \geq 2$ follows from our estimates and from computer experiments performed at CEN-Saclay by Professor Drouffe.

The bound for the measure of irrationality of $\pi/\sqrt{3}$ is indeed very simple and several people have observed that elementary methods give this bound. However, the bound varied depending on the author. The first result for $\pi/\sqrt{3}$ belongs to Danilov [38] who got the exponent 9.35... instead of 8.309... . In any case, the proposition 6.2 is the best result following from the 1-point Padé approximation to ln x . Recently I improved considerably the measure of irrationality of $\pi/\sqrt{3}$ using the Padé approximation to $\{1, \ln x, (\ln x)^2\}$ and new bounds for the denominators of the coefficients of Padé approximants. We get:

7. MULTIDIMENSIONAL INTERPOLATION IN $\mathbb{C}^n$. GROTHENDICK RESIDUE FORMULA AND LAGRANGE INTERPOLATION IN $\mathbb{C}^n$.

Let us turn now to the multidimensional interpolation problems in an attempt to generalize Padé approximation results of §1 (see[15]).

The situation in the n-dimensional space $P\mathbb{C}^n$ (or simply $\mathbb{C}^n$) is totally different from that in the one-dimensional case.

Let us start with the finite set S of points in $\mathbb{C}^n$. In order to start with the interpolation problem in $\mathbb{C}^n$, corresponding to a set S we need first of all to have some natural representation of S as a set of zeros of some algebraic expression.

For n=1 this is trivial: if $S=\{x_1,\ldots,x_n\}$, then $P_S(x)=\prod_{i=1}^n (x-x_i)$ is the unique (up to multiplicative constant) polynomial of degree n having x_i as zeros of multiplicity (≥ 1) counted as the time of occurrence of x_i at S.

This simple fact is the basis for all interpolation methods and formulae. However for $\mathbb{C}^k$: k>1 the situation is changed. As codimension of S in $\mathbb{C}^k$ is k, we can try to represent S only as an intersection of k hypersurfaces

$$P_1 = 0, \ldots, P_k = 0$$

for $P_i \in \mathbb{C}[x_1,\ldots,x_k]$: i=1,...,k. How does one find such P_i? Do they exist at all?

Let us take k=2, e.g., the two curves f=0, g=0 of degrees n, m. Then, by Bezout's theorem, if these curves have no common components, then the *number* of points of intersections (counted *with* multiplicities) is mn (in $\mathbf{PC}^2$).

There arises an immediate difference with the 1-dimensional situation:

a) only that set S can be represented as a set of *simple* intersections of two curves of degrees n,m for which $|S|=n\cdot m$;

b) Moreover only *some* sets S, $|S|=n\ m$, can be written in the form of simple intersections of curves of degrees n and m.

In order that mn given points S should be the intersections of two curves of orders m and n, with m>n, the coordinates of the mn must be connected by mn-3n+1 conditions; when m=n, they must be connected by n^2-3n+2 conditions.

E.g. 9 points, to be common to two cubic must have their coordinates subject to 2 conditions.

However it's possible to present *any* set S in C^n as an intersection of n hypersurfaces (though not only as simple intersections). We have

<u>*Theorem 7.1*</u> - *Let S be a finite set in* $\mathbf{C}^n$. *Then these are* n *hypersurfaces of degrees* $\leq |S|$:

$$P_1 = 0, \quad \ldots, P_n = 0 \quad : P_i \in \mathbf{C}[x_1,\ldots,x_n] : \quad i=1,\ldots,n,$$

such that the set S is the set-theoretic intersection of the hypersurfaces $P_1=0,\ldots, P_n=0$.

The proof (rather simple and based on the dimensional considerations mainly) was proposed as an answer to our question by J.P.Serre as an extension of the work of Mme Poitou.

We know now that the degrees of P_i can be dropped further: $d(P_i) \ll_n \sqrt[n]{|\bar{S}|}$ and we can take P_i in such a form that $P_1=0,\ldots,P_n=0$ have non-simple intersection only at *one* of the points of S.

Now instead of polynomials $P_S(x)$ for $\mathbf{C}^1$ we use as a basis set of interpolation polynomials the following expressions

$$P_1^{i_1}\ldots P_n^{i_n} : i_1\geq 0,\ldots,i_n\geq 0 .$$

The first non-trivial attemp to do something with the interpolation in two-dimensional case belongs to Angelesco (1916) where he considered the case of the intersections of two quadratic polynomials.

Now we can solve this problem completely. For this it should be noted that the polynomials $P_1(x_1,\ldots,x_n),\ldots,P_n(x_1,\ldots,x_n)$ are polynomials in the coordinates of the points of S. Now let's take Z as a 0-cycle (divisor of codim k in $P\mathbb{C}^k$) corresponding to an element $\vec{x}_1,\ldots,\vec{x}_m$ of S:

$$Z = \tilde{m}_1\vec{x}_1 + \ldots + \tilde{m}_m\cdot\vec{x}_m \quad .$$

Then Z can be represented for *some* values of $\tilde{m}_i$ corresponding to the multiplicities of the intersections of $P_1=0,\ldots,P_n=0$ at $\bar{x}_i$ as the intersection of hypersurfaces

$$R_1 = 0, \ \ldots \ , R_n = 0$$

where the multiplicity of the intersection of $R_1=0,\ldots,R_n=0$ in $\bar{x}_j$ is $\tilde{m}_j$ precisely and $R_1,\ldots,R_n$ are composed from $P_1^{i_1}\ldots P_n^{i_n}$.

Now instead of Hermite interpolation formula, having $S_0=\{\bar{x}_1^0,\ldots,\bar{x}_{m-1}^0\}$ fixed and $\bar{x}_m=\bar{x}$ varying we can write

$$Q(\bar{x}) = \frac{1}{(2\pi i)^n}\oint\frac{dz_1\wedge\ldots\wedge dz_n}{R_1\ldots R_n} \quad .$$

We'll present now some version of multidimensional the Residue Formula and its applications following our common work with D.V.Chudnovsky [15].

There exists a very nice residue formalism connected with the Grothendiek residue symbol.

Let U be the ball $\{x\in\mathbb{C}^n:|z|<\varepsilon\}$ and $f_1,\ldots,f_n\in\mathcal{O}(\bar{U})$ functions holomorphic in a neighborhood of the closure $\bar{U}$ of U. We assume that

$$D_i = (f_i) = \text{divisors of } f_i\text{: } i=1,\ldots,n$$

have the origin as their set-theoretic intersection,

$$f^{-1}(0) = \{\bar{0}\}$$

for

$$f = \left(f_1,\ldots,f_n\right) : U^* = U\setminus\{\bar{0}\}\to C^n\setminus\{\bar{0}\} \quad .$$

We are interested in residue associated with a meromorphic n-form

$$\omega = \frac{g(z)dz_1\wedge\ldots\wedge dz_n}{f_1(z)\ldots f_n(z)} \quad : g\in\mathcal{O}(\bar{U})$$

having polar divisor

$$D = D_1 + \ldots + D_n \quad .$$

In order to define the Grothendick residue symbol we take the cycl of integration

$$\Gamma = \{z : |f_i(z)| = \varepsilon\}$$

(with the orientation, say $d(\arg f_1)\wedge\ldots\wedge d(\arg f_n) \geq 0$).

Then the *residue* of ω at $\bar{0}$ is

$$\mathrm{Res}_{\{\bar{0}\}}\, \omega = \left(\frac{1}{2\pi i}\right)^r \int_\Gamma \omega \quad . \tag{7.1}$$

First of all, $\mathrm{Res}_{\{\bar{0}\}}\omega$ possesses all "normal" local properties:

Lemma 7.2 - *(Local properties of residues).*

1) In the generic case, when D_i *are smooth and meet transversely, i.e. Jacobian of* f

$$J_f(\bar{0}) = \frac{\partial(f_1,\ldots,f_n)}{\partial(z_1,\ldots,z_n)}(\bar{0}) \neq 0, \tag{7.2}$$

then

$$\mathrm{Res}_{\{\bar{0}\}}\omega = g(\bar{0})/J_f(\bar{0}). \tag{7.3}$$

2) (Transformation formula). Suppose that $f=(f_1,\ldots,f_n)$ *and* $g=(g_1,\ldots,g_n)$ *give holomorphic maps* $f,g:\bar{U}\to\mathbf{C}^n$ *with* $f^{-1}(0)=g^{-1}(0)=\{\bar{0}\}$. *Suppose that for ideals we have*

$$\{g_1,\ldots,g_n\} \subset \{f_1,\ldots,f_n\} ,$$

i.e.

$$g_i(z) = \sum_{j=1}^n a_{ij}(z)f_i(z)$$

for holomorphic matrix $A(z)=a_{ij}(z)$. *Then for* $h(z)\in\mathcal{O}(\bar{U})$ *we have:*

$$\mathrm{Res}_{\{\bar{0}\}}\left(\frac{h\,dz_1\wedge\ldots\wedge dz_n}{f_1\ldots f_n}\right) = \mathrm{Res}_{\{\bar{0}\}}\left(\frac{h\det A\,dz_1\wedge\ldots\wedge dz_n}{g_1\ldots g_n}\right)$$

Residues can be also used for an analytic formula of local intersection number of $f=0$ at $\bar{0}$. For this we define

$$(D_1,\ldots,D_n)_{\{\bar{0}\}} = \mathrm{Res}_{\{\bar{0}\}}\left(\frac{df_1}{f_1}\wedge\ldots\wedge\frac{df_n}{f_n}\right)$$

Then $(D_1,\ldots,D_n)_{\{\bar{0}\}}$ has indeed sense as the local intersection number of $f=0$:

a) For the local ring θ_0 at the origin and $I_f\subset\theta_0$ the ideal, generated by the f_i, we have

$$(D_1,\ldots,D_n)_{\{\bar{0}\}} = \dim_{\mathbf{C}} \theta_0/I_f \;;$$

b) $f : U^*\to\mathbf{C}^n\setminus\{\bar{0}\}$ has topological degree

$$(D_1,\ldots,D_n)_{\{\bar{0}\}}$$

All these assertions, in the form presented here, together with Global Residue Formula below belong to Ph.Griffiths.

Of course, the Global Residue Formula is just kind of expression we need in the Padé Approximations.

Let M be the n-dimensional compact complex manifold and ω a meromorphic differential form on M, whose polar divisor D can be expressed as a union $D=D_1\cup\ldots\cup D_n$ on n divisors D_i with the property that their intersection

$$Z = D_1\cap\ldots\cap D_n$$

is a finite set. Then we have

Lemma 7.3 - *(Global Residue Formula)*

$$\sum_{\bar{p}\in Z} \mathrm{Res}_{\{\bar{p}\}}\omega = 0 \quad . \qquad (7.4)$$

The most interesting applications proposed by Griffiths deal with $M=\mathbb{P}^n$. We assume that $D_1,\ldots,D_n$ are hypersurfaces of respective degrees $d_1,\ldots,d_n$ with intersections at isolated points P_ν and this intersection we present as a zero cycle

$$D_1\ldots D_n = \sum_\nu m_\nu P_\nu \ ,$$

where by Bezout's theorem

$$\sum_\nu m_\nu = d_1\ldots d_n .$$

We'll assume below simply that all P_ν lie in $\mathbb{C}^n\subset\mathbb{P}^n$ and that D_i is defined by

$$f_i(x_1,\ldots,x_n) = 0$$

for polynomials f_i of degree d_i.

The most general meromorphic n-form on $\mathbb{P}^n$ with polar divisor $D=D_1+$ $+\ldots+D_n$ has in $\mathbb{C}^n$ an expression

$$\omega = \frac{g(x)dx_1\cap\ldots\cap dx_n}{f_1(x)\ldots f_n(x)}$$

for a polynomial g(x). Here ω doesn't have the hyperplane at infinity as a component of its polar divisor when the degree of g(x) satisfy:

$$\deg(g) \leq (d_1+\ldots+d_n) - (n+1) \quad . \qquad (7.5)$$

Thus the Global Residue formula gives in this case (GENERALIZED JACOBI-KRONECKER FORMULA)

$$\sum_\nu \mathrm{Res}_{P_\nu}\left(\frac{g(x)dx_1\cap\ldots dx_n}{f_1(x)\ldots f_n(x)}\right) = 0 \quad . \qquad (7.6)$$

Why this is called the Jacobi-Kronecker formula? Because Jacobi and

Kronecker proved using only linear algebra the following important identity

$$\sum_{\nu} \frac{g(P_{\nu})}{J_f(P_{\nu})} = 0 \qquad (7.7)$$

if $\deg(g) \le \sum_{i=1}^{n} d_i - (n+1)$ and if D_i meet transversely at $d_1 \dots d_n$,

$$J_f = \frac{\partial(f_1,\dots,f_n)}{\partial(z_1,\dots,z_n)}$$

is the Jacobian of f. Of course (7.7) follows from (7.6) and results above.

This formula was already applied by I.Petrovsky (1936-1953) to 16th Hilbert Problem on real plane curves (n=2).

In particular, from these formulae we obtain an interesting multidimensional generalization of Hermite interpolation formula from D.V. and G.V.'s paper [15].

Corollary 7.4 - *Let* $f(z)$ *be holomorphic in* $\bar{U}$ *and we define in the neighborhood of* $\bar{0}$ *a new function* $F(\bar{x})$ *by*

$$F(\bar{x}) = \left(\frac{1}{2\pi i}\right)^n \int_{\Gamma} \frac{f(\bar{x}\bar{z})dz_1 \wedge \dots \wedge dz_n}{f_1(z)\dots f_n(z)}$$

Then $F(\bar{x})$ *vanishes at* $\bar{x}=\bar{0}$ *of order* $\ge \sum_{i=i}^{n} d(f_i) - n$.

We can write an explicit expression for $F(\bar{x})$ in terms of $\partial_z^k f$ in certain cases, e.g. when $f_1,\dots,f_n$ have only simple intersection $\partial^z f_i^!$ are powers of such f_i, etc. or in any case when the singularities of the intersections are known. In each of these cases $F(\bar{x})$ is a linear form from partial derivatives $\partial_z^{k_1,\dots,k_n} f(\bar{x}\bar{z})$ with coefficients being rational functions in partial derivatives in $f_i(\bar{x})$ at fixed $\bar{x}_0$.

Let's explain how from the generalized Jacobi-Kronecker formula it follows Lagrange interpolation formula.

Let n=1; $x_0,\dots,x_m$ are fixed and x is a variable. We set

$$f(z) = \prod_{i=0}^{m} (x-x_i)(z-x) .$$

Then the formula (4.7) can be written as

$$\sum_{i=0}^{m} g(x_i)/\psi(x_i)(x_i-x) + g(x)/\psi(x) = 0$$

for

$$\psi(x) = \prod_{i=0}^{m} (x-x_i) = f'(x)\big|_{z=x}$$

or

$$g(x) = \sum_{i=0}^{m} \frac{g(x_i)\psi(x)}{\psi(x_i)(x-x_i)}$$

for $d(g) \le m$.

Now (7.6) is a *natural* generalization of the Lagrange interpolation formula, if one of the $\bar{x}_\nu$ varies.

8. PROBLEMS IN TRANSCENDENTAL NUMBER THEORY. RELATIONS WITH SOLUTIONS OF LINEAR DIFFERENTIAL EQUATIONS.

8.1 -

We discuss now problems from Transcendental Number Theory. The major problem in this field is the problem of the determination of the arithmetic nature of the values of analytic functions satisfying algebraic differential equations. By the words "arithmetic nature" we understand irrationality, transcendencity and the determination of the measure of irrationality, of the measure of transcendencity, and, in the case of several numbers, of the measure of algebraic independence.

For the last hundred years (since Hermite) most attention was devoted to the study of *entire* or *meromorphic* functions in $\mathbb{C}$ satisfying algebraic differential equations.

Perhaps, the most general result in this direction is the

<u>*Schneider-Lang theorem 8.1*</u> - [35]. *Let* $f(z)$ *be a transcendental meromorphic function in* $\mathbb{C}$ *of finite order of growth* $\le\rho$. *Let* $f(z)$ *satisfy an algebraic differential equation over* $\overline{\mathbb{Q}}$:

$$P_1(z,f(z),\dots,f^{(q-1)}(z)) = 0 \quad : \quad q\ge 2 , \tag{8.1}$$

or

$$P_2(f(z),\dots,f^{(q-1)}(z)) = 0 , \qquad q\ge 2 . \tag{8.2}$$

Let $\mathbb{K}$ *be an algebraic number field and* $P_1(x_0,x_1,\dots,x_{q-1})$, $P_2(x_1,\dots,x_{q-1})$ *be polynomials from* $\mathbb{K}[\bar{x}]$.

1. If $f(z)$ *satisfies* (8.1), *then the set* $S^1_{\mathbb{K}}=\{z\in\mathbb{K}: f^{(k)}(z)\in\mathbb{K}$ for all $k\ge 0\}$ *has cardinality* $\le\rho\cdot[\mathbb{K}:\mathbb{Q}]$.

2. Let $q''=\deg \operatorname{tr} \mathbb{C}(f(z),\dots,f^{(q-2)}(z))\ge 2$ *and* $f(z)$ *satisfy* (8.2). *Then the set* $S^2_{\mathbb{K}} = \{z\in\mathbb{C}: f^{(k)}(z)\in\mathbb{K}$ for all $k\ge 0\}$ *has cardinality* $\le\rho\frac{q''}{q''-1}[\mathbb{K}:\mathbb{Q}]\le 2\rho[\mathbb{K}:\mathbb{Q}]$.

Only in case (8.1) can we add a little more:

Proposition 8.2 - [3], [11]. *Let* f(z) *be a meromorphic transcendental function of order* $\leq\rho$ *and* f(z) *satisfy* (8.1) *for* $P_1(\bar{x})\in\mathbb{K}[\bar{x}]$. *If* $S^1_{\mathbb{Q},\mathbb{K}}=\{z\in\bar{\mathbb{Q}}:f^{(k)}(z)\in\mathbb{K} \text{ for all } k\geq 0\}$, *then* $|S^1_{\mathbb{Q},\mathbb{K}}|\leq\rho[\mathbb{K}:\mathbb{Q}]$.

Unfortunately nothing more is known, about general meromorphic functions. There is however the long staying

Bombieri conjecture. Let f(z) *be a transcendental meromorphic function of order* $\leq\rho$ *satisfying an algebraic differential equation over* $\bar{\mathbb{Q}}$. *In the case of equation* (8.1), *the set*

$$S^1_{\bar{\mathbb{Q}}} = \{z\in\bar{\mathbb{Q}}:f^{(k)}(z)\in\bar{\mathbb{Q}} \text{ for all } k\geq 0\}$$

has cardinality $\leq\rho$. *In the case of* (8.2) *the set*

$$S^2_{\bar{\mathbb{Q}}} = \{z\in\mathbb{C}:f^{(k)}(z)\in\bar{\mathbb{Q}} \text{ for all } k\geq 0\}$$

is finite whenever f(z) *and* f'(z) *are algebraically independent over* $\mathbb{C}$ *and*

$$|S^2_{\bar{\mathbb{Q}}}| \leq 2\rho \quad (\text{or } |S^2_{\bar{\mathbb{Q}}}| \leq \frac{\rho\cdot q''}{q''-1}) .$$

However it may happen that the Bombieri conjecture will not give us any example of new transcendental numbers.

We propose the following Problem containing our suspicions:

Problem 8.3 - 1) Let f(z) satisfies all the assumptions of the Bombieri conjecture. If

$$|\{z\in\mathbb{C}:f^{(k)}(z)\in\bar{\mathbb{Q}} \text{ for all } k\geq 0\}| > \rho ,$$

then f(z) is reduced to z, e^z or $\wp(z)$.

2) Let f(z) is not reduced to z, e^z or $\wp(z)$ and let f(z) be meromorphic function of finite order of growth satisfying the equation $R(f,f',\dots,f^{(q-1)})=0$ for $R(x_0,\dots,x_{q-1})\in\mathbb{C}[x_0,\dots,x_{q-1}]$, $q\geq 3$. Let f(z) cannot be reduced by a change of a variable to a function satisfying an algebraic differential equation of order < q-1. Then $|S_{\bar{\mathbb{Q}}}|\leq 1$.

Of course, the best examples should be equations of the Painlevé type [16], [23]. E.g. Painlevé I-IV [16], [23], [36], are meromorphic functions, and we can study the arithmetic properties of their values at algebraic points. Nothing is known yet and we propose the following

Problem 8.4 - Let w(z) be a (meromorphic) solution of Painlevé I: $w''=6w^2+z$ defined in a way such that $w(z_0)$ and $w'(z_0)$ are algebraic numbers for an algebraic $z_0 \in \bar{\mathbb{Q}}$. Then for an algebraic $z_1 \in \bar{\mathbb{Q}}$, $z_1 \neq z_0$, either $w(z_1)$ or $w'(z_1)$ is transcendental.

It's possible even to conjecture that $w(z_1)$ and $w'(z_1)$ are algebraically independent for any $z_1 \in \bar{\mathbb{Q}}$, $z_1 \neq z_0$.

The same problem 8.4 can be formulated for any Painlevé I-IV. However for Painlevé V-VI (having critical points at $z=0$, $z=1$) the famous "Picard solution" of the case $\alpha=\beta=\gamma=\delta=0$, $w=\Lambda(C_1\omega_1+C_2\omega_2,z)$, $u=\int_0^{\Lambda(u,z)} dw \cdot (w(w-1)(w-z))^{-\frac{1}{2}}$, is the counterexample to a natural generalization of the Bombieri conjecture for non-meromorphic solutions.

Here is one evidence, confirming Problem 8.3.1). By Straus's theorem, if $f(z)$ is an entire function of order of growth $\leq \rho$ and if $|\{z \in \mathbb{C} : f^{(k)}(z) \in \mathbb{Z}$ for all $k \geq 0\}| > \rho$, then $f(z)$ is a finite sum of exponents.

8.2 -

In general we can ask: what entire or meromorphic functions $f(z)$ satisfy the algebraic differential equation

$$P(z, f(z), \ldots, f^{(q-1)}(z)) = 0 \quad . \tag{E}$$

Let's consider the case $q=2$ because only in this case does the Bombieri conjecture or problem 8.3 imply a precise transcendental statement on the transcendence of $f(\alpha)$ or $f'(\alpha)$ or $f'(\alpha)/f(\alpha)$ at algebraic $\alpha \in \bar{\mathbb{Q}}$.

First of all, let $q=2$ and $f(z)$ be a meromorphic or even a finite-valued function $f(z)$ satisfying (E). According to the Hermite-Picard theorem, $f(z)$ is reduced either to an elliptic function or is reduced to the solution of a Riccati equation . In the last case, $f(z)$ is reduced to $g'(z)/g(z)$ and $g(z)$ satisfies a linear differential equation of the second order with rational function coefficients. In other words, the only interesting case for $q=2$ is the equation

$$p(x)y'' + q(x)y' + r(x)y = 0 \tag{E_2}$$

where $p(x)$, $q(x)$, $r(x) \in \bar{\mathbb{Q}}[x]$ and

$$f(x) = y'(x)/y(x) \quad .$$

We assume, of course, that $y(x)$ is an entire function. If $y(x)$ is an E-function, e.g. the hypergeometric function $y={}_pF_q(x|a_1,\ldots,a_p;b_1,\ldots,b_q)$ [2] for $p \leq q$ and rational a_i, b_j, then by Siegel's theorem [29],

[30] f(x) is a transcendental number for $x\in\overline{Q}$, different from singularities of (E_2).

However, if y(x) is an entire function it need not be an E-function. It may be, say,

$$y(x) = {}_pF_q(x|a_1,\dots,a_p;b_1,\dots,b_q)$$

with an *algebraic* a_i, b_j.

We know one very general class of entire functions satisfying a *linear* differential equation (E) with polynomials coefficients: generalized hypergeometric ${}_pF_q$-functions with $p\le q$. However these functions may not be E-functions though they may satisfy a linear differential equation over $\mathbb{Q}[x]$.

Let's consider the most general function of the sort

$$f(x) = \sum_{n=0}^{\infty} x^n \prod_{j=1}^{n} \frac{P(j-1)}{Q(j)} \tag{8.3}$$

where $P(x), Q(x)\in\overline{\mathbb{Q}}[x]$. Then f(x) satisfies a linear differential equation

$$\{Q(\delta) - xP(\delta)\} f = 0 \tag{8.4}$$

for $\delta=x(d/dx)$. If f(x) is transcendental and entire then $d(P)<d(Q)$. Equation (8.4) is defined over Q if $P(x)\in Q[x]$, $Q(x)\in Q[x]$ and the zeros of Q(x) are different from negative integers.

Such a function (8.3) is simply a generalized hypergeometric function ${}_pF_q$ for $p\le q$. Similarly any entire function

$$f(z) = {}_pF_q\left(z\middle|{a_1,\dots,a_p \atop b_1,\dots,b_q}\right) \quad : \quad p\le q \qquad (a_i,b_j\in\overline{\mathbb{Q}})$$

can be reduced to a form (8.3) with $P(x),Q(x)\in Q[x]$ where $a_i: i=1,\dots,p$ and $b_j: j=1,\dots,q$ are roots of the polynomials of degree p and q, respectively, with algebraic integer coefficients.

Now for the functions (8.3) we constructed in §9 below the system of Padé approximants for any a_i, b_j or any $P(x), Q(x)$. This system of Padé approximations can be used in the proof of the linear independence of values of f(x) and it's derivatives at an algebraic point. Before in §2 and in §4 we presented such results even with the measure of linear independence for $d(P)=0$ (when $y(x)=f(x)$). Also we mentioned in §2 the Hurwitz theorem, that for $d(P)<d(Q)$ and $P(x),Q(x)\in\mathbb{K}[x]$, $\mathbb{K}=Q(\sqrt{-D})$, $D>0$ at least one of the numbers $f^{(i)}(x)/f(x): i=1,\dots,d(Q)-1$ is irrational over $\mathbb{K}$ for any $x\in\mathbb{K}$, $x\neq 0$.

Now we are able to prove the most general result in this direction, which we present in the following form

Theorem 8.5 - *Let* $\mathbb{K}$ *be a quadratic imaginary field and* $P(x), Q(x) \in \mathbb{K}[x]$, $d(P)<d(Q)$. *Let* $f(x)$ *be defined in* (8.3) *and* $f(x),\dots,f^{(\ell)}(x)$ *be linearly independent over* $\mathbb{K}[x]$: $\ell \le d(Q)-1$. *If* $\alpha \in \mathbb{K}$, $\alpha \ne 0$ *then* $f(\alpha),\dots,f^{(\ell)}(\alpha)$ *are linearly independent over* $\mathbb{K}$.

Moreover, if $1, f(x),\dots,f^{(\ell)}(x)$ *are linearly independent over* $\mathbb{K}[x]$, *then* $1, f(\alpha), \dots, f^{(\ell)}(\alpha)$ *are linearly independent over* $\mathbb{K}$.

Of course it is possible to obtain the bound for the measure of linear independence of $\{f(\alpha),\dots,f^{(\ell)}(\alpha)\}$ and of $\{1, f(\alpha),\dots,f^{(\ell)}(\alpha)\}$.

In any case it's clear from results of §§2,3,4,8 that the classes of E- and G-functions [29] studied up to now are not the most general for which the *present* analytic technique of transcendental numbers can be applied.

9. EFFECTIVE CONSTRUCTION OF A PADE APPROXIMATION TO SOLUTIONS OF LINEAR DIFFERENTIAL EQUATIONS IN TERMS OF RIEMANN DATA.

9.1 -

The Darboux-Bäcklund transformation for the solution of certain nonlinear equations has another incarnation that leads us immediately to a Riemann monodromy problem. Let's consider isomonodromy deformation equations [16] [36] [37] such as Painlevé VI [16] [23]. These isomonodromy deformation equations appear as the condition on the coefficients of a Fuchsian linear differential equation to have a fixed monodromy group. E.g. Painlevé VI corresponds to the equation of order 2 with four regular singularities. However the Fuchsian equation is not determined completely by singularities and the monodromy group. There remains an ambiguity because when the monodromy group is the same, local exponents may be different by integers (contiguous systems in Riemann's sense [28]). Here comes the Riemann theorem [28] on contiguous systems: any m+1 systems of contiguous functions, satisfying a Fuchsian linear differential equation of order m, are related by a linear relation with polynomial coefficients.

These *linear* relations between solutions of different linear equations with a fixed monodromy group, imply *nonlinear* relations between solutions of nonlinear isomonodromy deformation equations. These nonlinear relations are indeed special cases of the Darboux-Bäcklund transformation.

We refer the reader to the paper of Proff.Boiti and Pempinelli [36] for the description of the general Bäcklund transformation for Painlevé transcendents.

The method of contiguous functions and Riemann's results [28] enable us to make another miracle: to construct explicitly the Padé approximations to solutions of *any* linear differential equations. We present below in this chapter our construction of the remainder function in the Padé approximation problem for the solutions of Fuchsian linear differential equations. By a limiting procedure we can formally write the remainder function for the Padé approximation problem for any linear differential equation in terms of singularities and Stoke's data. The corresponding formulae will be presented elsewhere.

As a consequence of the present construction, we give the solution of a Padé approximation problem for arbitrary generalized hypergeometric functions.

9.2 -

We consider an arbitrary Fuchsian linear differential equation of order m

$$L_m y = 0 \tag{9.1}$$

where $L_m \in \mathbf{C}[d/dx, x]$ is a linear differential operator in d/dx with coefficients from $\mathbf{C}(x)$. Let $a_1,\ldots,a_n$ all be singularities of (9.1). Then choosing a certain fundamental solution $Y=Y(x)$ of (9.1), and considering in $\mathbf{PC}^1$ a circuit γ_i in the positive direction having a_i but none of a_j: $j\neq i$ inside, we obtain the following linear transformation,

$$Y(x) \mapsto Y(x)A_i \quad , \tag{9.2}$$

after continuing $Y(x)$ along the γ_i. Then $A_1,\ldots,A_n$ are called monodromy matrices of (9.1) corresponding to a choice of $Y(x)$.

If $\lambda_{j,k}: j=1,\ldots,m$ are eigenvalues of $A_k: k=1,\ldots,n$, then possible local exponents of $Y(x)$ at $x=a_k$ are $\mu_{j,k}$, where $\exp(2\pi\sqrt{-1}\,\mu_{j,k})=\lambda_{j,k}: k=1,\ldots,n; j=1,\ldots,m$. According to B.Riemann [28] we say that a system of functions $Y(x)$ belongs to the class

$$\left\{\begin{matrix} a_1,\dots,a_n \\ \hline A_1,\dots,A_n \\ \\ t\bar{\mu}_1,\dots,t\bar{\mu}_n \end{matrix}\right\} \quad : \bar{\mu}_k = (\mu_{j,k})_{j=1,\dots,m}$$

if $Y(x)$ satisfies (9.1) with monodromy substitutions A_k at $x=a_k$ (9.2) and $Y(x)$ has local exponents $\mu_{j,k}: j=1,\dots,m$ at $x=a_k: k=1,\dots,n$. For systems of functions $Y^{(i)}(x)$ having the same regular singularities, a_k and monodromy matrices $A_k: k=1,\dots n$ are called contiguous.

In Riemann's studies [28] it was shown that any m+1 systems of functions $Y^{(j)}(x) : j=1,\dots,m+1$ that are contiguous are linearly dependent over $\mathbb{C}[x]$. Moreover degrees of polynomials coefficients were determined in terms of local multiplicities of $Y^{(j)}(x)$.

We may assume, without loss of generality, that

$$a_1 = 0, \quad a_2 = \infty, \quad a_3 = 1 .$$

Also we can always assume that at $x=a_1=0$, $\mu_{1,1}=0$ and we have one branch of $Y(x)$ at $x=0$, $y_1(x)$, say, such that

$$y_1(x) \text{ is analytic at } x=0,\ y_1(0)\neq 0 .$$

Assume now that we have m arbitrary contiguous systems of functions, linearly independent over $\mathbb{C}[x]$, belonging to classes

$$Y^{(j)}(x) = \left\{\begin{matrix} a_1=0 & a_2=\infty & \dots & a_n \\ \hline A_1 & A_2 & \dots & A_n \\ \\ t\bar{\mu}_1^{(j)} & t\bar{\mu}_2^{(j)} & \dots & t\bar{\mu}_n^{(j)} \end{matrix}\right\} \quad : j=1,\dots,m \tag{9.3}$$

where $\bar{\mu}_k^{(j)}\in \mathbb{C}^m$, and

$$\bar{\mu}_k^{(j)} - \bar{\mu}_k^{(j')}\in \mathbb{Z} \qquad \text{for } j,j' = 1,\dots,m,\ k=1,\dots,n.$$

We choose branches $y^{(j)}(x) \overset{\text{def}}{=} f_j(x) : j=1,\dots,m$ from the system $Y^{(j)}(x)$ such that $f_j(x)$ is analytic at $x=0$, $f_j(0)\neq 0$.

Now we want to construct a Padé approximation system to functions $\{f_1(x),\dots,f_m(x)\}$ corresponding to the weights $\{n_1,\dots,n_m\}$. We consider the case of bounded $|n_i-N| : i=1,\dots,m$.

In these cases the expression for the remainder function can be obtained by considering a new, (m+1)-th system of contiguous functions of the form

$$Y_N(x) = \left\{ \begin{array}{ccc} \dfrac{0}{A_1} & \dfrac{\infty}{A_2} \;\cdots & \dfrac{a_n}{A_n} \\ \nu_{1,1}+mN & \nu_{1,2}-N & \nu_{1,n} \\ \nu_{2,1} & \nu_{2,2}-N \;\cdots & \nu_{2,n} \\ \vdots & \vdots & \vdots \\ \nu_{m,1} & \nu_{m,2}-N \;\cdots & \nu_{m,n} \end{array} \right\} \qquad (9.4)$$

where $\nu_{i,j}$ depend only on $\bar{\mu}_k^{(i)}$ and $|n_i-N|: i=1,\ldots,m,\ k=1,\ldots,n.$

If we now take the branch $F_N(x)$ of $Y_N(x)$ at $x=0$, corresponding to the exponent $\nu_{1,1}+mN$ at $x=0$, we obtain the remainder function for the Padé approximation problem for $\{f_1(x),\ldots,f_m(x)\}$:

$$F_N(x) = \sum_{i=1}^{m} P_i(x|f_1,\ldots,f_m;n_1,\ldots,n_m)f_i(x) \qquad (9.5)$$

where $P_i(x|f_1,\ldots,f_m;n_1,\ldots,n_m)$ are polynomials of degree $\leq n_i$: $i=1,\ldots,m$ and

$$\mathrm{ord}_{x=0}\, F_N(x) \geq \sum_{i=1}^{m} \{n_i+1\} - 1 \quad .$$

However, $|\mathrm{ord}_{x=0}F_N(x)-\sum_{i=1}^{m}\{n_i+1\}+1|$ is always bounded by an absolute constant (depending only on the exponents $\bar{\mu}_k^{(i)}$. For the case of the canonical system of contiguous functions, we have a perfect system of Padé approximations of type. I.

Theorem 9.1 - *Let us consider the following canonical system of contiguous functions:*

$$Y^{(i)}(x) = \left\{ \begin{array}{ccccc} \dfrac{0}{A_1} & \dfrac{\infty}{A_2} & \dfrac{1}{A_3} & \cdots & \dfrac{a_n}{A_n} \\ \mu_{1,1}^{(i)} & \mu_{1,2}^{(i)} & \mu_{1,3} & \cdots & \mu_{1,n} \\ \vdots & \vdots & \vdots & \cdots & \vdots \\ \mu_{m,1}^{(i)} & \mu_{m,2}^{(i)} & \mu_{m,3} & & \mu_{m,n} \end{array} \right\}$$

with $\bar{\mu}_k^{(i)} = \bar{\mu}_k : k>2$ *and* $\mu_{1,1}^{(i)} = 0$,

$$\mu_{\ell,1}^{(i)} = \mu_{\ell,1} - \delta_{\ell,i} \cdot (i-1) \; : \quad i=1,\dots,m; \; \ell=1,\dots,m$$

$$\mu_{\ell,2}^{(i)} = \mu_{\ell,2} + \delta_{\ell,i} \cdot (i-1) \; : \quad i=1,\dots,m; \; \ell=1,\dots,m \; .$$

We take the branch $f_i(x)$ *of* $Y^{(i)}(x)$ *at* $x=0$ *such that* $f_i(0)\neq o,\infty$. *Let* $N\geq 1$ *and let us consider a new system of contiguous function*

$$Y_N(x) = \left\{ \begin{array}{cccc} \underline{0} & \underline{\infty} \; \dots & \underline{a_k} \; \dots & \underline{a_n} \\ A_1 & A_2 & A_k & A_n \\ \nu_{1,1} & \nu_{1,2} & \mu_{1,k} & \mu_{1,n} \\ \vdots & \vdots & \vdots & \vdots \\ \nu_{m,1} & \nu_{m,2} \; \dots & \mu_{m,k} & \dots \mu_{m,n} \end{array} \right\}$$

where $\nu_{1,1}=0$, $\nu_{j,1}=\mu_{j,1}-mN$; $j=2,\dots,m$; $\nu_{j,2}=\mu_{j,2}+(m-1)N$; $j=1,\dots,m$. *Let* $f_N(x)$ *be a branch of* $Y_N(x)$ *corresponding to a local exponent* $\nu_{1,1}=0$ *at* $x=0$. *Then*

$$x^{(mN)-1} \cdot f_N(x) = \sum_{i=0}^{m} P_{i,N}(x) f_i(x)$$

where $P_{i,N}(x)$ *are polynomials of degree* $\leq N-1$ *and* $f_N(0)\neq 0$. *In the nondegenerate case, all* $P_{i,N}(x)$ *are of the exact degree* $N-1$.

9.3 -

Let us present now the remainder function for the generalized hypergeometric function for the Padé approximation problem.

As usual we work with ${}_qF_p$-hypergeometric functions [2] or §8. For complex numbers $a_1,\dots,a_q$; $b_1,\dots,b_p$ such that $b_1,\dots,b_p$ are different from non-positive integers we define

$${}_qF_p\left(x \left| \begin{matrix} a_1,\dots,a_q \\ b_1,\dots,b_p \end{matrix} \right. \right) = \sum_{n=0}^{\infty} \frac{(a_1)_n \cdots (a_q)_n}{(b_1)_n \cdots (b_p)_n n!} x^n \quad .$$

Remark. The function

$${}_{p+1}F_p\left(x \left| \begin{matrix} a_1,\dots,a_{p+1} \\ b_1,\dots,b_p \end{matrix} \right. \right) \qquad \text{(note } q=p+1\text{)}$$

satisfies a natural Fuchsian linear differential equation of order p+1. This is one of the few cases when the monodromy group has been computed. Here the singularities at $\{0,1,\infty\}$ only, are regular and the corresponding local multiplicities are

$$\begin{pmatrix} \frac{0}{} & \frac{\infty}{} & \frac{1}{} \\ 0 & a_1 & 0 \\ 1-b_1 & a_2 & 1 \\ \vdots & \vdots & \vdots \\ & & p-1 \\ 1-b_p & a_{p+1} & d \end{pmatrix}$$

for $d=\sum_{i=1}^{p} b_i - \sum_{j=1}^{p+1} a_j$.

Let $q\leq p+1$ and we fix parameters $a_1,\ldots,a_q$ and $b_1,\ldots,b_p$. Now we have the function

$$f_0(x) \overset{\text{def}}{=} {}_qF_p\left(x\left|\begin{matrix}a_1,\ldots,a_q\\ b_1,\ldots,b_p\end{matrix}\right.\right) .$$

Now we define p more functions contiguous to $f_0(x)$ in the following way

$$f_i(x) \overset{\text{def}}{=} {}_qF_p\left(x\left|\begin{matrix}a_1,\ldots,a_i+i,\ldots,a_q\\ b_1,\ldots,b_i+i,\ldots,b_p\end{matrix}\right.\right) \quad : \quad i=1,\ldots,p$$

(here the term a_i+i is present only if $i\leq q$). The function $f_0(x)$ satisfies the linear differential equation of the order p+1 with polynomial coefficients. Of course, any system of p+1 linearly independent contiguous functions (as $\{f_0^{(j)}(x):j=0,\ldots p\}$ or $\{f_j(x):j=0,\ldots,p\}$) are equivalent. For this reason we construct the system of Padé approximants to the functions $\{f_0(x),\ldots,f_p(x)\}$, with polynomials $P_{N,i}$ of degree N (the so-called diagonal case).

<u>*Theorem 9.2*</u> - *We define for* $N\geq 1$

$$R_N(x) = {}_qF_p\left(x\left|\begin{matrix}a_1+Np,\ldots,a_q+Np\\ b_1+(p+1)N,\ldots,b_p+(p+1)N\end{matrix}\right.\right)$$

Then we have

$$x^{(p+1)N-1} R_N(x) = \sum_{i=0}^{p} f_i(x)\cdot P_{N,i}(x) \quad ,$$

where $P_{N,i}(x)$ *are polynomials of degrees* $N-1:i=0,1,\ldots,p$. *Here, of course,* $R_N(0)=1$.

APPENDIX TO §9 - RESIDUE FORMULA AND NEW EXPRESSIONS FOR PADÉ APPROXIMATIONS. APPLICATIONS TO IRRATIONALITY PROOFS.

We can apply now residue formulae from §§1,7 to new Padé approximations to systems of hypergeometric functions.

E.g. we can write down the expression for the remainder function for the Padé approximation for the system of functions

$$1,\ {}_2F_1(1,\omega_i;\gamma;x)\ :\ i=1,\dots,m$$

for complex number $\omega_1,\dots,\omega_m$ such that $\omega_i-\omega_j$ are not integers for $i\neq j$.

We put

$$R_0\left(x;\begin{matrix}\omega_1,\dots,\omega_m\\ \kappa_1,\dots,\kappa_m\end{matrix}\right) = \frac{1}{2\pi i}\int_{C_0}\frac{{}_2F_1(1,s;\gamma;x)\,ds}{\Phi_0(s)}$$

for

$$\Phi_0(s) = \prod_{i=1}^{m}\prod_{\kappa=0}^{\kappa_i}(s+\kappa-\omega_i)\prod_{\kappa=0}^{\kappa_0}(s+\kappa)$$

and C_0 being a contour in an s-plane containing all zeros of $\Phi_0(s)$.

Then for $\kappa_0\geq\max(\kappa_1,\dots,\kappa_m)$ we have

$$R_0\left(x;\begin{matrix}\omega_1,\dots,\omega_m\\ \kappa_1,\dots,\kappa_m\end{matrix}\right) = \sum_{i=0}^{m} P_i(x)f_i(x)\ ,$$

$$f_0(x)=1,\quad f_i(x)={}_2F_1(1,\omega_i;\gamma;x)\ :\ i=1,\dots,m\ .$$

Here $P_i(x)$ are polynomials in x of degree $\kappa_i : i=0,1,\dots,m$. In the particular case of m=1 we reconstruct the famous Padé formulae for the Padé approximation for ${}_2F_1(1,b;c;x)$ in terms of Jacobi polynomials [2].

Our explicit construction of Padé approximants from §§1-4,5,7 and especially §9 are very useful and are the only tool in the irrationality proof and estimates of the measure of irrationality. We already gave many examples in §2 (theorem 2.2), §3 (theorem 3.1), §4 and §6 (theorems 6.1-6.3). In our papers [4]-[8] many other examples (e.g. connected with the elliptic function $\mathscr{P}(z)$) are treated. Let us give you just a few examples.

Corollary 9.3 - For rational integers p, q *we have*

$$|\pi - p/q| > |q|^{-19.88444333\dots}$$

for $|q|\geq q_0'$.

In the proof we use Padé approximants to $\{1,\log x,\dots,(\log x)^{m-1}\}$ at x=1 that we can get from §1 or §9. Denominator considerations are very important in the proof. The previous bound for π from [5] has 19.88... in the exponent.

In [6], [7], [15] we treated diophantine approximations to values of the polylogarithmic function $L_k(x)=\sum_{n=1}^{\infty}x^n/n^k\ :\ k=1,2,\dots$. For this we used another version of Padé approximants (of the so-called type II

[39]) constructed in [6], [7].

E.g. $L_2(^1/_q)$ is an irrational number, whenever q is an integer ≥ 14.

In later publications we'll consider application of the Padé approximation problem to numbers like $\zeta(k)$ for $k\geq 2$ (cf. Apery's results on $\zeta(2)$, $\zeta(3)$).

REFERENCES

[1] Baker Jr.,G.A.: Essentials of Padé approximants, Academic Press, 1975.

[2] Bateman, H. and Erdélyi, A.: Higher transcendental functions, McGraw-Hill, 3V,1953.

[3] Chudnovsky, G.V.: Proceedings ICM, Helsinki, pp.169-177, 1978.

[4] Chudnovsky, G.V.: C.R.Acad.Sci.Paris, 288A, pp.A-439-A-440.

[5] Chudnovsky, G.V.: C.R.Acad.Sci.Paris, 288A, pp.A-965-A-967.

[6] Chudnovsky, G.V.: C.R.Acad.Sci.Paris, 288A, pp.A-1001-A-1004.

[7] Chudnovsky, G.V.: J.Math.Pure Appl. (to appear) 1979.

[8] Chudnovsky, G.V.: C.R.Acad.Sci.Paris, 288A, pp.A-607-A-609.

[9] Chudnovsky, G.V.: C.R.Acad.Sci.Paris, 1979 (to appear).

[10] Chudnovsky, G.V.: 1979 Preprint IHES/M/79/26 (to be published in Lecture Notes in Mathematics, Springer, 1979).

[11] Chudnovsky, G.V.: Annals of Math., Princeton 109 (1979), n°2, pp.353-377.

[12] Chudnovsky, D.V. and Chudnovsky, G.V.: Il Nuovo Cimento, 40B (1977), n°2, 339-353.

[13] Chudnovsky, D.V. and Chudnovsky, G.V.: C.R.Acad.Sci.Paris 287A(1978), pp.573-576.

[14] Chudnovsky, D.V. and Chudnovsky, G.V.: Lettere al Nuovo Cimento, 25 (1979), pp. 263-265.

[15] Chudnovsky, D.V. and Chudnovsky, G.V.: Service de Physique Théorique, Preprint DPh.T79/115 (to be published in Lectures Notes in Mathematics, Springer).

[16] Chudnovsky, D.V.: Lecce lectures (see this volume).

[17] Chudnovsky, D.V.: Proceedings of Nat.Acad.Sci.USA, 75, n°9, 1978, pp.4082-4085.

[18] Darboux, G.: C.R.Acad.Sci.Paris 91(1879); C.R.Acad.Sci.Paris 94(1882), p.1456.

[19] Gelfand, A.O.: Transcendental and algebraic numbers, Dover, N.Y., 1960.

[20] Galočkin , A.: Math.Skornic, 95, pp.396-417, 1974.

[21] Galočkin, A.: Math.Zametki 8(1970), pp.19-27.

[22] Hermite, Ch.: C.R.Acad.Sci.Paris, 77(1873), pp.18-24, 74-79, 226.

[23] Ince, I.L.: Ordinary Differential Equations, Dover, 1959.

[24] Laguerre, : J.de Math., 1, p.135-165, 1885.

[25] Mahler, K.: Mathemat.Ann.168, pp.200-227, 1966.

[26] Osgood, Ch.F.: Tran.Am.Math.Soc.123(1966), pp.64-87.

[27] Padé, H.: Ann.Ecole Norm.Sup.(3) 16, pp.156-259, 1899.

[28] Riemann, B.: Ouvres Mathématiques, Albert Blanchard, Paris, pp.353-363, 1968.

[29] Siegel, C.L.: Abh.Preussichen Akad.Wissen.Phys.Math.Classe, n°1, 1929.

[30] Siegel, C.L.: Transcendental Numbers, Princeton University Press, 1949.

[31] Szëgo, G.: Orthogonal Polynomials, Providence, 1939.

[32] E.Grosswald; Bessel Polynomials, Lect.Not.Math., v.698, Springer-Verlag, 1978.

[33] A.J.van der Poorten,Math.Intelligencer, 1979.

[34] Khintchin. Continued fractions, Addison-Wesley, 1960.

[35] Lang, S.: Introduction to Transcendental Numbers, Addison-Wesley, 1966.

[36] Boiti,M. and Pempinelli, F.: Lecce lectures (this volume).

[37] D.V.Chudnovsky, C.R.Acad.Sci.Paris, 288A, 1979.

[38] V.Danilov: Math.zametki, v.24, n°4, 1978 (Russian).

[39] Jager, H.: A Multidimensional Generalization of the Padé Table, Drukkerij Holland N.V., Amsterdam, 1964.

THE SOLITON THEORY OF STRONG LANGMUIR TURBULENCE

D. ter Haar*

Instituut voor Theoretische Fysica

Princetonplein 5, P.O. Box 80006

3508 TA Utrecht, The Netherlands

*Permanent address:
Department of Theoretical Physics, 1 Keble Road, Oxford, England.

June 1979

We give a brief account of soliton theories of strong Langmuir turbulence, of the reasons for such theories and the difficulties encountered by them.

In this talk we shall consider an unmagnetized plasma in which the electron temperature T_e is much higher than the ion temperature T_i. In that case there are essentially three kinds of elementary excitations in the plasma: the longitudinal Langmuir oscillations (ℓ) with a dispersion relation

$$\omega^2 = \omega_{pe}^2 + 3k^2 v_{Te}^2 \tag{1}$$

where the Langmuir or plasma frequency (for electrons) is given by the equation

$$\omega_{pe}^2 = 4\pi n_e e^2/m_e \quad ; \tag{2}$$

transverse plasma oscillations (t) with a dispersion relation

$$\omega^2 = \omega_{pe}^2 + k^2c^2 \quad ; \tag{3}$$

and ion-acoustic oscillations with a dispersion relation

$$\Omega^2 = \frac{K^2 v_s^2}{1 + K^2 r_D^2} \quad . \tag{4}$$

In these equations k and K are wavenumbers, ω and Ω frequencies, v_{Te} the electron thermal velocity given by the equation

$$v_{Te}^2 = T_e/m_e \quad , \tag{5}$$

where we have expressed the temperature in energy units (which is equivalent to putting Boltzmann's constant equal to unity), m_e the electron mass, e the electron charge, c the velocity of light, v_s the ion-acoustic velocity given by the equation

$$v_s^2 = T_e/m_i \quad , \tag{6}$$

m_i the ion mass, and r_D the (electron) Debye radius given by the equation

$$r_D = v_{Te}/\omega_{pe} \quad . \tag{7}$$

If the total energy in these elementary excitations is sufficiently low, one can to a good approximation treat them as being independent and describe the physical situation of the plasma as corresponding to a linear superposition of elementary excitations. One can then develop a quasi-linear theory in which one takes into account interactions between the excitations, but still treating them as independent (or nearly independent) entities. In this way one can for that situation -the case of weak turbulence- derive the turbulence spectrum, that is, the energy density as function of wavenumbers. An excellent account of the theory of weak plasma turbulence can, for instance, be found in the book by Kaplan and Tsytovich (1973). For the case of weak Langmuir turbulence it turns out that the energy accumulates at very low wavenumbers, $k \sim k_o$, where k_o is much smaller than the Debye wavenumber k_D $(=1/r_D)$.

To get some idea of the orders of magnitude involved we remind ourselves first of all of the fact that for all plasmas of interest in fusion research or in astrophysics the ratio $\hbar\omega_{pe}/T_e$, where $\hbar$ is Planck's constant divided 2π, is exceedingly small (see, e.g., Thornhill and ter Haar, 1978, Table 1 on p.47) so that one can use classical statistics for the Langmuir oscillations. As each mode in that case will have an average energy T_e and as modes with wavenumbers larger than k_D will not be present, since they are heavily Landau damped, we expect that the total energy density W of the Langmuir waves in thermodynamic equilibrium will be given by the equation

$$W \approx k_D^3 T_e \quad . \tag{8}$$

This must be compared with the kinetic energy density, $n_e T_e$. The ratio $\overline{W}$ of those two energy densities will give us a measure for the strenght of the Langmuir turbulence; it is given by the relation

$$\overline{W} = W/n_e T_e \quad . \tag{9}$$

In the case of thermodynamic equilibrium we have

$$\overline{W} \approx k_D^3/n_e = 1/N_D \quad , \tag{10}$$

where the Debye number N_D, given by the equation

$$N_D = n_e r_D^3 \quad , \tag{11}$$

is a very large number for practically all plasmas of interest (see, e.g., Table 1 in Thornhill and ter Haar, 1978).

Another dimensionless parameter of interest for the present discussion characterizes the pumping power Q, that is, the rate at which energy is put into the plasma. We can write

$$Q = \eta n_e T_e \omega_{pe} \quad , \tag{12}$$

where η is the dimensionless pumping parameter.

If the energy pumped into the plasma is dissipated through electron collisions -as will be the case for weak turbulence- we have the balance equation

$$\nu_e W = Q \quad , \tag{13}$$

where ν_e is the collisional damping rate which is approximately given by the equation

$$\nu_e \approx \omega_{pe}/N_D \quad . \tag{14}$$

From equation (13), (14), (12), and (9) we thus find

$$\overline{W} = \eta N_D \quad , \tag{15}$$

and we see by comparing equations (15) and (10) that it needs only very small values of η,

$$\eta > N_D^{-2} \quad , \tag{16}$$

to get a turbulence level higher than that pertaining to thermodynamic equilibrium.

In the quasi-linear theory collisional damping is the only dissipation mechanism. However, if the level of turbulence, as measured by $\overline{W}$, increases, the Langmuir condensation, that is, the accumulation of energy at wavenumbers k_o much smaller than k_D will lead to new possibilities for dissipation and it is the study of these mechanism which is the subject of the soliton theory of Langmuir turbulence. Indeed, one can show that the kinetic equations which are used to derive the weak turbulence spectrum can be derived by using a random phase approximation, clearly indicating that as soon as coherent effects become important one most look elsewhere. Such a coherence is initiated through the modulational instability (M.I.) which sets in when the energy of the Langmuir oscillations accumulates at very low wavenumbers (Vedenov and Rudakov, 1965). One sees easily that a practically uniform plasma density is unstable: if through fluctuations locally the density is lowered, the local plasma frequency will be less than the plasma frequency in the vicinity and as the dielectric constant ε to a good approximation is given by the relation

$$\varepsilon = 1 - (\omega_{pe}^2/\omega^2) \quad , \tag{17}$$

plasma oscillations with the local Langmuir frequency will be trapped and through the ponderomotive force (vide infra) will tend to decrease the plasma density even further. The equations describing the M.I. were derived by Zakharov (1972). The simplest, though not the most rigorous, way to derive them in the one-dimensional case is to start from the full expression for the dielectric constant of the plasma in the form

$$\varepsilon(\omega,k) = 1 - \frac{\omega'^2_{pe}}{\omega^2} - \frac{3k^2v_{Te}^2}{\omega^2} \quad , \tag{18}$$

and from the equation

$$\varepsilon(\omega,k)\, E' = 0 \quad , \tag{19}$$

where E' is the electric field strenght and where ω'_{pe} is the local Langmuir frequency. Treating $\varepsilon(\omega,k)$ as an operator, ω and k as the operators $-i\partial/\partial t$ and $i\partial/\partial x$, and writing ω'^2_{pe} in the form

$$\omega'^2_{pe} = \omega^2_{pe} + 4\pi e^2\,\delta n/m_e \quad , \tag{20}$$

where ω_{pe} corresponds to the overall plasma density, which we assume to be uniform ($n_e = n_o$), while δn is the deviation of the plasma density n_o, we find the equation

$$-\frac{\partial^2 E'}{\partial t^2} - \omega^2_{pe} E' - \frac{4\pi e^2}{m_e}\,\delta n\, E' + 3v^2_{Te}\,\frac{\partial^2 E'}{\partial x^2} = 0 \quad . \tag{21}$$

As the main contribution to E' will come from Langmuir oscillations we can write

$$E' = \varepsilon\,\exp(-i\omega_{pe} t) \quad , \tag{22}$$

where ε is a slowly varying amplitude so that we have

$$\frac{\partial^2 E'}{\partial t^2} = \left[-\omega^2_{pe}\varepsilon - 2i\omega_{pe}\frac{\partial \varepsilon}{\partial t} + \frac{\partial^2 \varepsilon}{\partial t^2}\right] \exp(-i\omega_{pe} t)$$

$$\approx \left[-\omega^2_{pe}\varepsilon - 2i\omega_{pe}\frac{\partial \varepsilon}{\partial t}\right] \exp(-i\omega_{pe} t) \quad , \tag{23}$$

and hence

$$2i\omega_{pe}\frac{\partial \varepsilon}{\partial t} - \frac{4\pi e^2}{m_e}\,\delta n\,\varepsilon + 3v^2_{Te}\frac{\partial^2 \varepsilon}{\partial x^2} = 0 \quad . \tag{24}$$

Equation (24) still contains the quantity δn. If there would be no electric field present, δn would satisfy the sound wave equation

$$\frac{\partial^2 \delta n}{\partial t^2} - v^2_s\,\frac{\partial^2 \delta n}{\partial x^2} = 0 \quad . \tag{25}$$

However, if one takes into account the ponderomotive force (radiation pressure) of the high-frequency Langmuir field on the plasma, corresponding to a force per unit mass gives by the relation (see, e.g., Motz and Watson, 1967)

$$F_{pond} = -\frac{\partial}{\partial x}\,\frac{|\varepsilon|^2}{16\pi n_o m_i} \quad , \tag{26}$$

equation (25) must be replaced by the equation

$$\frac{\partial^2 \delta n}{\partial t^2} - v_s^2 \frac{\partial^2 \delta n}{\partial x^2} = \frac{\partial^2}{\partial x^2} \frac{|\varepsilon^2|}{16\pi m_i} \quad . \tag{27}$$

We note in passing that in the three-dimensional case equations (24) and (27) have the form

$$2i\omega_{pe} \frac{\partial \vec{\varepsilon}}{\partial t} - \frac{4\pi e^2}{m_e} \delta n\, \vec{\varepsilon} + 3v_{Te}^2 \nabla(\nabla \cdot \vec{\varepsilon}) - c^2 \left[\nabla \wedge (\nabla \wedge \vec{\varepsilon})\right] = 0 \ , \tag{28}$$

$$\frac{\partial^2 \delta n}{\partial t^2} - v_s^2 \nabla^2 \delta n = \nabla^2 \frac{|\vec{\varepsilon}|^2}{16\pi m_i} \quad . \tag{29}$$

We also note that equations (28) and (29) are non-linear, containing the terms involving $\delta n \vec{\varepsilon}$ and $|\vec{\varepsilon}|^2$, while in the linear limit they yield the dispersion relations (1), (3), and (4) -the latter in the limit as $Kr_D \to 0$ and the first two in the limit as $\omega - \omega_{pe} \ll \omega_{pe}$, where we have to take into account the fact that we have taken out of the $\vec{\varepsilon}$ oscillations the frequency ω_{pe} through relation (22).

For the discussion of equations (24) and (27) or of equations (28) and (29), it is convenient to introduce dimensionless variables through the relation

$$\vec{r}' = \frac{2}{3}\sqrt{\mu}\, \frac{\vec{r}}{r_D} \ , \quad t' = \frac{2}{3}\mu\omega_{pe} t \ , \quad \vec{E} = \frac{\vec{\varepsilon}}{8(\pi\mu n_o T_e/3)^{\frac{1}{2}}} \ ,$$
$$n = \frac{3}{4}\frac{\delta n}{\mu n_o} \ , \quad \alpha = \frac{c^2}{3v_{Te}^2} \ , \tag{30}$$

where

$$\mu = m_e/m_i \quad . \tag{31}$$

In terms of these variables equations (28), (29), and their one-dimensional counterparts are

$$i\frac{\partial \vec{E}}{\partial t} - \alpha\left[\nabla \wedge (\nabla \wedge \vec{E})\right] + \nabla(\nabla \cdot \vec{E}) - n\vec{E} = 0 \quad , \tag{32}$$

$$\frac{\partial^2 n}{\partial t^2} - \nabla^2 n = \nabla^2 |\vec{E}|^2 \quad , \tag{33}$$

and

$$i \frac{\partial E}{\partial t} + \frac{\partial^2 E}{\partial x^2} - nE = 0 \quad , \tag{34}$$

$$\frac{\partial^2 n}{\partial t^2} - \frac{\partial^2 n}{\partial x^2} = \frac{\partial^2 |E|^2}{\partial x^2} \quad , \tag{35}$$

where we have omitted the primes on $\vec{r}$ and t.

The turbulence parameter $\overline{W}$ follows from the fact that W in the present case equals $|\vec{\varepsilon}|^2/8\pi$ so that we have

$$\overline{W} = \frac{|\vec{\varepsilon}|^2}{8\pi n_o T_e} = \frac{8}{3} \mu \, |\vec{E}|^2 \quad . \tag{36}$$

Equations (34) and (35) have the following "soliton" solution:

$$E = E_o \operatorname{sech} \frac{E_o(x-Mt)}{\sqrt{2(1-M^2)}} \exp\left[\frac{1}{2} iMx - \frac{1}{4} iM^2 t - \frac{iE_o^2 t}{2\cdot(1-M^2)}\right] \quad , \tag{37}$$

$$n = -|E|^2/(1-M^2) \quad , \tag{38}$$

where M -the soliton velocity or Mach number- is an arbitrary parameter. The word soliton is put in quotes as the solution (37), (38) is not a soliton in the strict, aristocratic sense as defined by Scott, Chu, and McLaughlin (1973) and to some extent the fact that the Langmuir soliton, as we shall call the solutions (37), (38) all the same, can fuse and decay made it attractive to develop a theory describing developed Langmuir turbulence by the interaction of Langmuir solitons. Certainly, in a one-dimensional system these solitons show a rich phenomenology as shown by computer experiments (e.g., Degtyarev, Makhan'kov, and Rudakov, 1975) which has to a large extent been explained by analytical work by Thornhill (Gibbons, Thornhill, Wardrop, and ter Haar, 1977; Thornhill and ter Haar, 1978). The first to propose a soliton theory for strong Langmuir turbulence was Rudakov (1933; Kingsep, Rudakov, and Sudan, 1973). In favour of such a theory is the existence of experiments indicating the existence of density depressions filled with electric field energy; the most striking experiments to date are perhaps those by Antipov, Nezlin, Snezhkin, and Trubnikov (1978,1979), although one should emphasize that the identification of the observed density expressions, which are probably produced by the ponderomotive force due to the enhanced field energy density, with

Langmuir solitons still involves a certain amount of faith.

One can prove that the solution (37), (38) is in the one-dimensional case a stable configuration, but the corresponding three-dimensional plane wave soliton, while being a solution of equations (32) and (33) is not stable (see, for instance, Wardrop and ter Haar, 1979). In fact, one can prove by considering the energy of the system that any three-dimensional structure will gain energy by contracting, so that such structures will, in fact, collapse. Such a collapse has been found in computer experiments (see, e.g., Denavit, Pereira, and Sudan, 1964; Degtyarov, Zakharov, and Rudakov, 1975; Zakharov and Synakh, 1976).

Let us now study the development of the situation in a plasma in which Langmuir condensation has taken place. In such a plasma there will be Langmuir waves, corresponding to the following solution of equations (32) and (33):

$$\vec{E} = \vec{E}_o \exp\left[i(\vec{k}_o \cdot \vec{r}) - i\omega_o t\right] \quad , \qquad n=0 \ , \tag{39}$$

where ω_o and $\vec{k}_o$ satisfy the Langmuir dispersion relation (remember relation (22))

$$\omega_o = k_o^2 \tag{40}$$

and the relation expressing the fact that the Langmuir waves are longitudinal:

$$\left[\vec{k}_o \wedge \vec{E}_o\right] = 0 \tag{41}$$

In the situation we are considering the wavenumber k_o will be small and, in fact, it will satisfy the inequality

$$k_o \ll \sqrt{\mu} \quad . \tag{42}$$

One can now consider the stability of the solution (39). One finds (for details see Thornhill and ter Haar, 1978) that the growth rate γ_{MI} of the fastest growing perturbation depends on the amplitude of the original wave, or on the value of the turbulence parameter $\overline{W}$ (see equation (36)). If $\overline{W} \gg \mu$ -the hydrodynamic limit- the maximum growth rate equals

$$\gamma_{MI} \sim \mu^{\frac{1}{2}}\overline{W}^{\frac{1}{2}}\omega_{pe} \tag{43}$$

and corresponds to a wavenumber q_{MI} given by the relation

$$q_{MI} \sim \mu^{1/6} \overline{W}^{1/3} k_D \quad . \tag{44}$$

On the other hand, in the static limit when $W \ll \mu$ we have

$$\gamma_{MI} \sim \overline{W} \omega_{pe} \tag{45}$$

and

$$q_{MI} \sim \overline{W}^{\frac{1}{2}} k_D \quad . \tag{46}$$

In equation (43) to (46) wa have restored the dimensions.

The first thing to notice is that it follows from equation (30) that the largest negative value n can attain is $-3/4\mu$. Secondly, we see from equation (33) that $|\vec{E}|^2$ is of the same order of magnitude as $-n$ so that $|\vec{E}|^2$ has a maximum value of the order of μ^{-1} -which, of course, is a large number- and that $\overline{W}$, given by equation (36) will be at most of order unity -at least as long as the situation can be described by the Zakharov equations (32) and (33). Hence it follows from equations (44) and (46) that the wavenumber corresponding to the MI is well below k_D so that Landau damping will not come into operation as long as the MI is the dominant factor determining the shape of the density depressions or cavitons as they are often called.

The growth rates and corresponding wavenumbers given by equations (43) to (46) correspond to longitudinal perturbations. The corresponding quantities for transverse perturbations are, provided $\overline{W} \gg \mu/\alpha$, where α, which is given by equation (30), is a very large number (see Table 1 on p.47 of Thornhill and ter Haar, 1978), given by the equations

$$\gamma_{MI} \sim \mu^{\frac{1}{2}} \overline{W}^{\frac{1}{2}} \alpha^{-\frac{1}{2}} \omega_{pe} \tag{47}$$

and

$$q_{MI} \sim \mu^{1/6} \overline{W}^{1/3} \alpha^{-2/3} k_D \quad . \tag{48}$$

We see that in the transverse direction the growth rate is significantly decreased and the length scale considerably increased.

We should mention here that in the hydrodynamic case (i) the MI dominates over other processes such as ion-sound or electron collision damping and (ii) the MI will still be the dominant process, even if we are dealing with a broad-band spectrum rather than with the monochromatic wave (39). This latter statement is correct provided the band width of the frequency spectrum is much less than γ_{MI} and the spread in wave-number much less than q_{MI} so that the perturbation has neither space nor time to realize that the spectrum is not monochromatic. These conditions are satisfied for the spectrum which is produced in the Langmuir condensation.

We can now envisage what will happen, if energy is pumped into a plasma. The energy will first of all accumulate at very low wave numbers. This state is unstable and the MI will develop on a time-scale τ_{MI} and length-scale λ_{MI} which in the case when $\overline{W}_o > \mu$ ($\overline{W}_o$ being the energy density in the Langmuir condensate divided by $n_e T_e$) are given by equations (43) and (44):

$$\tau_{MI} \sim \mu^{-1/2}\, \overline{W}_o^{-1/2}\, \omega_{pe}^{-1} \quad , \tag{49}$$

$$\lambda_{MI} \sim \mu^{-1/6}\, \overline{W}_o^{-1/3}\, r_D \quad . \tag{50}$$

Bunching will proceed until a "soliton" is formed along the field; the transverse collapse is still negligible. The time- and length-scales τ_{sol} and λ_{sol} for this process follow from expression (37) and are

$$\tau_{sol} \sim \overline{W}_{sol}^{-1}\, \omega_{pe}^{-1} \quad , \tag{51}$$

$$\lambda_{sol} \sim \overline{W}_{sol}^{-1/2}\, r_D \quad , \tag{52}$$

which are, as we should have expected, just the expressions (45) and (46) obtained for the static regime.

We can find the turbulence parameter $\overline{W}_{sol}$ of the developed soliton field from the requirement that the soliton -which at the initial stage is one-dimensional- takes up all the energy of the MI whence

$$\overline{W}_{sol}\, \lambda_{sol} = \overline{W}_o\, \lambda_{MI} \quad , \tag{53}$$

or

$$\overline{W}_{sol} = \mu^{-1/3}\, \overline{W}_o^{4/3} \quad . \tag{54}$$

We thus find

$$\tau_{sol}/\tau_{MI} \sim (\mu/\overline{W}_o)^{5/6} < 1 \quad . \tag{55}$$

The transverse instabilities are likely to become important at a later stage. The transverse size R_o of the initial cavitons will thus be determined by the relation

$$R_o \sim 1/\Delta k \quad , \tag{56}$$

where Δk is the spread in wavenumber in the Langmuir condensate which will be of the order of the average wavenumber k_o of the condensate and thus satisfy inequality (42) so that

$$R_o \gg \mu^{-1/2} r_D \quad , \tag{57}$$

and hence

$$\lambda_{sol}/R_o \ll (\mu/\overline{W}_o)^{2/3} < 1 \tag{58}$$

The cavitons start out with a pancake shape with a soliton field along the small axis. As the field energy trapped in the caviton is conserved (conservation of plasmon number or of action) as long as the Zakharov equations hold, we have at a later stage when the cavitons are collapsing

$$\overline{W}_{sol}(t)\ \lambda_{sol}(t)\ R(t)^2 = \text{constant} \quad , \tag{59}$$

and using equation (52)

$$\lambda_{sol}(t)/\lambda_{sol}(0) = R(t)^2/R(0)^2 \quad , \tag{60}$$

so that the pancake structure will become more and more pronounced.

If it is, indeed, the MI which removes the energy from the Langmuir condensate into the cavitons, γ_{MI} will govern the dissipation of the condensate so that we have the balance equation

$$Q = \gamma_{MI} W_o \quad (= \gamma_{MI} \overline{W}_o n_e T_e) \quad , \tag{61}$$

or using equations (12) and (43)

$$\overline{W}_o = \mu^{-1/3}\ \eta^{2/3} \quad . \tag{62}$$

It would be interesting to measure the dependence of the turbulence energy on the pumping power to check whether relation (62) holds. If it did hold, it would be strong support for the soliton (caviton) theory of strong Langmuir turbulence.

During the collapse ion dynamics will become important so that the collapse time-scale τ_{coll} will be given by the equation

$$\tau_{coll} \sim R_o/v_s \quad . \tag{63}$$

This time-scale is much longer than the time-scale τ_{sol} for the relaxation to a soliton field along the small axis:

$$\tau_{coll}/\tau_{sol} \sim \frac{k_D}{k_o} \frac{\overline{W}_{sol}}{\mu^{1/2}} \gg 1 \quad . \tag{64}$$

However, we now run into difficulties, because τ_{coll} is also much longer than the time for the development of the MI:

$$\tau_{coll}/\tau_{MI} \sim \frac{k_D}{k_o} \overline{W}_o^{1/2} \gg 1 \quad . \tag{65}$$

where we have used inequality (42) and the fact that we have assumed that $\overline{W}_o > \mu$. This means that the collapse can not remove the energy as fast as the MI collects it.

We can put this differently. Let us assume that a fraction f_{cav} of the plasma is taken up by collapsing cavitons. Energy is collected by the MI at a rate γ_{MI} and removed by the collapse at a rate $1/\tau_{coll}$. For a consistent picture we thus need

$$\overline{W}_o \gamma_{MI} \sim f_{cav} \overline{W}_{sol}/\tau_{coll} \quad , \tag{66}$$

whence we find

$$f_{cav} \sim \frac{k_D}{k_o} \mu^{1/3} \overline{W}_o^{1/6} \gg 1 \quad . \tag{67}$$

One can possibly save the situation by relaxing condition (56) and assuming a much smaller initial transverse size, but be do not want to discuss this here any further.

REFERENCES

S.V. Antipov, M.V. Nezlin, E.N.Snezhkin, and A.S. Trubnikov: 1978 Sov. Phys.JETP 47, 506.

1979 Kurchatov Institute Preprint, IAE-3107.

L.M.Degtyarev, V.G.Makhan'kov, and L.I.Rudakov: 1975 Sov.Phys.JETP 40, 264.

L.M.Degtyarev, V.E.Zakharov, and L.I.Rudakov: 1975 Sov.Phys.JETP 41, 57.

J.Denavit, N.R.Pereira, and R.N.Sudan: 1974 Phys.Rev.Lett. 33, 1435.

J.Gibbons, S.G.Thornhill, M.J.Wardrop, and D.ter Haar: 1977 J.Plasma Phys. 17, 153.

S.A.Kaplan and V.N.Tsytovich: 1973 Plasma Astrophysics, Pergamon Press, Oxford.

A.S.Kingsep, L.I.Rudakov, and R.N.Sudan: 1973 Phys.Rev.Lett. 31, 1482.

H.Motz and C.J.H.Watson: 1967 Adv.Electron. 23, 153.

L.I.Rudakov: 1973 Sov.PHys.Doklady 17, 1166.

A.C.Scott, F.Y.F.Chu, and D.W.McLaughlin: 1973 Proc.IEEE 61, 1443.

S.G.Thornhill and D.ter Haar: 1978 Phys.Repts. 43, 43.

A.A.Vedenov and L.I.Rudakov: 1965 Sov.Phys.Doklady 9, 1073.

M.J.Wardrop and D.ter Haar: 1979 Physica Scripta, in course of pubblication.

V.E.Zakharov: 1972 Sov.Phys.JETP 35, 908.

V.E.Zakharov and V.S.Synakh: 1976 Sov.Phys.JETP 41, 465.

GEOMETRY OF BÄCKLUND TRANSFORMATIONS

F. A. E. Pirani*

Department of Mathematics,

University of Arizona

Tucson, Arizona

Introduction. In this lecture I suggest that a suitable setting for Bäcklund transformations is the theory of jet bundles. The transformations may be described by maps of bundles; those maps define connections in the sense of Ehresmann. The vanishing of the corresponding curvature yields the differential equation or equations of physical interest. The group which acts on the connection is not specified in advance and is not completely determined by the Bäcklund map.

The advantages of this approach are that it uses concepts and methods which are already familiar from differential geometry, that it allows generalization to space-times of dimension greater than 2, that it makes possible a unified treatment of Bäcklund transformation equations and inverse scattering equations, and that it gives a context in which to place the method of Estabrook and Wahlquist[10], from a study of which the present work first arose. What is reported here is joint work of M. Crampin, D.C.Robinson, W.F.Shadwick and myself, much of which has been published elsewhere[2-7-8-9]. This version differs in technical detail from the other published work, and is less didatic in style than the lecture as delivered. In content, it is roughly a summary of [8]. My par-

---

* On leave of absence from Department of Mathematics, King's College, London.

ticipation in the conference was supported in part by the Norman Foundation. The hospitality of Professors Boiti and Soliani and other members of the Istituto di Fisica, Lecce, is very happily acknowledged.

Example. The scattering problem for the sine-Gordon equation may be written in the form[1]

$$\frac{\partial}{\partial x}\begin{bmatrix} y^1 \\ y^2 \end{bmatrix} = \begin{bmatrix} \eta & -\frac{1}{2} z_x \\ \frac{1}{2} z_x & -\eta \end{bmatrix}\begin{bmatrix} y^1 \\ y^2 \end{bmatrix} ,$$

with eigenfunctions y^1 and y^2 whose time-dependence is given by

$$\frac{\partial}{\partial t}\begin{bmatrix} y^1 \\ y^2 \end{bmatrix} = \frac{1}{4}\eta^{-1}\begin{bmatrix} \cos z & \sin z \\ \sin z & -\cos z \end{bmatrix}\begin{bmatrix} y^1 \\ y^2 \end{bmatrix} .$$

Here η is a parameter, and suffixes denote partial derivatives of the function $z(x,t)$. Cross-differentiating these equations, one obtains

$$\frac{\partial}{\partial t}\frac{\partial}{\partial x}\begin{bmatrix} y^1 \\ y^2 \end{bmatrix} - \frac{\partial}{\partial x}\frac{\partial}{\partial t}\begin{bmatrix} y^1 \\ y^2 \end{bmatrix} = \frac{1}{2}(z_{xt} - \sin z)\begin{bmatrix} 0 & -1 \\ 1 & 0 \end{bmatrix}\begin{bmatrix} y^1 \\ y^2 \end{bmatrix} .$$

Thus the sine-Gordon equation (in characteristic coordinates)

$$z_{xt} - \sin z = 0$$

appears as the integrability condition for the scattering and time-dependence equations.

However, one might well, coming upon these equations for the first time, and unaware of their origin, rewrite them in the form

$$\frac{\partial y^A}{\partial x^b} + \Gamma^A_{D\,b}\, y^D = 0 ,$$

where y^A $(A=1,2)$ are again the eigenfunctions, $-\Gamma^A_{D\,1}$ and $-\Gamma^A_{D\,2}$ are the entries in the square matrices in the scattering and time-dependence equations respectively, x^1 and x^2 have been written in place of x and t $(b=1,2)$, and summation over repeated indices is understood[2]. The equations thus rewritten look exactly like equations of parallel transport, if the y^A are interpreted as components of a vector in a 2-dimensional vector space, copies of which have been attached to space-time at each point. Thus one is confronted with a vector bundle over space-time. From the further observation that the square matrices are trace-free, and hence lie in the Lie algebra SL(2,R), one may infer that the connection determining the parallel transport is an SL(2,R)-connection. Finally, the fact that the coefficients of connection depend on a function and its partial derivative suggest that the source for the connection might be sought in the theory of jet bundles. It turns out that this point of view has several advantages, including those enumerated in the Introduction. I therefore proceed, from this point of view, to describe a general context in which the example may be placed.

Notation and conventions. An elementary exposition of the relevant ideas of the theory of jet bundles and of the theory of connections has been published elsewhere[8]. Other accounts may be found in Robert Hermann's Interdisciplinary Mathematics[4], or (jet bundles, less elementary) in a paper of Kupershmidt[5], for example.

All manifolds, bundles and maps are supposed smooth (C^∞). Space-time is an m-dimensional manifold M, with local coordinates x^a. The field quantities, or dependent variables, are local fibre coordinates z^μ and y^A in bundles $\pi: E \to M$ and $\pi': E' \to M$ respectively. The appearance of two lots of field quantities, which by notation mirrors the situation in the example above, will be explained shortly.

The set of local sections of π is denoted $\Gamma(\pi)$, and the k-jet bundle of π is denoted $\pi_k: J^kE \to M$. The natural projection of the k-jet bundle on the ℓ-jet bundle, for $\ell \leq k$, is denoted $\pi^k_\ell : J^kE \to J^\ell E$. If $\gamma \in \Gamma(\pi)$, its k-jet, which is a local section of π_k, is denoted $j^k\gamma$. The module of p-forms on any manifold is denoted $\Lambda^p(\cdot)$, and the functions by $C^\infty(\cdot)$.

The contact module $\Omega^k(E)$ on J^kE is the $C^\infty(J^kE)$-module of Pfaffian forms which are annihilated by every k-jet:

$$\Omega^k(E) = \{\theta \in \Lambda^1(J^kE) \,|\, (j^k\gamma)^*\theta = 0 \quad \forall \gamma \in \Gamma(\pi)\} \quad .$$

Local coordinates on J^kE are written x^a(from M), z^μ(from fibres of π), $z^\mu_{a_1}, \ldots, z^\mu_{a_1 \ldots a_k}$ (tensor, not multi-index, notation). In these coordinates $\Omega^k(E)$ has a basis with presentation

$$\theta^\mu = dz^\mu - z^\mu_a dx^a \quad , \qquad \theta^\mu_a = dz^\mu_a - z^\mu_{ab} dx^b \quad , \quad \ldots$$

$$\theta^{\mu}_{a_1\ldots a_{k-1}} = dz^{\mu}_{a_1\ldots a_{k-1}} - z^{\mu}_{a_1\ldots a_k}\, dx^{a_k} \quad .$$

If $\phi : B \to M$ is any bundle, then the bundle induced from B by π' is denoted $\pi'^{*}(\phi) : \pi'(B) \to E'$, as in the diagram

$$\begin{array}{ccc} \pi'(B) & \longrightarrow & B \\ {\scriptstyle \pi'^{*}(\phi)}\downarrow & & \downarrow{\scriptstyle \phi} \\ E' & \xrightarrow[\pi']{} & M \end{array}$$

(for induced bundles, see for example Ref.3.

Construction of a Bäcklund map. The essential idea of a Bäcklund map is the definition of new field quantities by a system of first-order partial differential equations, the integrability conditions for which yield the original equations of interest[6]. To this end, consider the induced bundles

$$\pi'^{*}(\pi_k) \; : \; \pi'(J^kE) \to E'$$

and define projections

$$\tilde{\pi}_k = \pi' \circ \pi'^{*}(\pi_k) : \pi'(J^kE) \to M$$

and

$$\tilde{\pi}^k_\ell : \pi'(J^kE) \to \pi'(J^\ell E) \quad ,$$

the latter being such that

$$\begin{array}{ccc} \pi'(J^kE) & \xrightarrow{\pi'(\pi_k)} & J^kE \\ {\scriptstyle \tilde{\pi}^k_\ell}\downarrow & & \downarrow{\scriptstyle \pi^k_\ell} \\ \pi'(J^\ell E) & \xrightarrow[\pi'(\pi_\ell)]{} & J^\ell E \end{array}$$

commutes. The bundle $\pi'(J^kE)$ admits local coordinates $x^a, z^\mu, z^\mu_a, \ldots,$ $z^\mu_{a_1\ldots a_k}, y^B$ induced from the local coordinates on J^kE and E'.

In this language, a Bäcklund map is a map

$$\psi : \pi'^{*}(J^kE) \to J^1E'$$

over E'. For most purposes it is sufficient to consider Bäcklund maps which leave E' pointwise fixed. In this case the coordinate presentation of the Bäcklund map may be written

$$y_b^A = \psi_b^A\,(x^a, z^\mu, z_a^\mu, \ldots, z^\mu_{a_1\ldots a_k}, y^B).$$

A Pfaffian module $(\pi'(\pi_k))^*(\Omega^k(E))$ is induced on $\pi'^*(J^kE)$ from the contact module on J^kE, and a Pfaffian module $\psi^*(\Omega^1(E'))$ is induced on $\pi'^*(J^kE)$ from the contact module on J^1E'. The sum of these two is denoted $\tilde{\Omega}^k(\psi)$:

$$\tilde{\Omega}^k(\psi) := (\pi'(\pi_k))^*(\Omega^k(E)) + \psi^*(\Omega^1(E')) \quad .$$

The integrability conditions for the Bäcklund map ψ may be expressed very simply in terms of this object. If Ω is any collection of forms let $I(\Omega) = \{\Sigma\eta\wedge\theta \mid \theta\in\Omega,\ \eta \text{ any forms}\}$, and let $d\Omega = \{d\theta \mid \theta\in\Omega\}$. Then[8] the integrability conditions for ψ are

$$\tilde{\pi}_k^{k+1*}\psi^*\ d\Omega^1(E') \subset I(\tilde{\Omega}^{k+1}(\psi)) \quad .$$

In local coordinates, these conditions are merely

$$\tilde{D}_c^{k+1}\ \psi_b^A - \tilde{D}_b^{k+1}\ \psi_c^A = 0 \quad ,$$

where $\tilde{D}^{k+1}$ denotes the total derivative

$$\tilde{D}_c^{k+1} = \frac{\partial}{\partial x^c} + z^\mu_c\frac{\partial}{\partial z^\mu} + z^\mu_{cd}\frac{\partial}{\partial z^\mu_d} + \ldots + z^\mu_{cd_1\ldots d_k}\frac{\partial}{\partial z^\mu_{d_1\ldots d_k}} + \psi_c^A(x,z,\ldots)\frac{\partial}{\partial y^A} \,.$$

It is interesting, and possibly useful, to observe that a Bäcklund map defines a connection in the sense of Ehresmann, and that the integrability conditions represent the vanishing of the curvature of this connection. This construction may be carried out in at least two different ways, one defining a connection on $\pi':E'\to M$, the other, on $\pi'^*(\pi_\infty):\pi'(J^\infty E)\to E'$, where $\pi_\infty:J^\infty E\to M$ denotes the projective limit. Only the former, which is more likely to illuminate applications, will be described here.

Let γ be any section of $\pi:E\to M$, and let $\pi'(\gamma):E'\to\pi'(E)$, be the induced section of π'. Then $j^k\pi'(\gamma)$ is a section of $\pi'^*(\pi_k):\pi'(J^kE)\to E'$, which annihilates the module $(\pi'(\pi_k))^*(\Omega^k(E))$. Therefore the module $H^*(\gamma,\psi) := (j^k\pi'(\gamma))^*(\tilde{\Omega}^k(\psi))$ is simply $(j^k\pi'(\gamma))^*\psi^*(\Omega^1(E'))$; it has a basis given in local coordinates on E'

$$\theta^A = dy^A - \Gamma^A_a dx^a$$

where $\Gamma^A_a := \psi^A_a \circ j^k\pi'(\gamma)$. This module $H^*(\gamma,\psi)$ may be chosen as module of vertical forms defining an horizontal distribution $H(\gamma,\psi)$ on $\pi':E'\to M$. In all examples, every curve on M admits an horizontal lift, so that the distribution defines a connection. It will be assumed here that this is the case. In general there need be no Lie group G for which this connection is a G-connection. The condition for this is, in local coordinates, that Γ^A_a should be of the form $\Gamma^A_a(x,y) = \omega^\alpha_a(x)\, X^A_a(y)$, where as the notation indicates, the ω's are functions lifted from M, and the index α has some finite range. The vector fields $X^A_\alpha(y)\frac{\partial}{\partial y^A}$ will generate a subalgebra of the Lie algebra of π'-vertical vector fields, which will in general be infinite-dimensional[9], but in pratical examples this Lie algebra may be replaced by one of its finite-dimensional quotients.

Finally, it is not difficult to show that <u>the curvature of the connection defined by $H(\gamma,\psi)$ vanishes if and only if γ is a solution of the integrability conditions defined by ψ.</u>

REFERENCES

1. Ablowitz, M.J., Kaup, D.W., Newell, A.C. and Segur, H. 1973 Phys.Rev.Lett. 31, 125-127.
2. Crampin, M., Pirani, F.A.E. and Robinson, D.C. 1977 Lett.Math.Phys. 2, 15-19.
3. Godbillon, C. 1968 Géométrie différentielle et mecanique analytique Paris, Hermann.
4. Hermann, R., Interdisciplinary Mathematics Volumes I -, 1974, Brookline MA, Math Sci Press, especially volumes X and XII.
5. Kupersemidt, B., 1979 "Geometry of jet bundles and the structure of Lagrangian and Hamiltonian formalism", preprint.
6. Lamb, G.L., 1976 in Bäcklund transformations, (ed. R.M.Miura) 69-79 Berlin, Heidelberg, New York Springer Verlag - Lecture Notes in Mathematics 515, for a lucid elementary account.
7. Pirani, F.A.E. and Robinson, D.C., 1977, C.R.Acad.Sci. Paris 285, 581-583.
8. Pirani, F.A.E., Robinson, D.C., and Shadwick, W.F., 1979, "Local jet bundle formulation of Bäcklund transformations", in press, supplement to Letters in Math. Phys., No.1.
9. Shadwick, W.F., 1979 Dissertation, University of London.
10. Wahlquist, H.D. and Estabrook, F.B., 1975, J.Math.Phys. 16, 1-7 .

EXISTENCE OF SOLUTIONS AND SCATTERING THEORY FOR THE NON LINEAR SCHRÖDINGER EQUATION

J. Ginibre and G. Velo
Orsay Bologna

INTRODUCTION

In this lecture we shall report on some work done in the past two years [8] on the existence of solutions and on the scattering theory for a class of non linear Schrödinger equations

$$i\frac{du}{dt} = (-\Delta + m)u + f(u) \tag{0.1}$$

where Δ is the Laplace operator in $\mathbb{R}^n$, m is a real constant ad f is a non linear complex valued function. The equation (0.1), and especially the case where $f(u)=\lambda|u|^{p-1}u$, namely

$$i\frac{du}{dt} = (-\Delta + m)u + \lambda|u|^{p-1}u, \tag{0.2}$$

is interesting for several reasons. Firstly, it is widely used in several domains of applied physics [13] : the equation (0.2) with p=3 arises as an approximation of the Maxwell equations describing the propagation of a laser beam in a non linear medium and as the classical approximation to the field equation for a quantum mechanical many-body system with a δ-function two-body interaction. It also occurs in the Landau-Ginsburg theory of superconductivity. A second reason of interest lies in the fact that the equation (0.2), in the special case of dimension n=1 and with p=3, exhibits the same remarkable properties as the Korteweg-de Vries equation: for $\lambda<0$, it possesses a family of solitary wave solutions or solitons that are obtained by an arbitrary Galilei transformation (the equation is indeed Galilei invariant, see section 4 below) from the one-parameter family of solutions

$$u(x,t) = (-2/\lambda)^{1/2}\alpha \exp\left[i(\alpha^2-m)t\right]\left[\cosh(\alpha x)\right]^{-1} , \qquad (0.3)$$

where α is any real number. The equation also possesses multisoliton solutions that describe scattering processes where N incoming solitons collide and then emerge with unchanged shapes and speeds. Like the KdV equation, the equation (0.2) with n=1, p=3 can be "solved" by the inverse scattering method [16] in the abstract form given by Lax, and can be encompassed in the general algebraic framework of Gelfand et al. [7]. In view of this remarkable special case, it is therefore interesting to develop a general theory of scattering for the class of equations (0.1). A third reason of interest on (0.1) is that there exist very few non linear equations or systems for which one can develop a fairly complete theory of scattering, the most interesting other example being that of the non linear Klein-Gordon equation treated by Segal, Strauss and Morawetz-Strauss (see [14] for a review and a comprehensive bibliography).

Here we consider the following two problems:

(1) Existence of solutions of the Cauchy problem for the equation (0.1). More precisely we prove the existence and uniqueness of solutions of (0.1) with prescribed initial data $u(t_0)=u_0$ at some initial time t_0. This is done in two steps: we first prove the existence and uniqueness of local solutions, by which we mean solutions in a small time interval (in general depending on u_0), by a contraction method (section 1). We then extend this solution into a global solution, i.e. a solution defined for all times, by the use of suitable conservation laws (section 2).

(2) Asymptotic behaviour in time of the solutions of (0.1) and in particular scattering theory for the pair of equations that consists of (0.1) and the free Schrödinger equation

$$i\frac{du}{dt} = (-\Delta + m)u \qquad . \qquad (0.4)$$

We first prove the existence of a class of dispersive solutions to (0.1), namely solutions that spread out in time like the solutions of the free equation (0.4), by solving the Cauchy problem for (0.1) with initial time t_0 at or in a neighbourhood of infinity. These solutions tend to solutions of the free equation in a sense that can be specified as a by-product of the investigation. In terms of scattering theory, this step yields in particular the wave operators (section 3). We finally isolate a class of repulsive interactions for which we prove that all the solutions of (0.1) with suitable initial data are dispersive. The key of the proof is an approximate conservation law, which we call pseudoconformal, associated with a suitable one-parameter group of transformations that leave the free equation invariant. In terms of scattering theory, this property implies asymptotic completeness for the previously constructed wave operators (section 4).

Because of space-time limitations we present only the main and less technical results. All proofs and technical details are suppressed. They can be found in [8] . The assumptions on f and the functional framework will be specified as we proceed.

In addition to [8] several papers concerned with the previous problems have appeared: non linear Schrödinger equations with local interactions have been considered in [1], [3] , [4], [11], [12], and non linear Schrödinger equations with non local interactions have been considered in [2] , [5] , [6] , [9] . In particular the scattering theory for (0.2) and (0.4) in space dimension 3 has been studied in [11] by a different method based on dilatation invariance instead of pseudoconformal invariance.

1. EXISTENCE OF LOCAL SOLUTIONS

In order to study the Cauchy problem for the equation (0.1), we first recast it into the form of an integral equation that incorporates the initial data. Let $U(t)=\exp[it(\Delta-m)]$. Then the relevant integral equation is

$$u(t) = U(t-t_0)u_0-i\int_{t_0}^{t} d\tau U(t-\tau)f(u(\tau)) \equiv [A(t_0,u_0)u](t) \qquad (1.1)$$

or more concisely

$$u = A(t_0,u_0)u \qquad . \qquad (1.2)$$

We then look for a suitable Banach space X such that the following property holds. Let $T\geq 0$ and $I=[t_0-T,t_0+T]$. Let $\mathcal{X}(I)\equiv C(I,X)$ be the Banach space of continuous functions from I to X, equipped with the Sup.norm:

$$|u|_I = \operatorname*{Sup}_{t\in I} \|u(t)\|_X \tag{1.3}$$

and let $B(I,\rho)$ denote the ball of radius ρ in $X(I)$. Let $u_0 \in X$ be such that $U(\cdot - t_0)u_0 \in B(I,\rho)$ for some ρ. We want then to ensure that if T is sufficiently small, depending on ρ, say $T \leq T_0(\rho)$, then te non linear operator $A(t_0,u_0)$ maps $B(I,2\rho)$ into itself and is contracting in $B(I,2\rho)$, tipically:

$$|A(t_0,u_0)v_1 - A(t_0,u_0)v_2|_I \leq \tfrac{1}{2}|v_1 - v_2|_I \tag{1.4}$$

for all $v_1, v_2 \in B(I,2\rho)$. By the contraction mapping theorem, this implies the existence and uniqueness of solutions of (1.2) in $B(I,2\rho)$, and by a simple additional argument, the uniqueness in $X(I)$.

We now have to choose X, and correspondingly to make suitable assumptions on f. The most natural choice for X would be the Sobolev space $H^1(\mathbb{R}^n)$, which -as we shall see in the next section- is naturally associated with the conservation laws of the equation. We have been able to, make the previous scheme work with this choice only in dimension n=1. In higher dimensions, and this natural choice being excluded, there is still a large amount of freedom left in the choice of X and correspondingly of the assumptions on f. Of course larger spaces yield larger sets of admissible initial data and stronger uniqueness results, while smaller spaces yield stronger regularity properties of the solutions. For the purposes of scattering theory, we find it more interesting to choose X as large as possible. We present below two such choices, which are adequate for any $n \geq 2$ and for $n \leq 3$ respectively.

Our first choice is suitable for any $n \geq 2$. We assume f to satisfy the following conditions:

(H1) f is continuously differentiable and f(0)=0.

(H2) There exist real numbers p_1, p_2, r_0 and r such that

$$1 \leq p_1 \leq p_2 < (n+2)/(n-2) \tag{1.5}$$

$$2 \leq r_0 \leq r < 2n/(n-2) \tag{1.6}$$

$$r_0/\bar{r} \leq p_1 \leq p_2 \leq r/\bar{r}_0 \tag{1.7}$$

and for all $z \in \mathbb{C}$,

$$|f'(z)| \leq C(|z|^{p_1-1} + |z|^{p_2-1}) \quad . \tag{1.8}$$

Here and below the prime denotes the first derivative with respect to z or $\bar{z}$, $\bar{q}$ the index conjugate to q ($q^{-1}+\bar{q}^{-1}=1$), and C a non negative constant.

Corresponding to these assumptions, we choose $X=L^r \cap L^{r_0}$, where r and r_0 are chosen once for all as in (H2), and where we write L^q for $L^q(\mathbb{R}^n)$. Note that the bound with the higher (resp.lower) power in (1.8) restricts the behaviour of f' at large (resp. small) z, and therefore the local behaviour of f(u) in space (resp. the behaviour of f(u) at infinity in space). Also note that $X \supset H^1$ in this case.

Our second choice is suitable for $n \leq 3$. We assume f to satisfy:

(H1a) f is twice continuously differentiable and $f(0)=f'(0)=0$.

The corresponding choice for X is $X=H^1 \cap L^\infty$. Note that now $X \subset H^1$. For $n=1$, $X=H^1(\mathbb{R})$, since $H^1(\mathbb{R})$ $L^\infty(\mathbb{R})$.

We can now state the main result of this section.

Proposition 1. – *Let either:* $n \geq 2$, f *satisfy* (H1,2) *and* $X=L^r \cap L^{r_0}$, *or:* $n \leq 3$, f *satisfy* (H1a) *and* $X=H^1 \cap L^\infty$.

Then for any $\rho>0$, *there exists* $T_0(\rho)>0$ *depending only on* ρ *and* f, *such that for any* $t_0 \in \mathbb{R}$ *and any* $u_0 \in X$ *such that* $U(\cdot - t_0)u_0 \in B(I,\rho)$, *where* $I=[t_0-T_0(\rho), t_0+T_0(\rho)]$, *the equation* (1.2) *has a unique solution in* X(I). *This solution belongs to* $B(I,2\rho)$.

The proof of proposition 1 rests on the estimate (1.4) as described above. This estimate in turn rests on the fact that the free evolution U(t) maps $L^{\bar{q}}$ into L^q for any q, $\bar{q}$ with $q^{-1}+\bar{q}^{-1}=1$, $1 \leq \bar{q} \leq 2$, with

$$\|U(t)v\|_q \leq (4\pi|t|)^{n/q-n/2} \|v\|_{\bar{q}} \tag{1.9}$$

for any $v \in L^{\bar{q}}$, where $\|\cdot\|_q$ denotes the norm in L^q.

2. EXISTENCE OF GLOBAL SOLUTIONS

In order to extend the previous solutions to all times, one may try to proceed as follows. Let for simplicity $t_0=0$. Then, by proposition 1, one can construct a solution of the equation (1.1) up to some time T_0 depending on u_0. Taking now as initial data $u_1=u(T_0)$ at time T_0, one can extend this solution up to some time T_0+T_1, where T_1 depends on u_1, again by proposition 1. One can iterate this procedure and extend the solution from $t_n=T_0+...+T_{n-1}$ to $t_{n+1}=T_0+...+T_n$ by solving the Cauchy problem with initial data $u_n=u(t_n)$ at time t_n. What can happen however at this stage is that $\|u_n\|_X$ may tend to infinity as n does, in such a way that T_n tends to zero sufficiently fast to make the series $\sum T_n$ convergent. Then the solution just constructed may blow up in a finite time. The standard method to exclude such a phenomenon is to derive a priori estimates for suitable norms of arbitrary solutions of (1.1) with initial data u_0 in a suitable space. More precisely, one needs an a priori control of $\|U(t-s)u(s)\|_X$ for s and t in bounded intervals. Such a control is readily obtained from conservation laws associated with the equation. The simplest and most natural assumption is that the equation (0.1) is derived from a gauge invariant Lagrangian by the usual rules of variational calculus. We therefore assume:

(H3) The function $z\mapsto z^{-1}f(z)$ is real valued and depends only on $|z|$.

If f is continuous, the condition (H3) is equivalent to the requirement that there exist a C^1 function V from $\mathbb{R}^+$ to $\mathbb{R}$ such that if we define $V(z)\equiv V(|z|)$ for arbitrary $z\in\mathbb{C}$, then $f(z)=\partial V(z)/\partial\bar{z}$. Under condition (H3), the equation (0.1) is derived from the Lagrangian with density

$$\mathscr{L}(u) = \frac{i}{2}\left(\bar{u}\frac{\partial u}{\partial t} - u\frac{\partial \bar{u}}{\partial t}\right) - |\nabla u|^2 - V(u) \quad . \tag{2.1}$$

This Lagrangian is gauge and Galilei invariant. By the Noether theorem, the equation (0.1) has a number of conservation laws, most important of which is the conservation of the L^2-norm $\|u\|$ (we drop the subscript 2 for brevity) and of the energy

$$E(u) = \|\nabla u\|^2 + \int V(u)\,dx \quad . \tag{2.2}$$

The fact that $\|u\|$ and $E(u)$ are conserved quantities can be checked, at least formally, by direct computation using the equation (0.1). Converting this formal derivation into an actual proof for solutions of (1.1) in $X(I)$, where I is some interval of the real line $\mathbb{R}$, requires some care. First of all, with our first choice $X=L^r\cap L^{r_0}$, neither $\|u_0\|$ nor $E(u_0)$ are defined in general. We must therefore assume explicitly in this case that $u_0\in H^1$, and prove in a first step that under this assumption any solution u of (1.1) is such that $u(t)\in H^1$ for all t for which u is defined. Furthermore in both cases that we consider, we deal only with solutions of the integral equation (1.1) while the proof of the conservation laws remains manageable only when one uses the differential equation (0.1). The remedy is to regularize the equation with the help of a smooth positive function h with compact support and $\|h\|_1=1$. One also regularizes the initial data, so that the regularized equation in integral form becomes

$$u(t) = U(t-t_0)\ (h*u_0) - i\int_{t_0}^{t} d\tau U(t-\tau)(h*f(h*u(\tau)))$$

where * denote convolution.

Solutions of this equation also satisfy the differential equation in regularized form, and tend to solutions of the original equation when h tends to a δ-function. One can then prove the conservation laws for the solutions of the regularized equation and then remove the regularization by a limiting procedure.

The conservation laws of the L^2-norm and of the energy provide an a priori control of the H^1-norm of solutions of the equation (1.1) provided E(u) can control $\|\nabla u\|^2$, namely that for fixed $\|u\|$ and E(u), one cannot have $\|\nabla u\|^2 \longrightarrow +\infty$ and $\int V(u)dx \longrightarrow -\infty$. This requires an additional assumption on f, which is best expressed as the following lower bound on V:

(H4) There exists p_3 such that $1\leq p_3<1+4/n$ and for all $\rho\geq 0$,

$$V(\rho) \geq -C(\rho^2 + \rho^{p_3+1}) \tag{2.3}$$

for some non negative constant C.

This assumption is easily seen to imply that if $v\in H^1$ and b is defined by $p_3=1+4b/n$, so that $0\leq b<1$, then the following estimate holds

$$\|v\|^2_{H^1} \leq (1-b)^{-1} E(v) + C(\|v\|^2+\|v\|^{2+4nb/(1-b)}) \tag{2.4}$$

and therefore $\|v\|_{H^1}$ is controlled by $\|v\|$ and E(v).

Assumptions (H3) and (H4) suffice to control the H^1-norm of arbitrary solutions of the equation (1.1) with $u_0\in H^1$. With our first choice $X=L^r\cap L^{r_0}\supset H^1$, this is sufficient to control $\|U(t-s)u(s)\|_X$ uniformly in s and t, and therefore to extend local solutions to arbitrary times. With our second choice $X=H^1\cap L^\infty$, for n=2 or 3, we need an additional estimate on $\|U(t-s)u(s)\|_\infty$. This is obtained through the use of the Sobolev inequality

$$\| v \|_\infty \leq C \|\nabla v\|_q^{n/q} \|v\|_q^{1-n/q} \quad , \tag{2.5}$$

which holds for $n<q<2n/(n-2)$, and the estimate (1.9). An additional assumption on f is required, namely a power bound on the behaviour of f at infinity. (Note that in proposition 1, such a condition appears in (H2), while none is required in (H1a)). We assume:

(H2a) (If n=2 or 3). There exists p_2 such that $2\leq p_2<(n+2)/(n-2)$ and for all $z\in\mathbb{C}$

$$|f'(z)| \leq C(|z| + |z|^{p_2-1}) \ . \tag{2.6}$$

We can now state the main result of this section.

<u>*Proposition 2.*</u> *- Let either:* $n\geq 2$, f *satisfy* (H1,2,3,4), $X=L^r\cap L^{r_0}$, $t_0\in\mathbb{R}$ *and* $u_0\in H^1$, *or:* $n\leq 3$, f *satisfy* (H1a,2a,3,4), $X=H^1\cap L^\infty$, $t_0\in\mathbb{R}$ *and* $u_0\in X$ *be such that* $U(\cdot-t_0)u_0\in X(\mathbb{R})$.
Then *(1) The equation* (1.1) *has a unique solution in* $X(\mathbb{R})$.

(2) The solution u *is a bounded and continuous function from* $\mathbb{R}$ *to* H^1 *and satisfies for all* $t\in\mathbb{R}$ *the equalities*

$$\|u(t)\| = \|u_0\| \quad \text{and} \quad E(u(t)) = E(u_0) \ . \tag{2.7}$$

We conclude this section with two remarks. First, the assumption (H4) is close to necessary since a counter example of Glassey [10] (see also [15]) shows that for $V(\rho)\leq -C\rho^{p_3+1}$ with $p_3>1+4/n$, solutions of (1.1) blow up in H^1-norm in a finite time. In the borderline case $p_3=1+4/n$, one can still ensure the existence of global solutions provided $\|u_0\|$ is sufficiently small. Second, the assumptions made on f in proposition 2 cover the case of a single power (equation (0.2)) with $1\leq p<(n+2)/(n-2)$ if $\lambda>0$ and $1\leq p<1+4/n$ if $\lambda<0$.

3. DISPERSIVE SOLUTIONS AND WAVE OPERATORS

In this section, we study the solutions of the equation (1.1) that behave asymptotically like solutions of the free equation (0.4), namely like $U(t)u_+$ for some suitable u_+, when $t\to\infty$. These solutions we call dispersive, since the solutions of the free equation spread out when $t\to\infty$, as shown by the estimate (1.9). For this purpose, we define $\tilde{u}_0 = U(-t_0)u_0$ and rewrite (1.1,2) as follows

$$u(t) = U(t)\tilde{u}_0 - i\int_{t_0}^{t} d\tau U(t-\tau) f(u(\tau)) \equiv [\tilde{A}(t_0,\tilde{u}_0)u](t) \qquad (3.1)$$

or more concisely

$$u = \tilde{A}(t_0,\tilde{u}_0)u \quad . \qquad (3.2)$$

Let now u be a fixed dispersive solution and let formally t_0 tend to infinity. We then expect that $\tilde{u}_0 = U(-t_0)u(t_0)$ tends to a well-defined limit u_+:

$$u_+ = \lim_{s\to\infty} U(-s)u(s) \qquad (3.3)$$

and that u is a solution of the limiting equation

$$u = \tilde{A}(\infty,u_+)u \quad . \qquad (3.4)$$

This suggests that we look for dispersive solutions of (1.1) by solving (3.4), i.e. by solving the Cauchy problem with infinite initial time. In terms of scattering theory, this will yield the wave operator Ω_+ as the mapping $u_+ \to u(0)$. More generally, we shall look for solutions of the equation (3.2) for t_0 in a neighborhood of infinity, with the requirement that the method apply uniformly in t_0 for t_0 sufficiently large. We do this again by a contraction method. Let T>0, $I=[T,\infty]$. We look for a Banach space $X_0(I)$ of bounded continuous functions from I to X, such that the following property holds. Let $B_0(I,\rho)$ be the ball of radius ρ in $X_0(I)$, let $\tilde{u}_0$ be such that $U(\cdot)\tilde{u}_0 \in B_0(I,\rho)$ for some ρ. We want then to ensure that if T is sufficiently large, depending on ρ, say $T \geq T_1(\rho)$, then for all $t_0 \geq T$, the non linear operator $\tilde{A}(t_0,\tilde{u}_0)$ maps $B_0(I,2\rho)$ into itself and is contracting, uniformly in t_0. By the contraction mapping theorem, this implies the existence and uniqueness of solutions of (3.2) in $X_0(I)$.

We now have to choose $X_0(I)$ and correspondingly to make suitable assumptions on f in addition to those already made in section 1. The definition of $X_0(\cdot)$ must incorporate some decay of the generic element u(t) as a function of t when $t\to\infty$, in order that the integral in (3.1) be norm convergent in X uniformly in t_0. This decay should of course be tailored after that of the solutions of the free equation, as expressed by (1.9). It will imply as a by-product the convergence in (3.3) in some norm, and in particular in X-norm. The additional assumptions on f are of the same nature as those needed in linear scattering theory to prove the existence of the wave operators, namely they express that the interaction term decays sufficiently fast at large distance in space. They appear typically as lower bounds on p_1 in (1.8).

Again there is a large amount of freedom in the choice of $X_0(I)$ and of the corresponding assumptions on f that make the previous scheme work. We give below two possible choices that extend those made in section 1. We state the definition of $X_0(I)$ by giving the expression $|u|_{0_I}$ of the norm of a generic $u\in X_0(I)$.

Our first choice: $n\geq 2$, f satisfying (H1,2) and $X=L^r\cap L^{r_0}$ is extended by choosing

$$|u|_{0_I} = \sup_{t\in I}\left\{\mathrm{Max}\left((1+|t|)^{n/2-n/r}\|u(t)\|_r,\ \mathrm{id}(r\to r_0)\right)\right\} \tag{3.5}$$

and by making the additional assumption on f that p_1, which occurs in (H2), now satisfies

$$p_1 > 2 + 2/n - 2/r \quad . \tag{3.6}$$

Our second choice: $n\leq 3$, f satisfying (H1a) and $X=H^1\cap L^\infty$, is extended by picking some r, $r>2$ and $r<2n/(n-2)$ if $n=2$ or 3, by choosing

$$|u|_{0_I} = \sup_{t\in I}\left\{\mathrm{Max}\left(\|u(t)\|_{H^1}, (1+|t|)^{n/2-n/r}\mathrm{Max}(\|u(t)\|_r, \|u(t)\|_\infty)\right)\right\} \tag{3.7}$$

and by making the additional assumption on f:

(H2b) There exists p_1 such that

$$p_1 > \mathrm{Max}(2+2/n-2/r,\ 1+2r/n(r-2)) \tag{3.8}$$

and for all z with $|z|\leq 1$,

$$|f''(z)| \leq C|z|^{p_1-2} \tag{3.9}$$

where f" stands for any second derivative of f with respect to z and/or $\overline{z}$.

We remark here that the weakest condition on p_1 in (3.8), obtained by minimizing the r.h.s. with respect to r, is $p_1>1+(1+\sqrt{2n+1})/n$. Both this lower bound and that in (3.6) for the allowed values of r lie in the interval $(1+2/n,\ 1+4/n)$.

We can now state the results concerning the existence of the limit (3.3) and the Cauchy problem at infinity.

<u>*Proposition 3.*</u> - *Let either:* $n\geq 2$, f *satisfy* (H1,2) *and* (3.6) , $X=L^r\cap L^{r_0}$ *and* $X_0(\cdot)$ *be defined by* (3.5),

or: $n\leq 3$, f *satisfy* (H1a,2b), $X=H^1\cap L^\infty$ *and* $X_0(\cdot)$ *be defined by* (3.7).*Then*

(1) If $T\in\mathbb{R}$, $I=[T,\infty)$, $t_0\in[T,\infty]$, *and* $u\in X_0(I)$ *is a solution of the equation* (3.2), *then there exists* u_+ *such that* $U(\cdot)u_+\in X_0(I)$, $U(\cdot-s)u(s)$ *tends to* $U(\cdot)u_+$ *stron-*

gly in $X_0(I)$, *and* $U(-s)u(s)-u_+$ *tends to zero strongly in* X *when* $s\to\infty$.

(2) *For any* $\rho>0$, *there exists* $T_1(\rho)>0$ *depending only on* ρ *and* f *such that for any* $t_0\in[T_1(\rho),\infty]$ *and any* $\tilde{u}_0$ *such that* $U(\cdot)\tilde{u}_0\in B_0(I,\rho)$ *where* $I=[T_1(\rho),\infty)$, *the equation* (3.2) *has a unique solution in* $X_0(I)$. *This solution belogs to* $B_0(I,2\rho)$.

If we add to the assumptions of proposition 3 the assumptions that f satisfies (H3), and if necessary that $\tilde{u}_0\in H^1$, then we can extend the conservation laws of the L^2-norm and of the energy to infinite times:

<u>*Proposition 4.*</u> - *Let all the assumptions of proposition 3, part 1, be satisfied, and let in addition* f *satisfy* (H3) *and* $\tilde{u}_0\in H^1$. *Then* $u_+\in H^1$, u *is a bounded continuous function from* I *to* H^1, $U(-s)u(s)$ *tends to* u_+ *strongly in* H^1 *when* $s\to\infty$, *and* u *satisfies for all* $t\in I$ *the equalities*

$$\|u(t)\| = \|u_+\| \qquad \text{and} \qquad E(u(t)) = \|\nabla u_+\|^2 .$$

(Under the second set of assumptions of proposition 3, only the last equalities in proposition 4 are new. The continuity properties in H^1 are already contained in proposition 3, part 1, since $X\subset H^1$ in this case).

Finally, if we add to the assumptions of proposition 3, part 2, the assumptions that f satisfies (H3,4), and if necessary (H2a), that $U(\cdot)\tilde{u}_0\in X_0(\mathbb{R})$ and if necessary that $\tilde{u}_0\in H^1$, then we can extend the solution of the equation (3.2) described in the latter proposition to all times through the use of proposition 2. In particular, if we take $t_0=\infty$, we see that the wave operator $\Omega_+:u_+\to u(0)$ is well defined on the space of those $u_+\in H^1$ such that $U(\cdot)u_+\in X_0(\mathbb{R})$. One can also show that this operator is bounded and continuous in a suitable sense.

4. PSEUDOCONFORMAL CONSERVATION LAW AND ASYMPTOTIC COMPLETENESS FOR REPULSIVE INTERACTIONS

In this section, we describe and exploit an approximate conservation law of the equation (0.1) to prove that all solutions with suitable initial data are dispersive for a class of repulsive interactions. In order to make this conservation law arise in a natural way, we first digress on the invariance properties of the equation. Under the assumption (H3) on f, the equation (0.1) is invariant under a unitary projective representation of the Galilei group, or equivalently under a unitary representation of a (non trivial) extension of the Galilei group by the one-dimensional torus T_1 ($\equiv$gauge group). This representation is generated by the following transformations:

Gauge transformations: $[Q(\theta)u](x,t) = e^{i\theta} u(x,t)$

Space translations: $[E(a)u](x,t) = u(x-a,t)$, $a\in\mathbb{R}^n$,

Time translations: $[T(b)u](x,t) = u(x,t-b)$, $b\in\mathbb{R}$,

Space rotations, which act in an obvious way, and Pure Galilei transformations: $[G(v)u](x,t) = \exp\left[\frac{i}{2}\langle v,x\rangle - \frac{i}{4}v^2 t\right]u(x-vt,t)$, $v\in\mathbb{R}^n$.

The free equation (0.4) however, and more remarkably the equation (0.2) for the special value $p=1+4/n$, has a larger invariance group. It is invariant under space time dilations

$$[D(\alpha)u](x,t) = \exp(-\tfrac{n}{2}\alpha)\, u(xe^{-\alpha}, te^{-2\alpha}) \qquad (\alpha\in\mathbb{R})$$

and also under the antilinear antiunitary transformation J (with $J^2=1$):

$$[Ju](x,t) = t^{-n/2} \exp(ix^2/4t)\, \overline{u(x/t,1/t)} \quad .$$

If we now use J as an external automorphism for the present transformations, we find

$$J\,Q(\theta)\,J = Q(-\theta), \qquad J\,D(\alpha)\,J = D(-\alpha), \qquad J\,E(a)\,J = G(a)$$

and $J\,T(b)\,J = C(b)$, where $C(\cdot)$ is a new one parameter group of unitary transformations, which we call pseudoconformal, defined by

$$[C(w)u](x,t) = (1-wt)^{-n/2} \exp\left[\frac{-iwx^2}{4(1-wt)}\right] u\left(\frac{x}{1-wt},\frac{t}{1-wt}\right), \quad w\in\mathbb{R}. \qquad (4.1)$$

One checks easily that the linear span of the infinitesimal gauge, Galilei, dilation and pseudoconformal transformations is a Lie algebra.

By the Noether theorem, associated with the pseudoconformal invariance of the equation (0.4) or (0.2) with $p=1+4/n$, there is a conserved quantity which can be obtained by standard computations from the Lagrangian density (2.1) and the expression of the infinitesimal pseudoconformal generator. In the general case of the equation (0.1), which is not invariant, the conservation law is only approximate and contains a remainder term depending on f. The conservation law takes the integral form

$$\|xU(-t)u(t)\|^2 + 4t^2\int V(u(t))dx = \mathrm{id}(t\to s)+\int_s^t 4\tau d\tau\int W(u(\tau))dx \qquad (4.2)$$

where W is an auxiliary potential defined by $W(z)=W(|z|)$ for all $z\in\mathbb{C}$, and:

$$W(\rho) = (n+2)V(\rho) - \frac{n}{2}\rho V'(\rho) \qquad \text{for all } \rho\geq 0 . \qquad (4.3)$$

(Note that $W=0$ when $f(z)=\lambda z|z|^{4/n}$).

The conservation law (4.2) can of course be checked formally in differential form by direct computation using the equation (0.1). As in the case of the L^2-norm and of the energy, converting this formal derivation into an actual proof requires some care, and is best achieved by introducing a local regularization and a space cut off, later to be removed by a limiting procedure. In addition, the r.h.s. of (4.2) must be well defined already for u_0, which we therefore require to lie in the Hilbert space Σ with norm defined by

$$\|u\|_\Sigma^2 = \|u\|_{H^1}^2 + \|xu\|^2 \quad . \qquad (4.4)$$

For dispersive solutions of the equation (3.2) with initial data $\tilde{u}_0\in\Sigma$, and under a slight additional assumption on f, the conservation law (4.2) can be extended to infinite time. We collect the previous results in the following proposition:

<u>*Proposition 5.*</u> - *(1) Let either:* $n\geq 2$, f *satisfy* (H1,2,3) *and* $X=L^r\cap L^{r_0}$, *or:* $n\leq 3$, f *satisfy* (H1a,3) *and* $X=H^1\cap L^\infty$.

Let I *be an interval of* $\mathbb{R}$, $t_0\in I$, $\tilde{u}_0\in\Sigma$ *and let* $u\in X(I)$ *be solution of the equation* (3.2). *Then* u *is a continuous function from* I *to* Σ *and satisfies the equality* (4.2) *for any* s *and* t *in* I.

(2) Let either: $n\geq 2$, f *satisfy* (H1,2,3) *with* $p_1>1+4/n$, $X=L^r\cap L^{r_0}$ *and* $X_0(\cdot)$ *be defined by* (3.5), *or:* $n\leq 3$, f *satisfy* (H1a,2b,3) *with* $p_1>1+4/n$, $X=H^1\cap L^\infty$, *and* $X_0(\cdot)$ *be defined by* (3.7).

Let furthermore $T\in\mathbb{R}$, $I=[T,\infty)$, $t_0\in[T,\infty]$, $\tilde{u}_0\in\Sigma$, *let* $u\in X_0(I)$ *be solution of the equation* (3.2) *and let* u_+ *be defined as in proposition 3, part 1.*

Then u *is a continuous function from* I *to* Σ, $u_+\in\Sigma$, $U(-s)u(s)$ *tends to* u_+ *strongly in* Σ *when* $s\to\infty$, *and* u *satisfies for all* $t\in I$ *the equality*

$$\|xU(-t)u(t)\|^2 + 4t^2\int V(u(t))dx = \|xu_+\|^2-\int_t^\infty 4\tau d\tau\int W(u(\tau))dx \qquad (4.5)$$

where the last integral converges absolutely.

We now turn to the case of repulsive interactions in the sense of the following condition:

(H5) f satisfies (H3), and for all $\rho\geq 0$, $V(\rho)\geq 0$ and $W(\rho)\leq 0$, where W is defined by (4.3).

(This condition is satisfied by $f(u)=\lambda|u|^{p-1}u$ iff $\lambda\geq 0$ and $p\geq 1+4/n$. Note also that $(H5)\Rightarrow(H4)$).

Taking then s=0 in (4.2) shows that the l.h.s. is a decreasing function of $|t|$ and therefore that $\|xU(-t)u(t)\|\leq\|xu(0)\|$. When combined with the conservation of the L^2-norm and of the energy, this result shows that $U(-s)u(s)$ is uniformly bounded in Σ. By the easily proved estimate

$$\|U(t)v\|_q \leq C(1+|t|)^{n/q-n/2}\|v\|_\Sigma \qquad (4.6)$$

which holds for all $q\geq 2$, $q<2n/(n-2)$ if $n\geq 2$, and if necessary an additional estimate based on (2.5), this implies that all solutions of the equation (3.2) with initial data in Σ are dispersive in the sense that they belong to $X_0(\mathbb{R})$. More precisely:

Proposition 6. - *Let either:* $n\geq 2$, f *satisfy* (H1,2,3,5), $X_0(\cdot)$ *be defined by* (3.5), $t_0\in\mathbb{R}$ *and* $\tilde{u}_0\in\Sigma$, *or:* $n\leq 3$, f *satisfy* (H1a,2a,2b,3,5), $X_0(\cdot)$ *be defined by* (3.7), $t_0\in\mathbb{R}$, *and* $\tilde{u}_0\in\Sigma$ *be such that* $U(\cdot)\tilde{u}_0\in X_0(\mathbb{R})$.

Then the unique solution u *of the equation* (3.2) *in* $X(\mathbb{R})$ *(see proposition 2) actually belongs to* $X_0(\mathbb{R})$. *It satisfies the estimate*

$$\|U(t-s)u(s)\|_q \leq C(1+|t|)^{n/q-n/2}\left[\|xu(0)\|^2+\|u(0)\|^2+E(u(0))\right]^{\frac{1}{2}} \qquad (4.7)$$

for all s *and* t *in* $\mathbb{R}$ *and all* $q\geq 2$, $q<2n/(n-2)$ *if* $n\geq 2$.

In terms of scattering theory, the results of this and the previous section have the following implications for the wave operators. Let $\tilde{\Sigma}$ be the set of those $v\in\Sigma$ such that $U(\cdot)v\in X_0(\mathbb{R})$. (By the estimate (4.6), $\tilde{\Sigma}$ coincides with Σ under our first choice of assumptions, while it is strictly smaller under our second choice if n=2 or 3). Then if either $n\geq 2$ and f satisfies (H 1,2,3,5), or $n\leq 3$ and f satisfies (H1a,2a,2b,3,5) and if in addition $p_1>1+4/n$, the wave operator Ω_+ is a bijection from $\tilde{\Sigma}$ onto itself. In particular, it is asymptotically complete on $\tilde{\Sigma}$. Actually one can also show that Ω_+ and Ω_+^{-1} are bounded and continuous from $\tilde{\Sigma}$ onto itself. Of course the same properties also hold for the wave operators Ω_- and the S-matrix $S=\Omega_+^{-1}\Omega_-$.

REFERENCES

[1] Baillon, J.B., Cazenave, T., Figueira, M., C.R.Acad.Sc.Paris, 284, 869-872, (1977).

[2] Bove, A., Da Prato, G., Fano, G., Comm.Math.Phys., 37, 183-191, (1974), Comm.Math.Phys. 49, 25-33, (1976).

[3] Bresiz, H., Gallouet, T., Nonlinear Schrödinger evolution equations, preprint 1979.

[4] Cazenave, T., Haraux, A., C.R.Acad.Sc.Paris, 288, 253-256, (1979).

[5] Chadam, J.M., Glassey, R.T., J.Math.Phys., 16, 1122-1130, (1975).

[6] Davies, E.B., Some time dependent Hartree equations, preprint 1979.

[7] Gelfand, I.M., Dikii, L.A., Russian Math.Surveys, 30, 77-113, (1975), (Transl.from Usp.Mat.Nauk 30, 67-100, (1975)).

[8] Ginibre, J., Velo, G., J.Func.Anal., 32, 1-32, (1979); J.Func.Anal., 32, 33-71, (1979); Ann.Inst.H.Poincaré, 28, 287-316, (1978).

[9] Ginibre, J., Velo, G., C.R.Acad.Sc.Paris, 288, 683-685, (1979); Non linear Schrödinger equations with non local interaction, Mathematische Zeitschrift, in press.

[10] Glassey, R.T., J.Math.Phys. 18, 1794-1797, (1977).

[11] Lin, J.E., Strauss, W.A., J.Func.Anal., 30, 245-263, (1978).

[12] Pecher, H., Von Wahl, W., Time dependent non linear Schrödinger equations, Manuscripta Mathematica, in press.

[13] Scott, A.C., Chu, F.Y.F., McLaughlin, D.W., Proc.IEEE, 61, 1443-1483, (1973).

[14] Strauss, W.A., Non linear Scattering theory, in Scattering theory in Mathematical Physics, Lavita J.A. and Marchand J. eds., pp.53-78, D.Reidel Publ.Comp., Dordrecht, Holland (1974); Non linear invariant wave equations, in Invariant wave equations, Velo G. and Wightman A.S. eds., pp.197-249, Lecture Notes in Physics 73, Springer Verlag, Berlin-Heidelberg-New York (1978). Reed, M., Abstract non-linear wave equations, Lecture Notes in Mathematics 507, Springer Verlag, Berlin-Heidelberg-New York (1976).

[15] Strauss, W.A., The non linear Schrödinger equation, in Contemporary developments in continuum mechanics and partial differential equations, De La Penha G.M. and Medeiras L.A.J. eds., pp.452-465, North-Holland Publ.Comp., Amsterdam-New York-Oxford (1978).

[16] Zacharov, V.A., Shabat, A.B., Sov.Phys., JETP 34, 62-69 (1972), (Transl.from Zh.Exp.Teor.Fiz. 61, 118-134 (1971)).

A GEOMETRICAL APPROACH TO THE NONLINEAR SOLVABLE EQUATIONS*

F. Magri

Istituto di Matematica del Politecnico

Piazza Leonardo da Vinci 32 - 20133 Milano - Italy

ABSTRACT

A geometrical approach to the nonlinear solvable equations, based on the study of the "groups of motion" of special infinite-dimensional manifolds called "symplectic Kähler manifolds", is suggested. This approach is constructive, tensorial and simple in its ideas. It allows to recover the equations obtained through the socalled AKNS approach, together with some other examples. It leads to conjecture a possible "integrability condition" for infinite-dimensional systems, and to hope to be able to give a geometrical explanation of the so-called "spectral transform method".

1. INTRODUCTION

The purpose of this paper is to introduce a geometrical-operator point of view in the study of the nonlinear equations

$$\partial_t u^A(x^j,t) = K^A(u^B, \partial_j u^B, \ldots) \tag{1.1}$$

describing the evolution of the tensor-valued functions $u^A(x^j,t)$ of type (p,q) defined, at any instant of time t, in a fixed region Ω of $\mathbb{R}^n$, and

* This work has been sponsored in part by the G.N.F.M. of the C.N.R. .

obeying prescribed boundary conditions on $\partial\Omega$. Basic elements of this geometrical approach are the linear function space U of the field functions, regarded as functions of the space coordinates only, its dual space U^* and the differential operators defined on U and U^*. The spaces U and U^* are used to define the infinite-dimensional configuration space associated with the given evolution equations (Sec.2), while the differential operators are used to define suitable geometrical structures on this space (Sec.s 3, 4 and 5).

The leading idea of the present approach, as it evolved from the analysis of the mathematical structure of the main nonlinear solvable equations [1], is that such equations may be identified with the "groups of motion" of a special geometrical structure defined on the configuration space, and called a "symplectic Kähler manifold" (Sec.s 5 and 6).

However far this point of view may seem from the usual approaches to the integrable Hamiltonian equations, it can be shown that: 1) it completely recovers the classical theory of the finite-dimensional systems [2]; 2) it allows to construct classes of infinite-dimensional "integrable" systems and, in particular, the equations given by the so-called AKNS method (Sec.s 6 and 7); 3) it suggests a possible *integrability condition* for the infinite-dimensional systems (Sec.6); 4) it leads to a *method of integration* of the finite-dimensional systems which closely res embles the "spectral transform method", being based on the study of a suitable eigenvalue problem associated with the symplectic Kähler manifolds (there is, consequently, a reasonable hope to arrive to a better understanding of the last method from a geometrical point of view)[2]; 5) it is a *tensorial approach* , in the sense explained in Sec. 8.

The aim of this paper is mainly in giving a global view of such a geometrical approach, emphasizing the general lines rather than the particular points. For this reason and in order to make the paper not too large, the proves are not given in all the details. We apologize for this omission and we hope to be able to cover the missing details, together with some supplementary remarks, in a subsequent paper.

2. THE CONFIGURATION SPACE

According to the present approach, the study of the eq.s (1.1) is set up into the linear function space U of the field functions regarded as functions of the space coordinates only. Any set of such functions is referred to as a point of the space U, and it is simply denoted by u. The particular functions obeying the prescribed boundary conditions are regarded as forming a possibly nonlinear manifold in the space U, which is called the *configuration space* associated with the given evolution equa tions. It is endowed with a first elementary geometrical structure as follows.

Consider the field functions φ obeying homogeneous boundary conditions and observe that they define the variations

$$\delta u = \varepsilon \cdot \varphi \tag{2.1}$$

of the functions u. Such variations may be referred to as the infinitesimal displacements of the point u ad may, consequently, be regarded as forming the "tangent space" at this point. This remark suggests to associate with any point u the linear space of the order•d pairs of functions (u,φ), and to call it the *tangent space* at the considered point. It will be denoted by T_u. In the pairs (u,φ), the first function u, obeying the prescribed boundary conditions, is regarded as denoting the point of the configuration space we are dealing with; the second function φ, obeying homogeneous boundary conditions, is regarded as defining the tangent vector associated with it (see Fig. 1).

In the same way, consider the tensor-valued functions $\alpha_A(x^j,t)$ (hereafter abbreviated as α) of type (q,p), defined on the same region Ω of the functions u and obeying homogeneous boundary conditions on $\partial\Omega$. They form a second linear function space which also may be associated with the point u. This space, once put in duality with T_u by means of a separating bilinear form like, for example, the following one

$$\langle\alpha,\varphi\rangle = \frac{1}{2}\int_{\Omega} \{\bar{\alpha}_A(\underline{x})\varphi^A(\underline{x}) + \alpha_A(\underline{x})\bar{\varphi}^A(\underline{x})\}\sqrt{g}\, d^3\underline{x} \tag{2.2}$$

where the bar denotes the complex conjugate functions, g is the determinant of the Riemannian metric over Ω, and the summation convention on

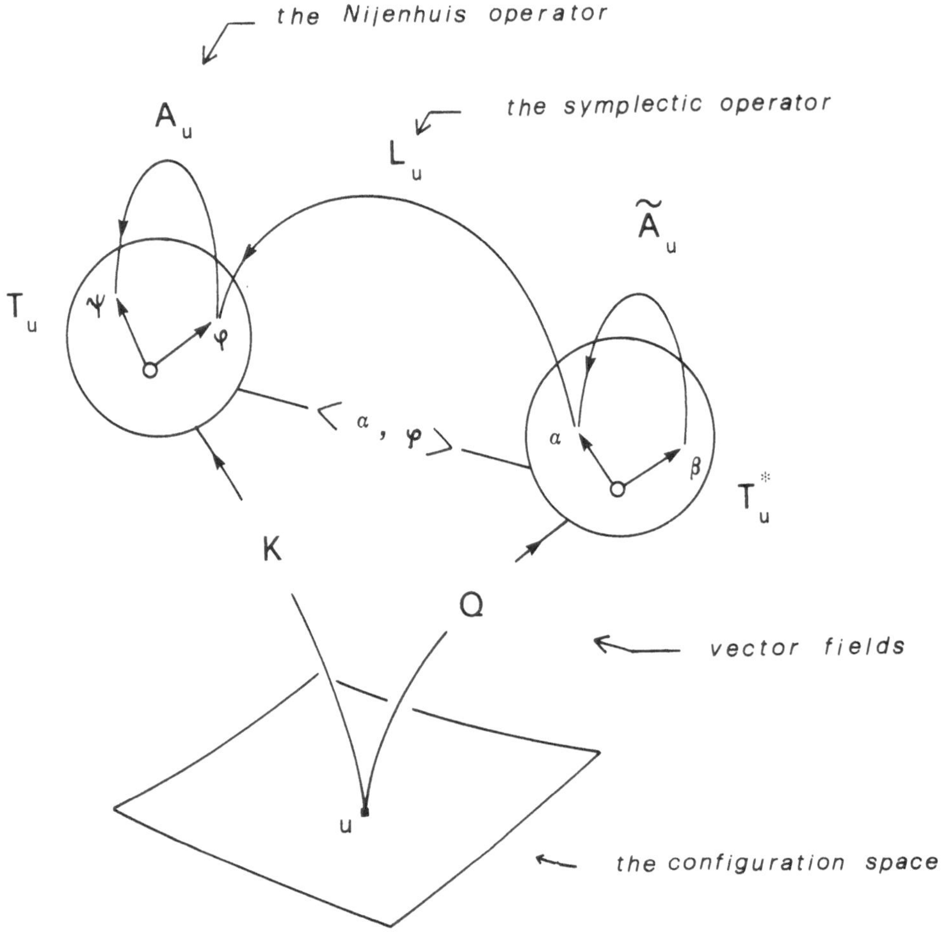

Example :

$K(u) = u_{xxx} + a u u_x$ $\qquad Q(u) = u_{xx} + \frac{a}{2} u^2$

$$A_u \varphi = \varphi_{xx} + \frac{2}{3} a u \varphi + \frac{a}{3} u_x \int_{-\infty}^{x} \varphi \, d\xi$$

$$L_u \alpha = \alpha_{xxx} + \frac{2}{3} a u \alpha_x + \frac{a}{3} u_x \alpha$$

Fig. 1: The basic geometrical scheme

the collective index A is applied, is called the *cotangent space* T^*_u at the point u.

By the introduction of the pair of spaces T_u and T^*_u, the configuration space associated with the given evolution equations is formally like the usual configuration space of the classical dynamical systems. It obeys, moreover, the same transformation laws of its classical model. To explain this point, consider a transformation

$$\bar{u}^A(x^j,t) = f^A(u^B, \partial_j u^B, \ldots) \tag{2.3}$$

of the field functions, possibly depending on the derivatives of such functions with respect to the space coordinates, and depict it as a point transformation

$$\bar{u} = F(u) \tag{2.4}$$

between the configuration space U and a second configuration space $\bar{U}$ (see Fig. 2). It is readily seen [3] that this transformation maps the vectors φ and α at the point u into the vectors $\bar{\varphi}$ and $\bar{\alpha}$ at the point $\bar{u}$ according to the relations

$$\bar{\varphi} = F'_u \cdot \varphi \tag{2.5}$$

and

$$\alpha = \tilde{F}'_u \cdot \bar{\alpha} \tag{2.6}$$

where F'_u is the Gateaux derivative of the operator F describing the transformation, and $\tilde{F}'_u$ is its adjoint operator with respect to the pair of bilinear forms defined over the spaces U and $\bar{U}$ (see the appendix). Such transformation laws simply extend the usual laws of the tensor calculus to an infinite-dimensional space. On them ultimately rests the legitimacy to call the φ and the α the tangent and the cotangent vectors at the point u. This example suggests that the use of the operator point of view may lead to a substantial enlargement of the field in which the methods of the classical tensor calculus may be applied, preserving however the formal properties and the conceptual contents of this calculus. The present work aim indeed to show the suitability of this form of "operator tensor calculus" in the study of the nonlinear evolution equations. To this end, however, we must first provide the configuration space with a suitable additional geometrical structure. This will be

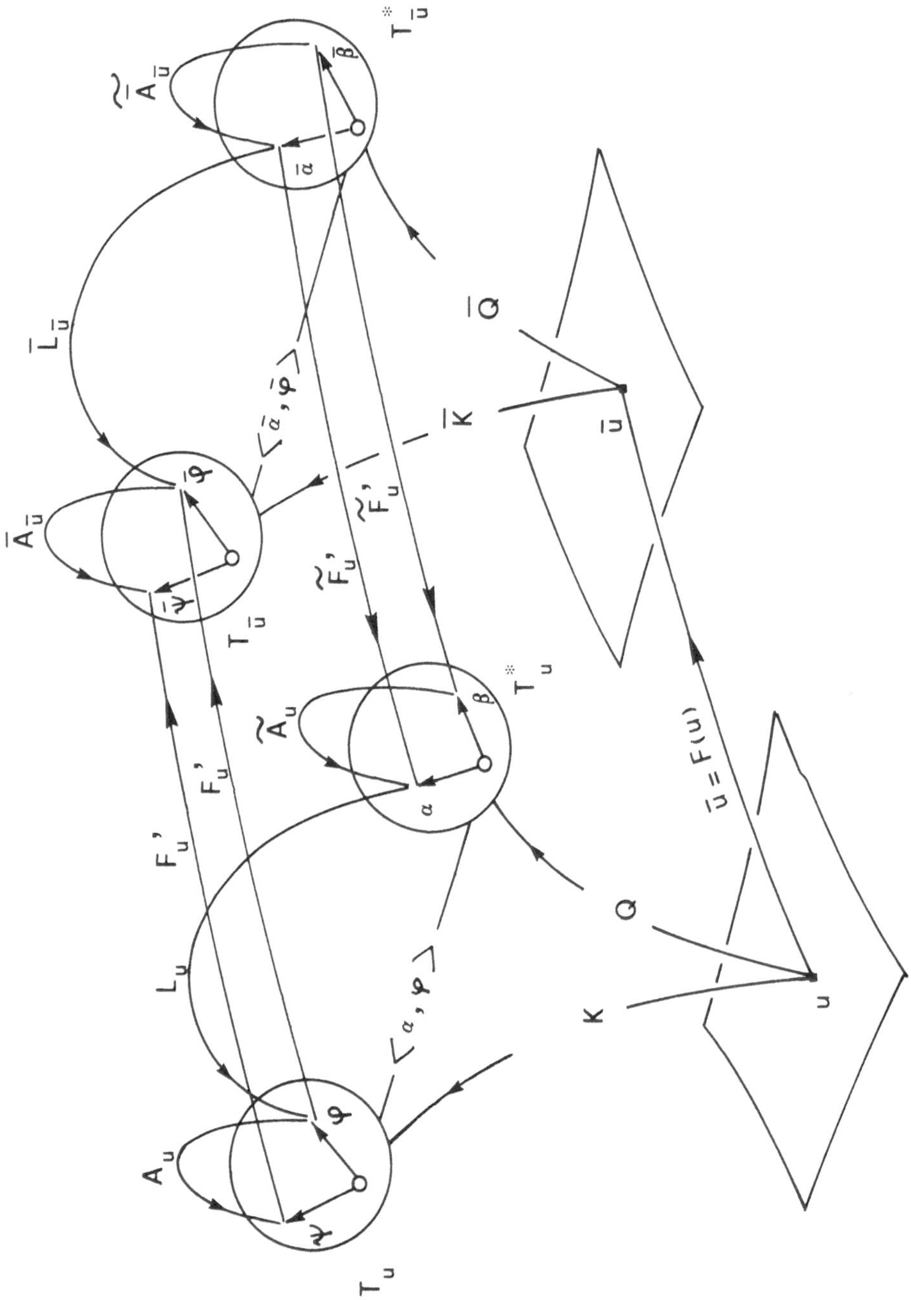

Fig. 2: The transformation scheme

done in the next three sections.

3. VECTOR FIELDS

Consider the given evolution equations which are now written in the abstract operator form

$$\frac{du}{dt} = K(u) \tag{3.1}$$

and recall that, under a transformation of the field functions (2.3), they change into new equations

$$\frac{d\bar{u}}{dt} = \bar{K}(\bar{u}) \tag{3.2}$$

given by [4]

$$\bar{K}(F(u)) = F'_u \cdot K(u) \quad . \tag{3.3}$$

From the present geometrical point of view, this law simply means that the evolution operator K obeys the same transformation law of the tangent vectors, so that it is possible to regard K as an operator associating a tangent vector with any point u of its domain (see Fig. 1). For this reason, we shall say that the given equations define a *controvariant vector field* on the configuration space U.

This example shows that it is possible to give simple geometrical representations of the differential operators defined on the space U, accounting for the way they change under a transformation of the field functions.

In this spirit, the operators Q obeying the transformation law (see Fig. 2)

$$Q(u) = \tilde{F}'_u \cdot \bar{Q}(F(u)) \tag{3.4}$$

will be regarded as associating in an intrinsic way a cotangent vector with any point u of their domain. They will be consequently referred to as the *covariant operators* on the space U.

Among such operators it is important to consider the special class of the operators for which the elementary circulation

$$\delta C \triangleq \langle Q(u), \delta u \rangle \quad , \tag{3.5}$$

relative to any infinitesimal displacement of the point u, coincides with the variation of a suitable functional I[u]. Such operators are called *potential operators* and are characterized by the following condition [5][6]

$$<Q'_u \cdot \varphi, \psi> = <Q'_u \cdot \psi, \varphi> \quad . \tag{3.6}$$

They will be referred to as the conservative covariant vector fields on U.

4. TENSOR OPERATORS

Pursuing the previous geometrical point of view, consider now a linear operator A_u like the operator

$$A_u \varphi = \varphi_{xx} + \partial_x \left(u \int_{-\infty}^{x} u \varphi d\xi \right) \tag{4.1}$$

which possibly may depend in a nonlinear way on the point u. Applied to any differential operator S_j, it gives another differential operator

$$S_{j+1}(u) = A_u \cdot S_j(u) \tag{4.2}$$

If, following the present point of view, we consider the operators S_j as controvariant vector fields on U, we must consistently regard A_u as a linear operator mapping the tangent space into itself. This remark compels us to enter into the theory of the tensor operators.

In this paper, we shall consider in particular two kinds of tensor operators, called the Nijenhuis operators and the symplectic operators.

The *Nijenhuis operators* are linear operator A_u mapping the tangent space into itself and obeying the following condition

$$A'_u(\varphi; A_u \psi) - A'_u(\psi; A_u \varphi) = A_u \left[A'_u(\varphi;\psi) - A'_u(\psi;\varphi) \right] \tag{4.3}$$

where A'_u is the Gateaux derivative of A_u with respect to u (see the appendix), and φ and ψ are any pair of tangent vectors. According to its geometrical meaning, any Nijenhuis operator is supposed to obey the transformation law (see Fig. 2)

$$\bar{A}_{F(u)} \cdot F'_u = F'_u \cdot A_u \tag{4.4}$$

showing that it is a tensor operator of type (1.1).

The *symplectic operators*, otherwise, are tensor operators mapping the cotangent space into the tangent space, and verifying the following two conditions

$$<\alpha, L_u \beta> = -<\beta, L_u \alpha> \tag{4.5}$$

$$<\alpha, L_u'(\beta; L_u \gamma)> + \ldots + \ldots = 0 \tag{4.6}$$

where the dots mean the cyclic permutation over α, β and γ. They obey the transformation law (see Fig. 2)

$$\bar{L}_{F(u)} = F_u' \cdot L_u \cdot \tilde{F}_u' \quad . \tag{4.7}$$

Leaving out a detailed study of the properties of the symplectic operators, which are on the other hand well-known, we limit ourselves to recall that they map the potential operators Q_j into the controvariant operators

$$S_j(u) \triangleq L_u \cdot Q_j(u) \tag{4.8}$$

which are called *Hamiltonian* (with respect to L_u); that the commutator [6, Sec.4]

$$[S_j, S_k](u) \triangleq S_{ju}' \cdot S_k(u) - S_{ku}' \cdot S_j(u) \tag{4.9}$$

of two Hamiltonian operators is again Hamiltonian, and that the corresponding functional

$$I_{jk}[u] \triangleq <Q_j(u), L_u Q_k(u)> \tag{4.10}$$

is called the *Poisson bracket* of the functionals associated with the operators S_j and S_k [1]. Here, on the contrary, we want to describe the less known properties of the Nijenhuis operators. They play an important part in the further study of the nonlinear solvable equations.

To describe the first property we need the concept of vector field leaving the Nijenhuis operator invariant. Given any controvariant vector field S, and any pair of infinitely near point u and $\bar{u}$ lying on a trajectory of this vector field

$$\bar{u} = u + \varepsilon S(u) \quad , \tag{4.11}$$

we can drag along the tangent vectors φ from the point u to the point $\bar{u}$

according to the relation (see Fig. 3)

$$\bar{\varphi} = \varphi + \varepsilon S_u' \cdot \varphi \tag{4.12}$$

The tensor operator A_u is then said to be invariant along the trajectories of the vector field S if the relation

$$\psi = A_u \cdot \varphi \tag{4.13}$$

at the point u, implies the same relation

$$\bar{\psi} = A_{\bar{u}} \cdot \bar{\varphi} \tag{4.14}$$

at the point $\bar{u}$. As is easily seen, the condition expressing this invariance is given by

$$A_u'(\varphi; S(u)) + A_u S_u' \varphi - S_u' A_u \varphi = 0 \tag{4.15}$$

which means that the "Lie derivative" of the tensor operator A_u along the trajectories of the vector field vanishes.

The first property of the Nijenhuis cperator we are dealing with concerns the vector fields leaving it invariant. According to this property the *operators*

$$S_j(u) \triangleq A_u^j \cdot S_1(u) \tag{4.16}$$

obtained from any given operator S_1 leaving A_u invariant by means of the Nijenhuis operator itself, *leave* A_u *invariant again and commute in pairs.*

The first part of this statement is a simple consequence of the following identity

$$A_u'(\varphi; S_{j+1}(u)) + A_u S_{j+1,u}' \varphi - S_{j+1,u}' A_u \varphi =$$

$$= A_u \left[A_u'(\varphi; S_j(u)) + A_u S_{ju}' \varphi - S_{ju}' A_u \varphi \right] \tag{4.17}$$

relating the Lie derivative of A_u along S_{j+1} to the Lie derivative along S_j. The second part follows, instead, from the identity

$$\left[A_u S_j, S_k \right] = A_u \left[S_j, S_k \right] \tag{4.18}$$

which holds for any pair of vector fields fulfilling the invariance condition (4.15). The iterative use of this identity, in fact, leads to the relation

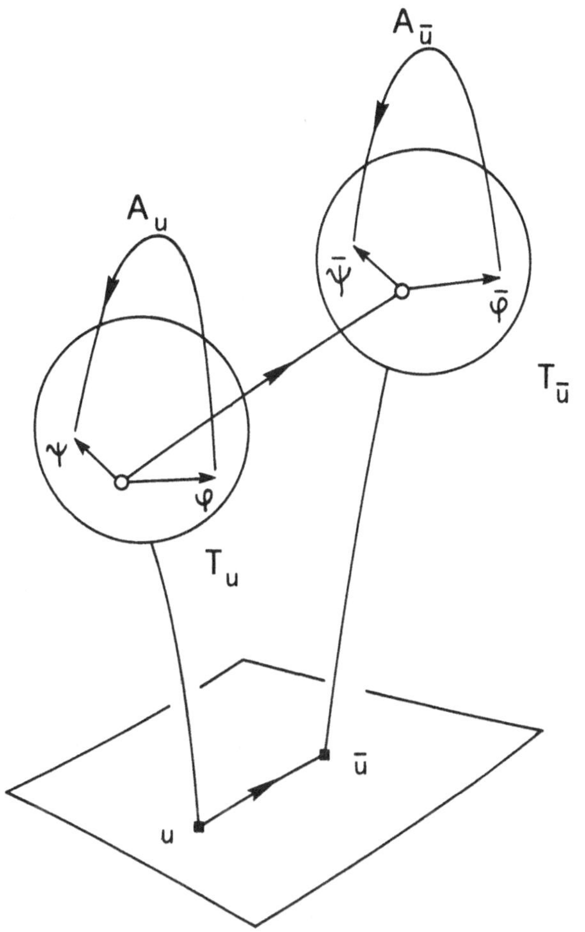

$$\bar{u} = u + \varepsilon K(u)$$
$$\bar{\varphi} = \varphi + \varepsilon K'_u \varphi$$

the invariarance condition is given by :

$$A'_u(\varphi; K(u)) + A_u K'_u \varphi - K'_u A_u \varphi = 0$$

Fig. 3: The Lie derivative of the Nijenhuis operator

$$\begin{aligned}[S_j, S_k] &= A_u\, [S_{j-1}, S_k] \\ &= A_u^2\, [S_{j-2}, S_k] \\ &\;\vdots \\ &= A_u^{j-k}\, [S_k, S_k] \\ &= 0 \end{aligned} \tag{4.19}$$

showing that the operators S_j are indeed commuting in pairs. The Nijenhuis operators are thus powerful tools for constructing abelian subalgebras of controvariant vector fields. This property gives a first reason for the present interest in the Nijenhuis operators.

To describe a second remarkable property of the Nijenhuis operators we must consider its adjoint operators $\tilde{A}_u$ defined by

$$<\tilde{A}_u \alpha, \varphi> \; = \; <\alpha, A_u \varphi> \tag{4.20}$$

It is a linear operator mapping the cotangent space T_u^* into itself (see Fig. 1), which makes a new covariant operator

$$Q_{j+1}(u) = \tilde{A}_u \cdot Q_j(u) \tag{4.21}$$

to correspond to any given covariant operator Q_j. It acts, in particular, on the potential operators, so that we can ask when the new operator Q_{j+1} is potential again.

The second property of the Nijenhuis operator we are dealing with gives an answer to the previous question. According to it, all the operators

$$Q_j(u) \triangleq \tilde{A}_u^j \cdot Q_1(u) \tag{4.22}$$

constructed by means of $\tilde{A}_u$ itself, are potential again if Q_1 is a given potential operator fulfilling the following condition

$$<\tilde{A}_u Q_u' \varphi, \psi> - <\tilde{A}_u Q_u' \psi, \varphi> + <Q(u), A_u'(\psi;\varphi) - A_u'(\varphi;\psi)> = 0 \tag{4.23}$$

The proof simply consists in verifying that the condition (4.23), assuring that Q_{j+1} is potential if Q_j is so, is iteratively fulfilled by all the operators Q_j given by eq. (4.22).

The operator $\tilde{A}_u$ allows thus to construct special families of potential operators exactly as the operator A_u allows to construct special subalgebras of controvariant vector fields. The former are important in the theory of the conservation laws, while the later are important in the theory of the symmetries of the nonlinear evolution equations.

5. A RELEVANT COMPOSITE STRUCTURE

So far, the Nijenhuis and the symplectic operators have been dealt with separately. Now, we study how to couple them in such a way that the geometrical structure defined by the Nijenhuis operator be compatible with the Hamiltonian structure defined by the symplectic operator.

To explain this point, let us consider Fig. 4. On the one side, we see the Nijenhuis operator A_u, defining a special class of controvariant vector fields S_j; on the other side, we see the adjoint operator $\tilde{A}_u$, defining a special class of covariant vector fields Q_j; in the middle, finally, we see the symplectic operator L_u. This picture immediately suggests to demand L_u should map the operators Q_j, associated with $\tilde{A}_u$, into the operators S_j associated with A_u. This is the compatibility condition we are requiring.

By explicitly writing the invariance condition (4.15), where S_j is now given by the eq. (4.8), and by exploiting the conditions (3.6) and (4.23) on the operators Q_j, we readily arrive at the following two conditions

1. $$A_u L_u = L_u \tilde{A}_u \tag{5.1}$$

2. $$\begin{aligned} &<\alpha, A_u'(L_u\beta;\varphi) - A_u'(\varphi;L_u\beta)> + \\ &\qquad + <\beta, A_u'(\varphi;L_u\alpha) + A_u L_u'(\alpha;\varphi) - L_u'(\alpha;A_u\varphi)> = 0 \end{aligned} \tag{5.2}$$

which will be called the *coupling conditions* between the Nijenhuis and the symplectic operators. Each pair of operators A_u and L_u fulfilling them will be said to form an *A-L pair*.

To clarify the meaning of the previous conditions by a comparison, let us consider the definition of the Kähler manifold [7]. A Kähler manifold is a composite structure defined, on the configuration space U, by a Nijenhuis operator A_u and by a Riemannian operator G_u obeying the

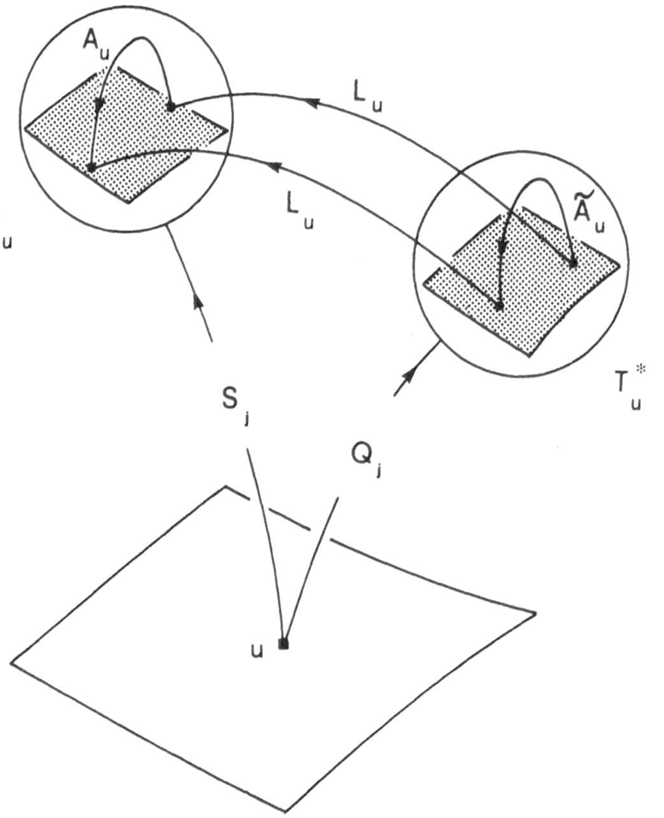

a symplectic Kähler manifold is a geometrical structure defined by a symplectic operator L_u *and by a Nijenhuis operator* A_u *fullfilling the following two coupling conditions:*

$$A_u L_u = L_u \widetilde{A}_u$$

$$\langle \alpha , A'_u(L_u\beta;\varphi) - A'_u(\varphi;L_u\beta)\rangle + \langle \beta , A'_u(\varphi;L_u\alpha) + A_u L'_u(\alpha;\varphi) + L'_u(\alpha;A_u\varphi)\rangle = 0$$

Fig. 4: A symplectic Kähler manifold

following conditions

1. $A_u^2 = -I$ (5.3)

2. $A_u G_u = -G_u \tilde{A}_u$ (5.4)

3. *the covariant derivative of* A_u *with respect to the Riemannian connection defined by* G_u *vanishes* (5.5)

Disregarding for the moment the first condition, are at first sight manifest the analogies between the conditions (5.4)-(5.5) and the present ones (5.1)-(5.2). The main difference is that in the present case the skew-symmetric operator L_u replaces the symmetric operator G_u. So, we can say that the geometrical structure defined by an A-L pair is the skew-symmetric counterpart of a Kähler manifold (this point of view will be confirmed in Sec.7). To stress this property, we tentatively call "*symplectic Kähler manifold*" the geometrical structure defined by an A-L pair.

6. INTEGRABLE HAMILTONIAN EQUATIONS

The reason to consider the symplectic Kähler manifolds is that they are intimately related with the theory of the integrable Hamiltonian equations. The point of contact between the two theories is given by the study of the "groups of motion" admitted by such manifolds, namely the study of the Hamiltonian vector field leaving A_u invariant. They are very special vector fields possessing many properties of the integrable Hamiltonian equations.

The relation we are dealing with is especially clear and complete in the case of the finite-dimensional symplectic Kähler manifolds. In this case, in fact, it can be said that the theory of the groups of motion of such manifolds *coincides* with the theory of the integrable Hamiltonian systems. This is due to the following theorem which is proved in [2]. Let U be a symplectic Kähler manifold of maximal rank (which means that the operator A_u is supposed to have only *double* eigenvalues). Then all the vector fields leaving this manifold invariant are completely integrable Hamiltonian vector fields and, conversely, every integrable Hamiltonian vector field leaves such a manifold invariant.

As a consequence of this identification the Nijenhuis operators enter, for the first time, as important objects into the theory of the Hamiltonian systems. This leads to a new method of integration of the Hamiltonian equations which closely resembles the "spectral transform method", as we shall see shortly.

No result of the same generality is so far known in the infinite-dimensional case. There are, however, many partial results which seem to point out the existence of the same kind of relation. A first indication is given by the following theorem, pointing out two general properties of the vector fields leaving a symplectic Kähler manifold invariant.

Consider any potential operator Q_1 fulfilling the condition (4.23) with respect to $\tilde{A}_u$, and let S_1 be the corresponding Hamiltonian operator given by

$$S_1(u) = L_u Q_1(u) \qquad . \tag{6.1}$$

Then the operators

$$S_j(u) \triangleq A_u^j \cdot S_1(u) \tag{6.2}$$

are *commuting Hamiltonian operators* leaving A_u invariant again, and the functionals [6, Sec.2]

$$I_j[u] = \int_{\lambda=0}^{\lambda=1} \langle Q_j(\lambda u), u\rangle \, d\lambda \tag{6.3}$$

associated with the potential operators

$$Q_j(u) \triangleq \tilde{A}_u^j \cdot Q_1(u) \tag{6.4}$$

are their *integrals which are in involution.*

The proof of the first part of this statement simply consists in observing that

$$\begin{aligned} S_j(u) &\overset{(6.1)}{=} A_u^j L_u Q_1(u) \\ &\overset{(5.1)}{=} L_u \tilde{A}_u^j Q_1(u) \\ &\overset{(6.4)}{=} L_u Q_j(u) \end{aligned} \tag{6.5}$$

and in recalling that the operators Q_j are potential and the operators S_j are commuting as a consequence of the properties of the Nijenhuis operators shown in Sec.4. The second part, concerning the integrals $I_j[u]$, is proved by explicitily computing the following Poisson bracket

$$\begin{aligned}\{I_j,I_k\} &\overset{(4.10)}{=} \langle Q_j(u),L_uQ_k(u)\rangle \\ &\overset{(6.3)}{=} \langle Q_j(u),L_u\tilde{A}_uQ_{k-1}(u)\rangle \\ &\overset{(5.1)}{=} \langle Q_j(u),A_uL_uQ_{k-1}(u)\rangle \\ &\overset{(4.20)}{=} \langle \tilde{A}_uQ_j(u),L_uQ_{k-1}(u)\rangle \\ &\overset{(6.3)}{=} \langle Q_{j+1}(u),L_uQ_{k-1}(u)\rangle \\ &\overset{(4.10)}{=} \{ I_{j+1},I_{k-1} \} \quad .\end{aligned} \tag{6.6}$$

From it we get

$$\{I_j,I_k\} = \{I_k,I_j\} \tag{6.7}$$

and then

$$\{I_j,I_k\} = 0 \tag{6.8}$$

showing that the functionals I_j are in involution. The same relation, written in the equivalent form

$$\langle Q_j(u),S_k(u)\rangle = 0 \tag{6.9}$$

on account of the eq.(6.5), proves finally that the functionals I_j are integrals of the Hamiltonian operators S_k [1, Sect.1], as it was stated.

The vector fields leaving the A-L pairs invariant reveal consequently, also in the infinite-dimensional case, the characteristic property of possessing a set of integrals which are in involution. This property makes them resemble the integrable Hamiltonian vector fields. The problem which naturally arises is to know when they have a "sufficient" number of integrals to be considered integrable (the classical criterion of Liouville becoming obviously meaningless in the infinite-dimensional case). The experience of the finite-dimensional case suggests then to make the *conjecture* that the previous vector fields may be considered integrable if the Nijenhuis operator A_u has only double eigenvalues. Since such eigenvalues are integrals of the vector fields we are considering, this condition substantially amounts to demand that such fields admit a number of integrals equal to "half" the dimension of the space. From this point of view, the present conjecture appears as a convenient form of the classical Liouville condition, preserving its meaning also in an infinite number of dimensions.

A second indication of the relation between the theory of the symplectic Kähler manifolds and the infinite-dimensional integrable systems is that an A-L pair is actually associated with the main nonlinear solvable equations which have been so far considered in the literature (see the next Section and the following Fig. 5).

A third indication is suggested by the comparison between the "spectral transform method" used in the study of the infinite-dimensional equations [8], and the method of integration developed in the framework of the theory of the symplectic Kähler manifolds [2]. This method comes out as a consequence of the change of point of view produced by the geometrical approach, which leads to give preeminence to the Nijenhuis operator A_u rather than to the integrals which are in involution. The leading idea is to use the Nijenhuis operator to define a privileged system of coordinates on the base manifold.

To explain this point, consider at any point u the eigenvectors of the adjoint operator $\tilde{A}_u$ of the Nijenhuis operator A_u (see Fig.6). They form a local basis in the cotangent space. Two eigenvectors α_k and β_k are by assumption associated with any eigenvalue λ_k (we are considering a symplectic Kähler manifold of maximal rank). It turns out that such eigenvectors can be always chosen so that they are the gradients of two suitable functions. The first function is the eigenvalue itself [2;Sec.4]

$$\alpha_k \triangleq \text{grad}\, \lambda_k(u) \qquad ; \tag{6.10}$$

the second function μ_k

$$\beta_k \triangleq \text{grad}\, \mu_k(u) \tag{6.11}$$

must be instead computed by a quadrature [2,Sec.3]. This means that the eigenvectors of $\tilde{A}_u$ at the different points of the configuration space join together themselves to form a network of coordinate lines on the base manifold.

A set of 2n coordinates $(\lambda_1 \ldots \lambda_n, \mu_1 \ldots \mu_n)$ is thus associated with any point u of the manifold. When this point evolves according to the

equation ↴			L_u ↴	A_u ↴	$\langle \alpha, \varphi \rangle$ ↴
kdv	$u_t + u_{xxx} + auu_x = 0$	:	∂_x	$\partial_{xx} + 2au + au_x \int_{-\infty}^{x}$	$\int_{-\infty}^{\infty} \alpha \varphi$
modified kdv	$u_t + u_{xxx} + au^2 u_x = 0$	:	∂_x	$\partial_{xx} + {}^2\!/_3 a \left(u \int_{-\infty}^{x} u \quad \right)_x$	"
Harry Dym	$u_t = (u^{-1/2})_{xxx}$	:	∂_{xxx}	$\left(u^{-1/2} \int_{-\infty}^{x} u^{-1/2} \quad \right)_{xxx}$	"
sine-Gordon	$u_t = \int_{-\infty}^{x} \sin u$	:	$\int_{-\infty}^{x}$	$\partial_{xx} + u_x \int_{-\infty}^{x} u_\xi \partial_\xi$	"
n. l. Schrödinger	$iu_t + u_{xx} + a\bar{u}u^2 = 0$	:	$i\cdot$	$i\cdot\left[\partial_x + au \int_{-\infty}^{x} \bar{u}(\cdot) + u\overline{(\cdot)} \right]$	$\int_{-\infty}^{\infty} \alpha\bar{\varphi} + \bar{\alpha}\varphi$
d. n. l. Schrödinger	$iu_t + u_{xx} + ia(\bar{u}u^2)_x = 0$	:	∂_x	$\partial_x\left[i - au \int_{-\infty}^{x} \bar{u}(\cdot) + u\overline{(\cdot)} \right]$	"
AkNS		:	$\begin{pmatrix} 0 & 1 \\ -1 & 0 \end{pmatrix}$	$i\cdot\begin{pmatrix} -\partial_x + u_1 \int_{-\infty}^{x} u_2 & u_1 \int_{-\infty}^{x} u_1 \\ -u_2 \int_{-\infty}^{x} u_2 & \partial_x - u_2 \int_{-\infty}^{x} u_1 \end{pmatrix}$	$\int_{-\infty}^{\infty} \alpha_1 \varphi_1 + \alpha_2 \varphi_2$

Fig. 5: Some A-L pairs associated with nonlinear solvable equations

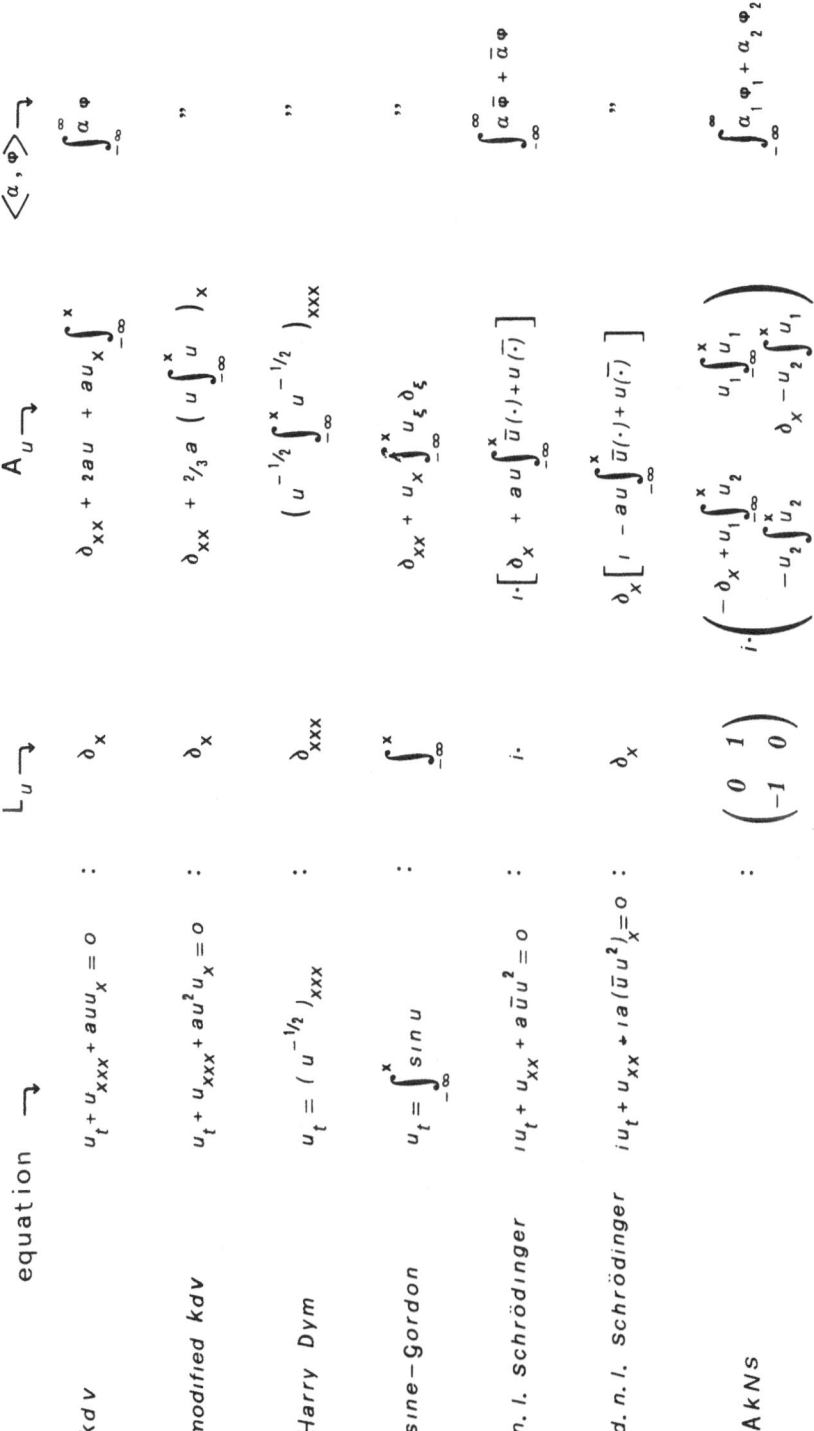

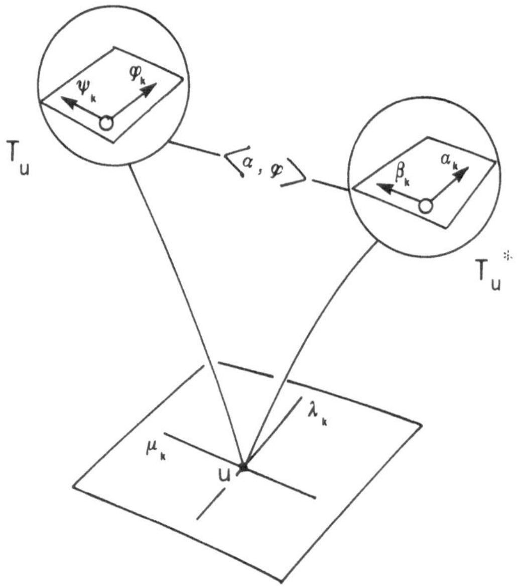

$$\tilde{A}_u \alpha_k = \lambda_k \alpha_k \qquad \alpha_k = grad\ \lambda_k$$
$$\tilde{A}_u \beta_k = \lambda_k \beta_k \qquad \beta_k = grad\ \mu_k$$

the eigenvectors of the Nijenhuis operator at the different points of the configuration space define a network of coordinate lines. In the new coordinates (λ_k, μ_k) the vector fields leaving A_u *invariant become integrable by quadrature alone*

(ref. 2, sect.s 3 and 6)

Fig. 6: A privileged system of coordinates

equation

$$\frac{du}{dt} = K(u) \tag{6.12}$$

its coordinates evolve according to the equations

$$\begin{aligned} \dot{\lambda}_k &= \langle \alpha_k, K(u) \rangle \\ \dot{\mu}_k &= \langle \beta_k, K(u) \rangle \quad . \end{aligned} \tag{6.13}$$

If the vector field K, in particular, leaves A_u invariant such equations reduce themselves to the following ones [2, Sec.6]

$$\begin{aligned} \dot{\lambda}_k &= 0 \\ \dot{\mu}_k &= \phi_k(\lambda_k, \mu_k) \end{aligned} \tag{6.14}$$

which may be solved simply by quadratures. The privileged system of coordinates (λ_k, μ_k) defined by the Nijenhuis operator reduces consequently to the quadratures all the vector fields leaving A_u invariant.

This situation may be compared with that encountered in the theory of the spectral transform method. To be definite, let us consider the case of the KdV equation

$$u_t + 6uu_x - u_{xxx} = 0 \quad . \tag{6.15}$$

As is well-known, the procedure for solving this equation consists in associating with it the Schrödinger eigenvalue problem

$$\psi_{xx} - u\psi = -\lambda^2\psi \tag{6.16}$$

and in considering as new coordinates the so-called "scattering data", including the eigenvalues λ and the reflection coefficients R [9]. When u evolves according to the KdV equation, they evolve according to the equations [9, p.236]

$$\begin{aligned} \dot{\lambda} &= 0 \\ 2i\lambda\dot{R} &= \int_{-\infty}^{+\infty} \psi^2 \cdot u_t \, d\xi \end{aligned} \tag{6.17}$$

To see the analogies with the method based on the theory of the symplectic Kähler manifolds it suffices then to observe that the eigenvalues λ of the Schrödinger equation are still eigenvalues of the Nijenhuis operator

$$A_u\varphi = -\varphi_{xx} + 4u\varphi + 2u_x\int_{-\infty}^{x}\varphi\, d\xi \qquad (6.18)$$

associated with the KdV equation (see the next section), and that the square of the Schrödinger eigenfunction ψ is exactly the corresponding eigenfunction of $\tilde{A}_u$. It follows that, with the notation used in this paper, the two eq.s(6.17) read

$$\dot{\lambda} = 0$$
$$2i\lambda\dot{R} = \langle\beta, K(u)\rangle \qquad (6.19)$$

namely they have the same form we would expect on the basis of the theory of the symplectic Kähler manifolds. This strongly suggests that the theory of the infinite-dimensional symplectic Kähler manifolds may give a simple geometrical explanation of the spectral transform method on the basis of the properties of the Nijenhuis operators. Further researches on this point are presently in progress.

7. A METHOD FOR CONSTRUCTING INTEGRABLE HAMILTONIAN EQUATIONS

As it was shown in the previous section, any A-L pair defines a simple iterative process for constructing hierarchies of commuting Hamiltonian equations. This process has been summarized for convenience in Fig. 7. It only requires to choose, at the beginning, a suitable pair of operators Ω_1 and S_1 and the next apply the Nijenhuis operator. In particular, in all the theories which are invariant under the space translation as the starting operators S_1 it can be chosen the generator of the space translation itself

$$S_1(u) = u_x \quad , \qquad (7.1)$$

while the operator Ω_1 can be obtained from S_1 by means of the symplectic operator L_u.

According to this result, the problem of constructing integrable Hamiltonian equations may be reduced to the problem of constructing suitable A-L pairs. This problem may be dealt with as follows.

Choose a symplectic operator L_u as simple as possible (in particular

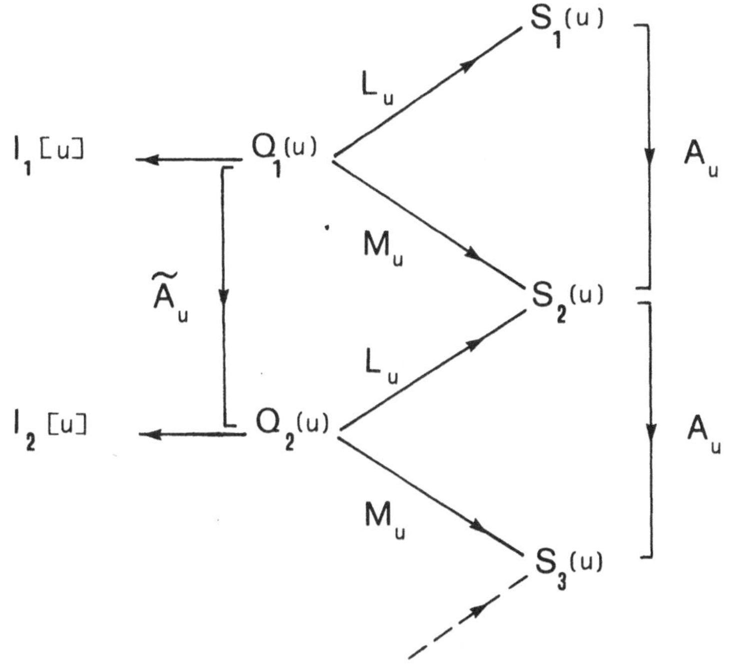

$$[S_i, S_k] = 0 \qquad \{I_i, I_k\} = 0$$

any A-L *pair gives rise to a simple iterative process for constructing " integrable " Hamiltonian equations*

(see also ref. 1 , sect. 3)

Fig. 7: The A-L pairs and the "integrable" Hamiltonian equations.

admitting an inverse operator L_u^{-1}), and consider the operators ℓ_u given by

$$\ell_u = \Omega'_u - \tilde{\Omega}'_u \tag{7.2}$$

where Ω is any covariant operator. Try then to find Ω so that the operator

$$A_u \triangleq L_u \ell_u \tag{7.3}$$

is a Nijenhuis operator. In the affirmative case, the operators A_u and L_u make an A-L pair. The reason is that the operator A_u given by eq.(7.3) is the more general operator fulfilling the coupling conditions (5.1) and (5.2). So, if it is also a Nijenhuis operator, it gives rise to an A-L pair.

By this recipe, the construction of an A-L pair is reduced to the problem of solving the single Nijenhuis condition (4.3) in the single unknown operator ℓ_u. This problem can be most easily dealt with by suitably fixing the form of the operator ℓ_u up to some undetermined coefficients. In this way, the Nijenhuis condition splits into a system of algebraic conditions on the undetermined coefficients. By solving them we are able to find the A-L pairs. This procedure will be made clear by the following examples.

Let $u=(u_1,\dots,u_n)$ be a vector-valued function defined on the real axis, and let L_u and ℓ_u be given by

$$L_u\alpha = E\alpha \tag{7.4}$$

$$\ell_u\varphi = A\varphi_x + \sum_{1}^{m}{}_{kj}\, a_{kj} B_k u \int_{-\infty}^{x} \langle B_j u,\varphi\rangle d\xi \tag{7.5}$$

where

$$a_{jk} = a_{kj} \tag{7.6}$$

and where A, B_k, E are n×n matrices with constant entries which are only required to obey the following symmetry conditions

$$\langle\alpha,E\beta\rangle = -\langle\beta,E\alpha\rangle\,,\ \langle A\varphi,\psi\rangle = \langle A\psi,\varphi\rangle\,,\ \langle B_k\varphi,\psi\rangle = \langle B_k\psi,\varphi\rangle \tag{7.7}$$

in order to assure that L_u is a symplectic operator and that ℓ_u may be given by the eq.(7.1) with a suitable operator Ω. The meaning of the crochet $\langle\alpha,\varphi\rangle$ will be precised shortly. The corresponding operator A_u is then

$$A_u\varphi = EA\varphi_x + \sum a_{jk}EB_k u\int_{-\infty}^{x} \langle B_j u,\varphi\rangle d\xi \tag{7.8}$$

By inserting this expression into the Nijenhuis condition (4.3), by using the symmetry condition (7.7) and with a simple integration by parts (where the boundary terms are disregarded, since φ and ψ are supposed to obey homogeneous boundary conditions) the following conditions on the unknown matrices A, B_k, E are then obtained

$$B_j E B_\ell = B_\ell E B_j \tag{7.9}$$

$$B_k EA = AEB_k \tag{7.10}$$

$$\sum_1^m a_{kj}\langle B_j EA\varphi,\psi\rangle\langle B_k u,\chi\rangle + \ldots + \ldots = 0 \tag{7.11}$$

where the dots mean the ciclic permutation over φ, ψ and χ.

Any solution of these conditions gives an A-L pair and then a hierarchy of integrable equations. As a first example, consider the following solution (for n=m=1)

$$E=i \qquad A=B=1 \qquad \langle\alpha,\varphi\rangle = \alpha\bar{\varphi} + \bar{\alpha}\varphi \quad . \tag{7.12}$$

In this case the Nijenhuis operator A_u turns out to be given by

$$A_u\varphi = i\left[\varphi_x + u\int_{-\infty}^{x}(u\bar{\varphi} + \bar{u}\varphi)\, d\xi\right] \tag{7.13}$$

and the first equation obtained is the *nonlinear Schrödinger equation*

$$u_t = i(u_{xx} + u^2\bar{u}) \tag{7.14}$$

A second example (for n=2 and m=1) is given by

$$E = \begin{pmatrix} 0 & 1 \\ -1 & 0 \end{pmatrix} \qquad -A = B^2 = \begin{pmatrix} 0 & i \\ i & 0 \end{pmatrix}$$
$$\langle\alpha,\varphi\rangle = \alpha_1\varphi_1 + \alpha_2\varphi_2 \tag{7.15}$$

which yields the following Nijenhuis operator

$$A_u = i\begin{pmatrix} -\partial_x + u_1\int_{-\infty}^{x} u_2 & u_1\int_{-\infty}^{x} u_1 \\ -u_2\int_{-\infty}^{x} u_2 & \partial_x - u_2\int_{-\infty}^{x} u_1 \end{pmatrix} \tag{7.16}$$

generating the hierarchy of the two-components AKNS equations [8, p.372]. Other examples are easily obtained in the same way, but they will not be reported here for brevity. Rather, let us show how the Nijenhuis operators associated with the KdV equation and with the modified-KdV equation are obtained from the previous one (7.16). This will be usefull to show a general procedure allowing to obtain new Nijenhuis operators from known ones. This procedure rests on the observation that the powers A_u^j of any Nijenhuis operator are themselves Nijenhuis operators. Consider then the square

$$A_u^2 = -\begin{pmatrix} \partial_{xx} - \left[u_1\int_{-\infty}^{x} u_2\partial_x + \partial_x u_1\int_{-\infty}^{x} u_2\right], & u_1\int_{-\infty}^{x} u_1\partial_x - \partial_x u_1\int_{-\infty}^{x} u_1 \\ u_2\int_{-\infty}^{x} u_2\partial_x - \partial_x u_2\int_{-\infty}^{x} u_2 & , \partial_{xx} - \left[u_2\int_{-\infty}^{x} u_1\partial_x + \partial_x u_2\int_{-\infty}^{x} u_1\right] \end{pmatrix} \tag{7.17}$$

of the operator (7.16), and observe that it admits the *invariant submanifolds* defined by

$$u_2 = -1 \tag{7.18}$$

and by

$$u_2 = -u_1 \quad . \tag{7.19}$$

By taking then the restriction of A_u^2 on such manifolds, the following two operators

$$A_u\varphi = \varphi_{xx} + 2u\varphi + u_x\int_{-\infty}^{x}\varphi d\xi \tag{7.20}$$

and

$$A_u\varphi = \varphi_{xx} + 2\partial_x\left(u\int_{-\infty}^{x} u\varphi d\xi\right) \tag{7.21}$$

are readily obtained. They are the Nijenhuis operators we were looking for (see table 5).

To give, finally, just one example of a class of equations which have been not previously considered in the literature, let us consider the following operators

$$\begin{aligned} L_u\alpha &= A\alpha_x \\ \ell_u\varphi &= B\varphi + Cu\int_{-\infty}^{x} \langle a,\varphi\rangle d\xi + a\int_{-\infty}^{x} \langle Cu,\varphi\rangle d\xi \end{aligned} \tag{7.22}$$

where $a=(a_1,\dots,a_n)$ and the matrices A, B, C are required to obey the symmetry conditions

$$<\alpha, A\beta> = <\beta, A\alpha>, \quad <B\varphi,\psi> = -<B\psi,\varphi>, \quad <C\varphi,\psi> = <C\psi,\varphi> \tag{7.23}$$

In this case, the Nijenhuis condition may be solved explicitely whatever the number n of components of u may be. The solution is

$$C = A^{-1} \qquad B_{hk} = a_h b_k - b_h a_k \tag{7.24}$$

where the b_k are n arbitrary constants. The corresponding evolution equations are

$$K(u) = ABu_{xx} + \partial_x\left[u<a,u> + \tfrac{1}{2}Aa<Cu,u>\right] \tag{7.25}$$

According with the results of Sec.6, they possess an infinite number of symmetries and of conservation laws which are in involution.

The previous examples aim to give an idea of the possibilities of obtaining integrable equations according to the theory of the A-L pairs. Only one procedure has been pointed out, however, without to mention other methods accounting for different aspects of the theory of the symplectic Kähler manifolds. We hope to be able to discuss them in a subsequent paper. In particular, the reader is referred to [2] for a discussion of the A-L pairs for finite-dimensional systems.

Remark 1. Many of the previous Nijenhuis operators were, of course, already well-known in the literature, where they appear in many different contexts. It is, however, interesting to observe that it has never been previously recognized that they were Nijenhuis operators. From this point of view, the present geometrical approach seems to be able to actually improve our understanding of the theory of the nonlinear solvable equations.

Remark 2. The previous construction of the A-L pairs was based on the observation that the coupling conditions (5.1) and (5.2) imply that the operator

$$\ell_u \triangleq L_u^{-1}A_u \tag{7.26}$$

verifies the conditions

$$<\ell_u\varphi,\psi> = -<\ell_u\psi,\varphi> \tag{7.27}$$

$$<\ell_u'(\varphi;\psi),\chi> + \ldots + \ldots = 0 \tag{7.28}$$

which mean that the bilinear form $<\ell_u \varphi, \psi>$ is a closed 2-form defined on the configuration space. Now, it is remarkable that a similar result holds in the theory of the Kähler manifolds, where it is shown that the operator

$$\ell_u = G_u^{-1} \cdot A_u \tag{7.29}$$

obeys itself the conditions (7.26) and (7.27) [7]. This fact clearly confirms the existence of a close relation between the symplectic Kähler manifolds and the usual Kähler manifolds, as it was formally stated in Sec.5.

Remark 3. Another result concerning the symplectic Kähler manifolds (which has not been proved in this paper) says that the operator

$$M_u \triangleq A_u \cdot L_u \tag{7.30}$$

is itself a symplectic operator, as L_u. This remark is usefull to relate the theory of the symplectic Kähler manifolds with the theory of the twofold Hamiltonian equationswhich has been suggested in [1].

8. CONCLUDING REMARKS

The present approach arised as an attempt of extending to the nonlinear evolutions equations the idea, originally due to Volterra, of combining together the nice ideas of the classical differential geometry with the powerful computational techniques of the nonlinear operator theory, in order to exploit the advantages of both such theories in the study of the nonlinear differential equations..

The main concept arrived at by this approach is that of "symplectic Kähler manifold". The main thesis is that the nonlinear solvable equations define the "groups of motion" of such a manifold, and that the spectral transform is a suitable change of coordinates defined by the Kählerian structure. This thesis, however, has been completely proved only in the finite-dimensional case [2].

The geometrical-operator scheme developed in this paper provides a unified point of view in dealing with symmetries and conservation laws of the nonlinear evolution equations [1]. It is sufficiently general to cover the main nonlinear solvable equations previously considered in the literature, suggesting at the same time some new equations. It leads to the discovery that the operators A_u defining the different classes of nonlinear solvable equations are Nijenhuis operators.

It is, finally, a tensorial scheme. We are able to reconstruct the scheme after an arbitrary change of the field functions, while there is not a systematic way to do this in the inverse method or in the Lax approach. To give an example, consider the modified-KdV equation

$$u_t = u_{xxx} + \tfrac{3}{8}u^2 u_x \tag{8.1}$$

and observe that the transformation

$$u = v_x + a\sin(v/2) \tag{8.2}$$

changes it into the equation

$$v_t = v_{xxx} + \tfrac{1}{8}v_x^3 + \tfrac{3}{8}a^2 v_x \sin^2(v/2) \tag{8.3}$$

according with the transformation law (3.3).

Both such equations have been separately studied in the literature, by repeating every time the analysis of their properties. The tensorial character of the present scheme allows instead to obtain the properties of the new equation by simply translating the properties of the modified-KdV equation by means of the transformation laws previously considered. It is well-known, for example, that the modified-KdV equation possesses an infinite number of symmetries which are in involution, generated by the following Nijenhuis operator

$$A_u\varphi = \varphi_{xx} + \tfrac{1}{4}\partial_x\left(u\int_{-\infty}^{x} u\varphi\, d\xi\right) \quad . \tag{8.4}$$

The same conclusion is then arrived at for the eq.(8.3) without any further computation. Moreover, the Nijenhuis operator

$$A_v\varphi = \varphi_{xx} + \frac{v_x}{4}\int_{-\infty}^{x} v_x\varphi\, d\xi + \frac{a^2}{4}\left(\varphi\sin^2\frac{v}{2} + \frac{v_x}{4}\int_{-\infty}^{x}\varphi\sin v\, d\xi\right) \tag{8.5}$$

generating the symmetries and the conservation laws of the new equation is readily obtained from (8.4), by using the transformation law (4.4). Wishing then to stress the invariance of the properties of the two equations in spite of the difference in their forms, it can even be said that the two equations (8.1) and (8.3) are the same equation written in two different system of coordinates. The tensorial character of the present approach may thus aid to gain a more unified point of view in the study of the nonlinear evolution equations.

APPENDIX

The Gateaux derivative of an operator $F:U\to\overline{U}$ may be denoted by F'_u and is defined by

$$F'_u\cdot\varphi = \frac{d}{d\varepsilon}F(u+\varepsilon\varphi)\Big|_{\varepsilon=0} \tag{A 1}$$

so that to the first-order in ε it is

$$F(u+\varepsilon\varphi) = F(u) + \varepsilon F'_u\cdot\varphi \quad . \tag{A 2}$$

Its adjoint operator $\tilde{F}'_u$, relative to the prefixed bilinear forms $<\alpha,\varphi>_U$ and $<\overline{\alpha},\overline{\varphi}>_{\overline{U}}$ is defined by

$$<\overline{\alpha},F'_u\varphi>_{\overline{U}} = <\tilde{F}'_u\overline{\alpha},\varphi> \quad . \tag{A 3}$$

The Gateaux derivative of the operator A_u, finally, is defined by

$$A'_u(\varphi;\psi) = \frac{d}{d\varepsilon}A_{u+\varepsilon\psi}\varphi\Big|_{\varepsilon=0} \quad . \tag{A 4}$$

So, for example, if A_u is given by

$$A_u\varphi = \varphi_{xx} + \partial_x\left(u\int_{-\infty}^{x}u\varphi d\xi\right) \quad , \tag{A 5}$$

the Gateaux derivative turns out to be

$$A'_u(\varphi;\psi) = \partial_x\left(\psi\int_{-\infty}^{x}u\varphi d\xi + u\int_{-\infty}^{x}\psi\varphi d\xi\right) \tag{A 6}$$

and we can easily verify that the operator (A 5) obeys the Nijenhuis condition (4.3).

REFERENCES

[1] F.Magri, J.Math.Phys. 19 (1978), 1156.

[2] F.Magri, Some remarks on the Integrable Hamiltonian Systems, submitted for the pubblication on the Nuovo Cimento B (1980).

[3] F.Magri, Proceedings of the "Joint IUTAM-IMU Symposium-Group Theoretical Methods in Mechanics", (Novosibirsk, August 25-29, 1978), N.H.Ibraghinov and L.V.Ovsiannjkov Ed.s, p.178.

[4] F.Magri, J.Math.Phys. 18 (1977), 1405.

[5] M.N.Vainberg, *Variational Methods for the Study of Nonlinear Operators*, Holden-Day, San Francisco, 1964.

[6] F.Magri, Ann.Phys. 99 (1976), 196.

[7] K.Yano, *Differential Geometry on Complex and Almost Complex Spaces*, Pergamon Press, Oxford, 1965, Ch.4.

[8] J.Moser (Ed.), *Dynamical Systems: Theory and Applications*, Lect.Notes in Phys. 38 (Springer, Berlin, 1975).

[9] F.Calogero, Nuovo Cimento B (1976), 229.

PROPERTIES OF A SPECIAL FUNCTION RELATED TO SELF-SIMILAR SOLUTIONS OF CERTAIN NONLINEAR WAVE EQUATIONS

M. Leo, R.A. Leo and G. Soliani
Istituto di Fisica dell'Università di Lecce,
73100 Lecce, Italy

ABSTRACT

Some basic properties of a new special function pertinent to the study of certain nonlinear wave equations are presented.

1. INTRODUCTION

The aim of this lecture, based on two recent papers[1-2] by M. Leo, R.A. Leo and G. Soliani, is to show some basic properties of the function

$$\Psi(\alpha,\beta,\gamma;x) = \int_x^\infty dt\, t^{\alpha-1}\left[1-\left(1-\frac{e^{-t}}{t^\beta}\right)^\gamma\right] \tag{1.1}$$

where α, β and γ are real parameters.

For the proofs of the majority of the theorems presented here, we refer the reader to Refs. 1 and 2.

The study of the function $\Psi(\alpha,\beta,\gamma;x)$ is interesting for two reasons;

i) it is related to solutions of self-similar type of the nonlinear wave equation

$$u_{tx} = a\,(e^{-u} + \beta u^{\beta-1}) \quad , \tag{1.2}$$

where a and β are constants;

Talk given by G.Soliani at the meeting on "Nonlinear Evolution Equations and Dynamical Systems", held in Lecce, 20-23 June 1979.

ii) it covers certain old (and new) special functions for particular choices of the parameters α, β and γ. For example, for $\gamma=1$ one obtains

$$\Psi(\alpha,\beta,1;x) = \Gamma(\alpha-\beta,x) \tag{1.3}$$

where

$$\Gamma(\alpha-\beta,x) = \int_x^\infty dt\ t^{\alpha-\beta-1}\ e^{-t} \tag{1.4}$$

is the (complementary) incomplete Gamma function.

We remark that, generally, exact self-similar solutions for nonlinear evolution equations cannot be obtained in terms of known functions, unlike what happens for, say, the Korteweg-de Vries and the Sine-Gordon equations[3]. However, such exact solutions may be found sometimes provided that it is possible to define suitably new functions.

For simplicity's sake, here we have assumed that α, β, γ and x are real, the latter being nonnegative.

Furthermore we have restricted ourselves to consider only real values of $\Psi(\alpha,\beta,\gamma;x)$. This implies that $e^{-t}<t^{\beta}$ for any t such that $x\leq t<+\infty$. We observe that the integral (1.1) can also be extended to the interval $(0,+\infty)$ open on the left, provided that $-e<\beta<0$ and $\alpha>-|\beta|$ (see Proposition 1.1 of Ref.1).

2. LINK BETWEEN THE Ψ-FUNCTION AND SELF-SIMILAR SOLUTIONS OF THE EQUATION $u_{tx}= a(e^{-u} + \beta u^{\beta-1})$.

We shall show that there exist a link between the Ψ-function and self-similar solutions of (1.2). To this end, we put into (1.2) $u=u(\xi)$, where $\xi=x+vt$ (v=constant). Then (1.2) transforms into the ordinary differential equation (*reduced form* of Eq.(1.2)):

$$\frac{1}{2} v\, u_\xi^2 = -a\, e^{-u} + a\, u^{\beta} + c \quad . \tag{2.1}$$

c being an integration constant. Limiting ourselves to choose $c=0$, (2.1) yields

$$\lambda(\xi-\xi_0) = \pm\int_{u_0}^{u} dt\ t^{-\beta/2}\left(1-\frac{e^{-t}}{t^{\beta}}\right)^{-\frac{1}{2}} \quad , \tag{2.2}$$

where $u_o=u(\xi_o)$, $\lambda=(\frac{v}{2a})^{-1/2}$ and ξ_o is a constant.

Let us consider two cases: $\beta\neq 2$ and $\beta=2$. In the first case, from (2.2) we have

$$\lambda\xi = \frac{2}{2-\beta}\, u^{\frac{2-\beta}{2}} + \Psi(\frac{2-\beta}{2},\beta,-\frac{1}{2};u) + \text{const} \quad , \tag{2.3}$$

while in the second case we get

$$\lambda\xi = \ln u + \Psi(0,2,-\frac{1}{2};u) + \text{const} \quad . \tag{2.4}$$

Of course, the self-similar solution of (1.2) can be obtained, in principle, from (2.3) or (2.4) by means of an inversion operation.

As an example, let us deal with (1.2) with $\beta=0$. Equation (1.2) becomes Liouville's equation[5-6]:

$$u_{tx} = a\,e^{-u} \tag{2.5}$$

from which we have

$$\frac{1}{2} v\, u_\xi^2 = -ae^{-u} + \text{const} \quad . \tag{2.6}$$

Choosing for simplicity the constant on the right of (2.6) equal to a, (2.6) yields

$$\frac{1}{2a} v\, u_\xi^2 = -e^{-u} + 1 \quad , \tag{2.7}$$

which gives

$$\lambda\xi = u + \Psi(1,0,-1/2;u) + \text{const} \quad . \tag{2.8}$$

In this case, the Ψ-function can be explicitly determined in terms of elementary functions, that is

$$\Psi(1,0,-\frac{1}{2};u) = -2\ln\left[1-(1-e^{-u})^{\frac{1}{2}}\right] - 2u - 2\ln 2 \quad . \tag{2.9}$$

By means of simple calculations, one obtains the following known expression of u in terms of ξ:

$$u = \ln \frac{\left[1+2A\,e^{\lambda(\xi-\xi_o)}\right]^2}{4A\,e^{\lambda(\xi-\xi_o)}\left[1+e^{\lambda(\xi-\xi_o)}\right]} \quad , \tag{2.10}$$

where A is a constant.

3. A SERIES REPRESENTATION

Consider the binomial expansion:

$$\left(1 - \frac{e^{-t}}{t^{\beta}}\right)^{\gamma} = 1 + \frac{1}{\Gamma(-\gamma)} \sum_{n=1}^{\infty} \frac{\Gamma(n-\gamma)}{n!} \frac{e^{-nt}}{t^{n\beta}} , \qquad (3.1)$$

where the series on the right is uniformly convergent for $t \geq x$, x being any fixed number such that $e^{-x} < x^{\beta}$.

Thus from (3.1) we deduce that

$$\Psi(\alpha,\beta,\gamma;x) = -\frac{1}{\Gamma(-\gamma)} \sum_{n=1}^{\infty} \frac{\Gamma(n-\gamma)}{n!} \int_x^{\infty} dt\, t^{\alpha-n\beta-1} e^{-nt} . \qquad (3.2)$$

Using now the integral representation:

$$\mu^{\nu}\Gamma(\nu,\mu x) = \int_x^{\infty} dt\, t^{\nu-1} e^{-\mu t} \qquad (3.3)$$

$(x>0,\ \mathrm{Re}\,\mu>0)$, we obtain the series representation:

$$\Psi(\alpha,\beta,\gamma;x) = -\frac{1}{\Gamma(-\gamma)} \sum_{n=1}^{\infty} \frac{\Gamma(n-\gamma)}{n!} n^{n\beta-\alpha} \Gamma(\alpha-n\beta,nx) . \qquad (3.4)$$

This formula will be employed later.

4. A RECURRENCE RELATION

Another useful property involving the Ψ-function is given by the following recurrence relation:

$$\left(1 - \frac{\gamma\beta}{\alpha}\right) \Psi(\alpha,\beta,\gamma;x) = -\frac{1}{\alpha} x^{\alpha}\left[1 - \left(1 - \frac{e^{-x}}{x^{\beta}}\right)^{\gamma}\right] + \qquad (4.1)$$

$$+ \frac{\gamma}{\alpha}\left[\Psi(\alpha+1,\beta,\gamma;x) - \Psi(\alpha+1,\beta,\gamma-1;x) - \beta\Psi(\alpha,\beta,\gamma-1;x)\right] ,$$

for $\alpha \neq 0$.

Notice that for $\gamma=1$, the relation (4.1) specializes to the well known recurrence relation:

$$\Gamma(\alpha-\beta+1,x) = x^{\alpha-\beta-1} e^{-x} + (\alpha-\beta)\Gamma(\alpha-\beta,x)$$

for the incomplete Gamma function.

5. SOME SPECIAL CASES

In the following, we shall denote by n any positive integer.

a) Case $\gamma = n$.

From (3.4) we have that the Ψ-function can be expressed as a finite sum of incomplete Γ-functions, that is:

$$\Psi(\alpha,\beta,n;x) = \sum_{k=1}^{n} (-1)^{k+1} \binom{n}{k} k^{\beta k-\alpha} \Gamma(\alpha-\beta k,kx) \quad . \tag{5.1}$$

b) Case $\gamma = -1$, $\alpha = n+1$, $\beta = 0$.

From (1.1) we obtain:

$$\Psi(\alpha,\beta,-1;x) = -\int_x^\infty dt \, \frac{t^{\alpha-1}}{e^t t^\beta - 1} \quad . \tag{5.2}$$

Setting in (5.2) $\alpha=n+1$ and $\beta=0$, we get

$$\Psi(n+1,0,-1;x) = -D(n,x) \quad , \tag{5.3}$$

where

$$D(n,x) = \int_x^\infty dt \, \frac{t^n}{e^t-1} \tag{5.4}$$

is a function introduced by Debye (see Ref.7).

For $x=0$ and $n \geq 1$, the function given by (5.3) becomes

$$\Psi(n+1,0,-1;0) = -D(n,0) = -\int_0^\infty dt \, \frac{t^n}{e^t-1} = -n! \, \zeta(n+1) \tag{5.5}$$

where $\zeta(z)$ is the Riemann zeta function.

More generally, from (5.2) we deduce that

$$\Psi(\alpha,0,-1;0) = -\int_0^\infty dt \, \frac{t^{\alpha-1}}{e^t-1} = -\Gamma(\alpha)\,\zeta(\alpha) \quad , \tag{5.6}$$

for $\alpha > 1$.

At this stage we shall call *generalized Debye function*, the integral

$$\int_x^\infty dt \, \frac{t^{\alpha-1}}{e^t-1} \tag{5.7}$$

which appears on the right of (5.2) for $\beta=0$. Using the symbol $D(\alpha-1,x)$ to denote (5.7), we have

$$\Psi(\alpha,0,-1;x) = - D(\alpha,x) \tag{5.8}$$

from (5.2).

Remark 5.1- We note that one is able to evaluate the sum of the series

$$\sum_{n=1}^{\infty} \frac{\Gamma(n-\gamma)}{n!} n^{n\beta-\alpha} \Gamma(\alpha-n\beta,nx) \tag{5.9}$$

for any $\gamma=-n$, in terms of a combination of incomplete Debye functions and other known functions. Moreover, putting $\gamma=-1$ in (5.9) and taking into account (5.8), we get

$$D(\alpha-1,x) = \sum_{n=1}^{\infty} \frac{\Gamma(\alpha,nx)}{n^{\alpha}} \tag{5.10}$$

for each $x>0$.

In view of (5.6) and (5.8), we find now the well -known expansion for the Riemann zeta function:

$$\zeta(\alpha) \equiv - \frac{1}{\Gamma(\alpha)} \Psi(\alpha,0,-1;0) = \sum_{n=1}^{\infty} \frac{1}{n^{\alpha}} \tag{5.11}$$

for $\alpha>1$.

c) Case $\gamma=-1/2$, $\beta=0$.

For $\beta=0$ and $\gamma=-1/2$ the integral on the right of (1.1) exist for any $\alpha>1/2$ when $x=0$. Using then the series expansion (3.1), we have

$$\Psi(\alpha,0,-1/2;0) = -\Gamma(\alpha) \sum_{n=1}^{\infty} \frac{(2n-1)!!}{(2n)!!} \frac{1}{n^{\alpha}} \tag{5.12}$$

If we now define the function

$$Z(\alpha) = - \frac{1}{\Gamma(\alpha)} \Psi(\alpha,0,-1/2;0) \quad , \tag{5.13}$$

the relation (5.12) yields

$$Z(\alpha) = \sum_{n=1}^{\infty} \frac{(2n-1)!!}{(2n)!!} \frac{1}{n^{\alpha}} \tag{5.14}$$

for $\alpha>1/2$.

d) Case $\beta=0$, $\alpha>\max(0,-\gamma)$ $(\gamma\neq 0,1,2,...)$.

Both the series on the right of (5.11) and (5.14) can be considered as special cases of the more general series

$$\sum_{n=1}^{\infty} \frac{\Gamma(n-\gamma)}{\Gamma(-\gamma)n!} \frac{1}{n^{\alpha}} \quad , \tag{5.15}$$

which converges for any $\alpha>\max(0,-\gamma)$ (see Ref.1).

From the expression (3.4) we infer that the sum of the series (5.15) is given by the function $-\frac{1}{\Gamma(\alpha)}\Psi(\alpha,0,\gamma;0)$.

6. A FUNCTIONAL RELATION FOR THE POLYGAMMA FUNCTIONS REDERIVED

Certain properties of the Ψ-function can be fruitfully used in order to rederive some relations concerning known special functions. As an example, let us consider the polygamma functions:

$$\psi^{(n)}(x) = \frac{d^{n+1}}{dx^{n+1}} \ln \Gamma(x) \quad , \tag{6.1}$$

where $n=1,2,3,\ldots$ and $x\neq 0,-1,-2,\ldots$.

One can prove the following

Proposition 6.1 - *If* m *is a non-negative integer, then*

$$\psi^{(n)}(m+1) = (-1)^n n! \left\{ -\zeta(n+1) + \sum_{j=1}^{m} \frac{1}{j^{n+1}} \right\} \quad , \tag{6.2}$$

where $\zeta(n+1)$ *is the zeta Riemann function.*

The proof of (6.2) will be obtained with the help of a few theorems, which have been demonstrated in Refs.1 and 2. Specifically:

Theorem 6.2 - *Let* α, β *and* γ *be (real) arbitrary parameters and* n *any positive integer. Then the following transform holds:*

$$\begin{aligned}\int_x^{\infty} dt\, e^{-nt}\, \Psi(\alpha,\beta,\gamma;t) &= \frac{1}{n} e^{-nx}\, \Psi(\alpha,\beta,\gamma;x) \\ &- \frac{1}{n}\sum_{k=0}^{n} (-1)^k \binom{n}{k} \Psi(\alpha+n\beta,\beta,\gamma+k;x) \\ &+ \frac{1}{n}\sum_{k=1}^{n}\sum_{j=1}^{k} (-1)^{k+j+1} \binom{n}{k}\binom{k}{j} j^{(j-n)\beta-\alpha}\, \Gamma(-(j-n)\beta+\alpha, jx) \quad ,\end{aligned} \tag{6.3}$$

for $x>0$ *such that* $e^{-x}<x^{\beta}$.

Theorem 6.3 - *Let* α *and* β *be such that* $-e<\beta<0$ *and* $\alpha>-|\beta|$. *Then*

$$\Psi(\alpha,\beta,\gamma;0) = \int_0^\infty dt\ \Psi(\alpha-1,\beta,\gamma;t) \tag{6.4}$$

for any value of γ.

At this stage, let us consider the integral representation:

$$\psi^{(n)}(m+1) = (-1)^{n+1}\int_0^\infty dt\ \frac{t^n e^{-(m+1)t}}{1-e^{-t}} \quad . \tag{6.5}$$

Since

$$\frac{d}{dt}\Psi(\alpha,0,-1;t) = \frac{t^{\alpha-1}e^{-t}}{1-e^{-t}} \quad , \tag{6.6}$$

from (6.5) we have (for $\alpha=n+1$):

$$\psi^{(n)}(m+1) = (-1)^{n+1}\int_0^\infty dt\ e^{-mt}\frac{d}{dt}\Psi(n+1,0,-1;t) \quad . \tag{6.7}$$

Integrating by parts, (6.7) yields

$$\psi^{(n)}(m+1) = (-1)^{n+1}\Big[-\Psi(n+1,0,-1;0) + m\int_0^\infty dt\ e^{-mt}\ \Psi(n+1,0,-1;t)\Big] \quad . \tag{6.8}$$

In virtue of theorems 6.2 and 6.3 and using the recurrence relation (4.1), the integral on the right of (6.8) reads

$$m\int_0^\infty dt\ e^{-mt}\ \Psi(n+1,0,-1;t) = \sum_{k=1}^{m}(-1)^k\binom{m}{k}\frac{n}{k}\Psi(n,0,k;0) \tag{6.9}$$

Since

$$\Psi(n,0,k;0) = (n-1)!\sum_{j=1}^{k}(-1)^{j+1}\binom{k}{j}\frac{1}{j^n} \quad , \tag{6.10}$$

the integral (6.9) becomes

$$m\int_0^\infty dt\ e^{-mt}\ \Psi(n+1,0,-1;t) = n!\sum_{k=1}^{m}\frac{(-1)^k}{k}\binom{m}{k}\sum_{j=1}^{k}(-1)^{j+1}\binom{k}{j}\frac{1}{j^n} \quad . \tag{6.11}$$

By interchanging the summation in (6.11) and using the identity

$$\frac{1}{k}\binom{k}{j} = \frac{1}{j}\binom{k-1}{j-1} \quad , \tag{6.12}$$

we are led to the expression

$$m\int_0^\infty dt\, e^{-mt}\, \Psi(n+1,0,-1;t) = -n!\sum_{j=1}^{m}\frac{(-1)^{j+1}}{j^{k+1}}\sum_{k=j}^{m}(-1)^{k+1}\binom{m}{k}\binom{k-1}{j-1} \quad . \tag{6.13}$$

Recall now that (see Ref.2) if j and m are any pair of positive integers such that $1\le j\le m$, then

$$\sum_{k=j}^{m}(-1)^{k+1}\binom{m}{k}\binom{k-1}{j-1} = (-1)^{j+1} \quad . \tag{6.14}$$

Using the result (6.14), equation (6.13) can be written as

$$m\int_0^\infty dt\, e^{-mt}\, \Psi(n+1,0,-1;t) = -n!\sum_{j=1}^{m}\frac{1}{j^{n+1}} \quad . \tag{6.15}$$

Then, making the substitution (6.15) into (6.8), with the help of (5.11) we finally obtain the relation (6.2).

7. ASYMPTOTIC BEHAVIOUR FOR LARGE x

Let us recall the generalization of Poincaré's definition of an asymptotic expansion as given in Ref.8, viz:

<u>*Definition 7.1*</u> - *A sequence* $\{\phi_s(x)\}$ *of functions such that*

$$\frac{\phi_{s+1}(x)}{\phi_s(x)} \longrightarrow 0 \tag{7.1}$$

for $x\to+\infty$ *and any* $s=0,1,2,\ldots$ *is called an* <u>*asymptotic sequence*</u> *or* <u>*scale*</u> *as* $x\to+\infty$. *Consider now a scale* $\{\phi_s(x)\}$ *as* $x\to+\infty$, *and let* $f(x)$, $f_n(x)$ $(n=0,1,2,\ldots)$ *be functions such that*

$$\left|\frac{f(x) - \sum_{s=0}^{N-1} f_s(x)}{\phi_N(x)}\right| \tag{7.2}$$

is bounded for $x\to+\infty$ *and for every nonnegative integer* N.

Then $\sum_{s=0}^{\infty} f_s(x)$ *is said to be a* <u>*generalized asymptotic expansion*</u> *with respect to the scale* $\{\phi_s(x)\}$, *and one writes*

$$f(x) \sim \sum_{s=0}^{\infty} f_s(x) \quad ; \qquad \{\phi_s(x)\} \text{ as } x\to+\infty .$$

In Ref.1 we have proved the following:

Theorem 7.1 - *The function defined by (1.1) admits the asymptotic behaviour*

$$\Psi(\alpha,\beta,\gamma;x) \sim x^{\alpha-\beta-1} e^{-x} \sum_{s=0}^{\infty} A_s(x) \frac{1}{x^s} \quad , \tag{7.3}$$

for $x\to+\infty$, *in the generalized sense of Poincaré, with respect to the scale* $\{\frac{1}{x^s}\}$, *where*

$$A_s(x) = (-1)^{s+1} \frac{1}{\Gamma(-\gamma)} \sum_{m=0}^{\infty} \frac{\Gamma(m+1-\gamma)\Gamma(s-\alpha+(m+1)\beta+1)e^{-mx}}{(m+1)!(m+1)^{s+1}\Gamma(-\alpha+(m+1)\beta+1)x^{m\beta}} \quad , \tag{7.4}$$

the series on the right of (7.4) being uniformly convergent for any $x>\bar{x}$, *where* $\bar{x}$ *is such that the inequality* $e^{-x}<x^{\beta}$ *is verified.*

Theorem 7.1 can be applied to some interesting special cases. For example:

a) Asymptotic expansion of the incomplete Γ*-function.*

Setting $\gamma=1$ in (7.4) we obtain

$$A_s(x) = (-1)^s (1-\alpha+\beta)_s \tag{7.5}$$

where

$$(1-\alpha+\beta)_s = \frac{\Gamma(s+1-\alpha+\beta)}{\Gamma(1-\alpha+\beta)} . \tag{7.6}$$

Then (7.3) becomes

$$\Psi(\alpha,\beta,1;x) \equiv \Gamma(\alpha-\beta,x) \sim x^{\alpha-\beta-1} e^{-x} \sum_{s=0}^{\infty} (-1)^s \frac{(1-\alpha+\beta)_s}{x^s} \quad , \tag{7.7}$$

that is the known asymptotic expansion for the incomplete Γ-function for fixed $(\alpha-\beta)$ and large x (see Ref.9).

b) Asymptotic expansion for the incomplete Debye function.

Putting $\gamma=-1$ and $\beta=0$ into (7.4) we obtain

$$A_s(x) = (-1)^{s+1} \frac{\Gamma(s-\alpha+1)}{\Gamma(1-\alpha)} \sum_{m=0}^{\infty} \frac{e^{-mx}}{(m+1)^{s+1}} . \tag{7.8}$$

Then recalling the function[10]

$$\Phi(z,s,v) = \sum_{m=o}^{\infty} \frac{z^m}{(m+v)^s} , \tag{7.9}$$

defined for $|z|<1$, $v\neq 0,-1,-2,\ldots$, the series on the right hand side of (7.8) can be expressed by $\Phi(e^{-x},s+1,1)$, i.e.,

$$A_s(x) = (-1)^{s+1} (1-\alpha)_s \Phi(e^{-x},s+1,1) . \tag{7.10}$$

Using this result and the integral representation:

$$\Phi(z,s,v) = \frac{1}{\Gamma(s)} \int_o^{\infty} dt \frac{t^{s-1} e^{-vt}}{1-z e^{-t}} , \quad \mathrm{Re}\, v > 0 , \tag{7.11}$$

from (7.3) and (5.8) we obtain

$$D(\alpha-1,x) \sim x^{\alpha-1} e^{-x} \sum_{s=o}^{\infty} (-1)^s \frac{(1-\alpha)_s}{s! x^s} \int_o^{\infty} dt \frac{t^s e^{-t}}{1-e^{-(x+t)}} , \tag{7.12}$$

for fixed α and large values of $x>0$.

8. SOME FUTURE OBJECTIVES

To conclude this lecture, we shall sketch briefly some further developments involving the Ψ-function.

Specifically:

i) One could try to extend to the complex plane all the results obtained for $\Psi(\alpha,\beta,\gamma;x)$ on the real line. Remember that we have introduced a series, namely (see (5.15)):

$$\sum_{n=1}^{\infty} \frac{\Gamma(n-\gamma)}{\Gamma(-\gamma)n!} \frac{1}{n^{\alpha}} ,$$

which can be considered as a kind of generalization of the Riemann series $\sum_{n=1}^{\infty} \frac{1}{n^{\alpha}}$. Therefore, for the function defined by the series (5.15) it should be interesting (but of course very difficult) to tackle the problem of finding the distribution of zeros in the α-complex plane (for any fixed γ).

ii) On the basis of the properties of the Ψ-function, one might investigate whether this can be interpreted in the light of Group Theory, as is possible for a wide class of known functions (see, for instance, Ref.11).

REFERENCES

1 M.Leo, R.A.Leo and G.Soliani, "On a special function related to a class of certain nonlinear wave equations", 1979, to be submitted for publication.

2 M.Leo, R.A.Leo and G.Soliani, "Some theorems concerning a new special function", *Quaderni dell'Istituto di Matematica dell'Università di Lecce*, 1979.

3 A.C.Scott, F.Y.F.Chu and D.W.Mc Laughlin, "The soliton: a new concept in applied science", *Proceedings IEEE* 61 (1973), pp.1443-1483.

4 V.G.Makhankov, "Dynamics of classical solitons (in non-integrable systems)", *Physics Reports* 35 (1978), pp.1-128.

5 W.F.Ames, "Nonlinear ordinary differential equations in transport processes", Academic Press, New York, 1968, p.42 and p.101.

6 H.T.Davis, "Introduction to nonlinear differential and integral equations", Dover Publications, New York, 1962, p.20.

7 M.Abramowitz and I.A.Stegun, "Handbook of mathematical functions", Dover, New York, 1965, p.998.

8 F.W.J. Olver, "Asymptotic and special functions", Academic Press, New York and London, 1974, p.25.

9 F.G.Tricomi, "Funzioni ipergeometriche confluenti", Edizioni Cremonese, Roma, 1954, p.174.

10 I.S.Gradshteyn and I.M.Ryzhik, "Table of integrals, series and products", Academic Press, N.Y., 1965, p.1075.

11 N.Ja.Vilenkin, "Fonctions spéciales et théorie de la représentation des groupes", Dunod, Paris, 1969.

VORTEX MOTIONS AND CONFORMAL MAPPINGS

Jacob Burbea

Department of Mathematics,

University of Pittsburgh

Pittsburgh-Pennsylvania 15260

ABSTRACT

An evolution equation describing vortex motions of invariant curves is established. Exact solutions of this equation, generalizing those of Kirchoff and Moore and Saffman, are found. A non linear dispersion relation extending a classical result of Lamb is demonstrated. Other results are proven.

1. INTRODUCTION

Analytic methods for determining stationary states of motions of fluid with zero vorticity except in the interior of certain Jordan curves are presently not available. An exception is Kirchoff's (1876) solution for the elliptic region $E=\{(x,y):x^2/a^2+y^2/b^2<1\}$. Here, E is a region with constant vorticity ω_0 while in the exterior of E the fluid is irrotational and is at rest at infinity. This region rotates without change of shape with a constant angular velocity $\Omega=\omega_0 ab/(a+b)^2$. The first extensive investigation of this stade of motion, called the "Kirchoff vortex", was conducted by Hill (1884). The stability analysis of this motion was given by Love (1893). These and other related results may be found in the monograph of Lamb [5, pp.230-233] and in Love's paper [6]. A generalization of this vortex motion was recently given by Moore and Saffman

[7], where a strain was imposed on the Kirchoff vortex.

In the previously described works the whole analysis depend heavily, and crucially, on the use of elliptic coordinates and skilfully matching, on the elliptic boundary, the interior and the exterior stream functions. It is therefore not surprising that one is not able analytically, by appealing only to the above methods, to include or exclude the possibility of a steady periodic solution of non-elliptic form. However, Lamb [5,p. 231] noted that for a nearly circular region whose boundary is given by $r=r_0+\varepsilon\cos m\theta$ rotates, when $\varepsilon<<1$ with an angular velocity $\Omega_m=(\omega_0/2)(m-1)/m$, where ω_0 is the vorticity and $m\geq 2$ is the wave number. This shows that a circular vortex with radius r_0 and constant vorticity ω_0 is stable and moreover that such a vortex supports infinitesimal azimuthal waves with the linear dispersion relation $\nu_m=m\Omega_m=(1/2)\omega_0(m-1)$. When $m=2$, the disturbed region is an ellipse rotating about its center with an angular velocity $\omega_0/4$. This, of course, is consistent with Kirchoff's exact solution. When $m>2$, the disturbed region is an m-fold symmetric hypotrochoid.

In spite of the fact that the solution, for $m>2$, of these motions is approximate it nevertheless provides a significant motivation. Indeed, Zabusky [8] in discussing this problem was led to conjecture that if the amplitude ε were made small-but-finite, these azimuthal waves would have nonlinear-dispersive or perhaps soliton-like properties. This paper, amongst other things, provides a proof for the first part of Zabusky's conjecture (see Theorem 6 for more details). Quite recently, Deem and Zabusky [3] employed a "contour dynamics" algorithm to find numerically a class of stable, rotating and uniformly translating solutions to the Euler equations in two dimensions. Their investigation is based on a nonlinear integro-differential equation for the boundary of the vortex that contains the rotational period or the translating velocity as a bifurcation parameter.

The above mentioned equation is, by all means, of non-elementary type and, in fact, a very complicated one. It is expressed in terms of polar coordinates and it is difficult to solve it analytically even if the boundary is elliptic. Moreover, the equation in its present form is not suitable for dealing with the more involved problem concerning the stability analysis of the steady state solutions.

Our main contributions in this paper are as follows: we employ methods from the theory of functions and conformal mapping to derive an effective evolution equation for motions with piecewise constant vorticity. This equation, in case of a simple vortex, is expressed via the conformal mapping ϕ of the exterior of the unit disk onto the exterior of the vortex where, the angular velocity Ω, the strain rate, of order $m\geq 2$, $2m^{-1}e$ and the shear rate, of order $m\geq 2$, $2m^{-1}\alpha$ serve as bifurcation parameters. With this equation we are able to find exact solutions of the form $\phi(e^{i\theta})=a(e^{i\theta}+b^{-i(m-1)\theta})$, where $a\neq 0$ is any complex number and where $b=b(t)$ is a complex valued function of the time t, subject to the condition that $|b|\leq(m-1)^{-1}$, $m\geq 2$. The case $m=2$ corresponds to an ellipse and we recover the Kirchoff solution and generalized the solutions of [7]. Even in this special case of 2-fold symmetry we obtain infinitely many more solutions then those found by Moore and Saffman [7]. For example, when $\Omega=0$ the ellipse $(m=2)$ is a steady state solution if and only if

$$|\omega_0|^{-1}(e^2+\alpha^2)^{\frac{1}{2}} \leq (\sqrt{5}-2)^{\frac{1}{2}}(3-\sqrt{5})/2(\sqrt{5}-1) \cong 0.15 .$$

This result is in agreement with that of [7] where the special case of $\alpha=0$ is considered. When $m\geq 3$ the resulting curve is an m-fold symmetric hypotrochoid and it is an exact solution of the evolution equation when strain and shear of order m are present while rotation is absent. We also show the existence of periodic solutions which bifurcate the trivial solution of the circle (i.e., when $b=0$). Moreover, the non linear dispersive relation generalizing the linearized solutions of Lamb [5, p.231] is established. Other related results are proven.

Our evolution equation yields a very effective and a short way in dealing with the problem of the stability of a steady state solution. However, this along with other relevant results will be elaborated in our forthcoming paper [2]. Also, the intimate connection of vortex motions with the theory of univalent functions and the classical Fredholm eigenvalue problem will be discussed elsewhere.

2. EULER EQUATIONS

In this section we briefly review a fairly general problem in two-dimensional fluid dynamics. The velocity of the fluid is denoted by $\vec{u}=(u,v)$ with $u=u(x,y)$, $v=v(x,y)$, (x,y) $\mathbb{R}^2$. We shall assume that the fluid is incompressible which means that

$$\operatorname{div} \vec{u} = \partial_x u + \partial_y v = 0 .$$

This shows the existence of a stream function $\psi=\psi(x,y)$ with

$$\partial_x \psi = v , \qquad \partial_y \psi = -u \tag{2.1}$$

Here

$$\operatorname{curl} \vec{u} = (0,0,\partial_x v - \partial_y u) = (0,0,\Delta\psi)$$

and the vorticity of the flow is $\xi \equiv \Delta\psi$. The vorticity must satisfy

$$d_t \xi \equiv \partial_t \xi + u\partial_x \xi + v\partial_y \xi = 0 \tag{2.2}$$

where ∂_t denotes differentiation with respect to the time t.

Let $D=D(t)$ be a region of vorticity at time t. This means that D is a plane region whose area Lebesgue measure is finite and

$$\Delta\psi \equiv \xi = \chi_{\overline{D}}\omega - f \tag{2.3}$$

where $f=f(x,y)$ is a C^1-function in $\mathbb{R}^2$, $\omega=\omega(x,y)$ is a C^1-function in D and $\chi_{\overline{D}}$ is the characteristic function of $\overline{D}$, the closure of D.

The customary definition of a region of vorticity or a vortex is, of course, when $f\equiv 0$ but we will find it more convenient to let f be an arbitrary function taylored for the specific problems we shall be considering.

In general the region D is not connected, for sake of simplicity, however, we shall assume that the boundary $C=\partial D$ is of class C^1. This means that each boundary component is described by $F(x,y,t)=0$, where $F=F(x,y,t)$ is a defining function for the boundary components and is of class C^1 in a neighborhood containing the boundary. As in (2.2) we must also assume that on the boundary of each vortex (or each component of D) we have

$$d_t F \equiv \partial_t F + u\partial_x F + v\partial_y F = 0 , \tag{2.4}$$

which means that the vortices always contain the same particles.

Equations (2.1)-(2.3) constitute the familiar Euler equations in two-dimensions and they are written here as a Hamiltonian system in terms of the vorticity and stream function . To solve this system one must take into consideration equation (2.4) and also that the pressure is continuous when crossing the boundary of the vortex. It is, of course, implicitly assumed here that we have a continuous motion, that is, $\vec{u}$ is continuous in crossing ∂D.

When dealing with these equation it is convenient to introduce some complex notation. For $z=x+iy\in\mathbb{C}$ we use

$$\partial_z = \tfrac{1}{2}(\partial_x - i\partial_y) , \qquad \partial_{\bar{z}} = \tfrac{1}{2}(\partial_x + i\partial_y)$$

and thus

$$\Delta = 4\partial_z \partial_{\bar{z}}$$

For $\zeta\in\mathbb{C}$ we write

$$\psi_0(\zeta) = \frac{1}{2\pi}\int_D \omega(z)\log|z-\zeta|d\sigma(z) , \qquad (2.5)$$

where $d\sigma(z)=dxdy$ is the area Lebesgue measure. This solves Poisson's equation $\Delta\psi_0(z)=\chi_{\bar{D}}(z)\omega(z)$, $z\in\mathbb{C}$, and, of course, $\partial_z\psi_0(\infty)=0$. The velocity induced by this stream function is given by $u_0+iv_0=2i\partial_{\bar{z}}\psi_0$ and is at rest at infinity. Clearly, ψ_0 is of class C^1 across ∂D as well as in the entire complex plane $\mathbb{C}$. It is also well known that (2.5) gives (up to a constant) the unique solution of a continuous motion which is rotational in the interior of D with vorticity $\omega(z)$, $z\in D$, irrotational in the exterior of D and which is at rest at infinity.

Let E_D be the exterior of D and let $A=A(D)=\int_D d\sigma(z)$ be the area of D. Plainly, ψ_0 is the real part of a multi-valued holomorphic function in E_D while $\partial_z\psi_0$ is a single-valued holomorphic function in E_D. The circulation around D is by definition

$$\kappa = \frac{1}{2\pi}\int_D \omega(z)d\sigma(z) .$$

We shall assume that $\omega=\omega(z)$ and $f=f(z)$ are independent of t which imply that the circulation is constant on each vortex and that (2.2) reduces to

$$u\partial_x\xi + v\partial_y\xi = 0 \qquad . \qquad (2.2')$$

With this assumption the pressure becomes continuous in crossing the boundary of the vortex. This continuity is secured by virtue of Bernoulli's theorem for, here the motion is continuous and the circulation is constant on each vortex (see Lamb [5, p.246]).

In the special case that $f\equiv 2\Omega$, where Ω is a real constant, and when ω is a piecewise constant whe have the further reduction that equation (2.2') is identically satisfied. Moreover, if D_j denotes a component of D then $\omega(z)\equiv\omega_j$ for all $z\in D_j$ with ω_j being a real constant and thus

$$\kappa = \frac{1}{2\pi}\sum_j \omega_j A_j \quad ,$$

where $A_j = A(D_j)$ is the area of the component D_j whose vorticity is the constant $\omega_j - 2\Omega_j$.

We close this section with the following interesting distortion theorem:

Theorem 1. - *Let* $D=\cup D_j$ *be a region of constant vorticity* ω_j *on each component* D_j *and let* $A_j=A(D_j)$ *be the area of* D_j. *Assume that the vortex* D *is in a fluid which is at rest at infinity with a flow field* $\vec{u}=(u,v)$. *Then*

$$|\vec{u}|^2 = u^2 + v^2 \le \frac{A}{4\pi}\sum_j \omega_j^2 \qquad ; \qquad A=\sum_j A_j \ ,$$

everywhere. Equality, at one point, holds if and only if D *is a disk whose radius is* $\sqrt{A/\pi}$ *with the above point being on the circumference of the disk.*

Proof. Under the above assumptions we must have

$$i(u-iv) = 2\partial_z \psi(z)$$

with

$$\psi(z) = \psi_0(z) + B = \frac{1}{2\pi}\sum_j \omega_j \int_{D_j} \log|\zeta - z|\, d\sigma(\zeta) + B \quad ,$$

where B is a real constant. Here, differentiation is allowed even if $z\in D$ and therefore

$$i(u-iv) = -\frac{1}{2\pi}\sum_j \omega_j \int_{D_j} \frac{1}{\zeta - z}\, d\sigma(\zeta) \quad .$$

Consequently

$$|\vec{u}| \le \frac{1}{2\pi}\sum_j |\omega_j|\,|\mu_j(z)| \qquad ; \quad \mu_j(z) \equiv \int_{D_j} \frac{1}{\zeta - z}\, d\sigma(\zeta) \quad .$$

By a theorem of Ahlfors-Beurling [1], $|\mu_j(z)| \le \sqrt{\pi A_j}$ for every $z\in\mathbb{C}$ with equality, at one point z, if and only if D_j is a disk whose radius is $\sqrt{A_j/\pi}$ and z must be on the boundary of this disk. Hence

$$|\vec{u}|^2 \le \frac{1}{4\pi}\left(\sum_j |\omega_j|\, A_j^{\frac{1}{2}}\right)^2 \le \frac{1}{4\pi}\left(\sum_j \omega_j^2\right)\left(\sum_j A_j\right)$$

which is the required inequality. The statement about equality follows now in an obvious manner.

3. THE EVOLUTION EQUATION

Having made the previous assumptions, namely the continuity of the motion and the time independence of the vorticity the problem becomes purely kinematic for, the dynamical conditions are identically satisfied. Indeed, the Euler equations are now reduced to (2.1),(2.2') ,(2.3) and (2.4).

Equation (2.2') is equivalent to

$$-\frac{\partial(\psi,\xi)}{\partial(x,y)} = 2i\,\frac{\partial(\psi,\xi)}{\partial(z,\bar{z})} = -4\,\mathrm{Im}\left[\partial_z\psi\,\partial_{\bar{z}}\xi\right] = 0$$

or

$$\mathrm{Im}\left[\partial_z\psi\,\partial_{\bar{z}}f\right] = \chi_{\bar{D}}(z)\,\mathrm{Im}\left[\partial_z\psi\,\partial_{\bar{z}}\omega\right] . \qquad (3.1)$$

Equation (2.4) means that the normal component of the velocity of the boundary of the vortex is equal to the normal component of the fluid on the boundary of the vortex. Let $z=x+iy$, $z=z(s,t)$ (s is the arc lenght) be the parametrization of the boundary of the vortex $C=C(t)$. The velocity of the boundary and the (outer) normal to the boundary are given by

$$\vec{u}_b = (\dot{x},\dot{y}) , \qquad \vec{n} = (y',-x') ,$$

where $\dot{x}=dx/dt$ and $x'=dx/ds$. Equation (2.4) is therefore equivalent to

$$\vec{u}_b\cdot\vec{n} = \vec{u}\cdot\vec{n} ,$$

on the boundary. However,

$$\vec{u}_b\cdot\vec{n} = -\mathrm{Re}\left[i\dot{\bar{z}}z'\right]$$

and by (2.1),

$$\vec{u}\cdot\vec{n} = -\frac{\partial\psi}{\partial s} = -2\mathrm{Re}\left[\partial_z\psi z'\right] .$$

Therefore,

$$\tfrac{1}{2}\mathrm{Re}\left[i\dot{\bar{z}}z'\right] = \mathrm{Re}\left[\partial_z\psi z'\right] \qquad (3.2)$$

on the boundary C.

This is an evolution equation on the parametrization $z=z(s,t)$ of the boundary $C(t)$ of the vortex $D(t)=UD_j$ whose vorticity is $\chi_{\bar{D}}\omega-f$. This evolution equation must be satisfied on each boundary component of C.

A time independent solution of this equation is called a steady state solution. It is a solution of the steady state equation

$$\mathrm{Re}\left[\partial_z \psi z'\right] = 0 \quad , \qquad z=C, \tag{3.3}$$

or that $\psi=k_j$ on C_j, where k_j is a real constant depending on the boundary component C_j.

The problem is now reduced to solving (3.2) or (3.3) subject to (3.1). However, we shall employ a further reduction by assuming piecewise constant vorticity. More specifically, we choose $f=2\Omega$ where Ω is a real constant and we shall assume that $\omega=\omega_j$, ω_j being a real constant, on each component D_j. In this case equation (3.1) is automatically satisfied and we are left only with equation (3.2) or (3.3).

Another reduction which will employed here is by insisting that the stream function is of the form

$$\psi(z) = \psi_0(z)-\tfrac{1}{2}\Omega|z|^2+\frac{2}{m^2}e\,\mathrm{Re}(z^m)+\frac{2}{m^2}\alpha\,\mathrm{Im}(z^m)+v_1\mathrm{Im}z-v_2\mathrm{Re}z+B , \tag{3.4}$$

where $m\geq 2$ is an integer, $\Omega, e, \alpha, v_1, v_2$ and B are real constants, and ψ_0 is as in (2.5). This choice is clearly consistent with our vorticity $\xi=\chi_{\bar{D}}\omega-2\Omega$, where $\omega=\omega_j$ on each D_j. The harmonic terms in (3.4) determine the velocity of the fluid at infinity. The physical meaning of the "bifurcation parameters" appearing in (3.4) are as follows: Ω is the angular velocity of the boundary, $2m^{-1}e$ is a strain rate of order m, $2m^{-1}\alpha$ is a shear rate of order m, v_1 is a translation velocity along the x-axis, v_2 is a translation velocity along the y-axis while the constant B has no significant meaning (see also the work of Moore and Saffman [7] for the case m=2).

The non trivial significance of the translation velocity (v_1,v_2) is only when D is not connected. Also, the presence of each individual term due to Ω,e or α is essentially the same. It is however of interest, as far as closed form solutions are concerned, to have a unified approach to the totality of all terms. A somewhat similar approach for the special case m=2 was also considered in [7].

In this paper we shall confine ourselves to a simple vortex D with vorticity $\omega=\omega_0$ where ω_0 is a real constant. Therefore, D is a Jordan domain whose vorticity is $\omega_0-2\Omega$ while the vorticity of it's exterior is

-2Ω. In this case there is no loss of generality in assuming $(v_1,v_2)=0$ and therefore

$$\psi(z) = \psi_0(z) - \tfrac{1}{2}\Omega|z|^2 + \frac{2}{m^2} e\,\mathrm{Re}(z^m) + \frac{2}{m^2}\alpha\,\mathrm{Im}(z^m) + B \qquad (3.5)$$

with

$$\psi_0(z) = \frac{\omega_0}{2\pi}\int_D \log|\zeta-z|\,d\sigma(\zeta) \ . \qquad (3.6)$$

From (3.5) we obtain

$$\partial_z\psi(z) = \partial_z\psi_0(z) - \tfrac{1}{2}\Omega\bar{z} + \frac{e-i\alpha}{m} z^{m-1} \qquad (3.7)$$

We now evaluate $\partial_z\psi_0$. Due to the removable singularity of the integral in (3.6) we have

$$\partial_z\psi_0(z) = -\frac{\omega_0}{4\pi}\int_D \frac{1}{\zeta-z}\,d\sigma(\zeta)$$

which holds for all $z\in\mathbb{C}$. A use of Green's formula shows that

$$\partial_z\psi_0(z) = \frac{\omega_0}{4}\left[\bar{z} - \frac{1}{2\pi i}\int_C \frac{\bar{\zeta}}{\zeta-z}\,d\zeta\right] \quad , \ z\in\bar{D} \qquad (3.8)$$

and

$$\partial_z\psi_0(z) = -\frac{\omega_0}{4}\,\frac{1}{2\pi i}\int_C \frac{\bar{\zeta}}{\zeta-z}\,d\zeta \quad , \ z\in E_D \ . \qquad (3.9)$$

Noting the Plemelj saltus theorem, expressions (3.8) and (3.9) exhibit the well-known fact that $\partial_z\psi_0$ is continuous across the boundary $C=\partial D$.

From (3.7) and (3.8) we have, for $z\in\partial D$,

$$\partial_s\psi(z) = 2\mathrm{Re}\left[z'\partial_z\psi\right] = \frac{\omega_0}{2}\mathrm{Re}\left\{z'\left[\lambda\bar{z} + \frac{2}{m}\mu z^{m-1} - \frac{1}{2\pi i}\int_C \frac{\bar{\zeta}}{\zeta-z}\,d\zeta\right]\right\} \qquad (3.10)$$

where

$$\lambda = 1 - \frac{2\Omega}{\omega_0} \quad , \qquad \mu = \frac{2}{\omega_0}(e-i\alpha) \ . \qquad (3.11)$$

Integration of (3.10) yields

$$\psi(z) = \frac{\omega_0}{2}\left[\frac{\lambda}{2}|z|^2 + \frac{2}{m^2}\mathrm{Re}(\mu z^m) + \mathrm{Re}\,S(z)\right] + B \ , \ z\in\bar{D} \qquad (3.12)$$

where B is a constant and

$$S(z) = \frac{1}{2\pi i}\int_C \bar{\zeta}\log\left(1-\frac{z}{\zeta}\right)d\zeta \quad , \ z\in\bar{D} \ . \qquad (3.13)$$

Here, we have assumed, for simplicity, that $0\in D$.

Similarly,

$$\psi(z) = \frac{\omega_0}{2}\left[-\frac{\Omega}{\omega_0}|z|^2 + \frac{2}{m^2}\mathrm{Re}(\mu z^m) + \kappa\log|z| + \mathrm{Re}T(z)\right] + B, \quad z\in E_D, \qquad (3.14)$$

where $\kappa = A/2\pi$ is the circulation and

$$T(z) = \frac{1}{2\pi i}\int_C \bar{\zeta}\log\left(1-\frac{\zeta}{z}\right)d\zeta \quad , \quad z\in E_D \quad . \tag{3.15}$$

The evolution equation (3.2), with the aid of (3.10), may now be written as

$$\mathrm{Re}[z' i \dot{\bar{z}}] = \frac{\omega_0}{2}\,\mathrm{Re}\left\{z'\left[\lambda\bar{z} + \frac{2}{m}\mu z^{m-1} - H(z)\right]\right\} , \qquad z\in C , \tag{3.16}$$

where

$$H(z) = \frac{1}{2\pi i}\int_C \frac{\bar{\zeta}}{\zeta - z}\,d\zeta \quad . \tag{3.17}$$

It is to be noted here that all singular integrals are taken in the principal value sense. The steady state equation has, by virtue of (3.12) and (3.16), the following form

$$\tfrac{1}{2}\lambda|z|^2 + \frac{2}{m^2}\mathrm{Re}(\mu z^m) + \mathrm{Re}\,S(z) = \mathrm{const.} , \qquad z\in C \quad . \tag{3.18}$$

It is convenient to note the following expressions:

$$H(z) = \lim_{n\to\infty} H_n(z) \quad , \quad H_n(z) = \sum_{k=1}^{n} r_k z^{k-1} \quad , \tag{3.19}$$

and

$$S(z) = \lim_{n\to\infty} S_n(z) \quad , \quad S_n(z) = -\sum_{k=1}^{n} k^{-1} r_k z^k , \tag{3.20}$$

where

$$r_k = \frac{1}{2\pi i}\int_C \bar{\zeta}\zeta^{-k}\,d\zeta \quad ; \qquad k=1,2,\ldots \quad . \tag{3.21}$$

4. EXACT SOLUTIONS

We assume that the Jordan domain D contains the origin as an interior point. The boundary C will be parametrized by means of $z=a\phi(\omega)$, $\omega\in\Gamma=\{\omega : |\omega|=1\}$, where $a\neq 0$ and

$$\phi(\omega) = \omega + \sum_{n=1}^{\infty} b_n \omega^{-n} \quad . \tag{4.1}$$

We may suppose that $a\phi(\omega)$ maps conformally the exterior of the unit disk Δ, $E_\Delta=\{\omega : |\omega|>1\}$ onto the exterior of D, E_D. In this case $a=g'(\infty)\neq 0$, $g(\omega)=a\phi(\omega)$, and thus $|a|$ is the *mapping radius* or the *capacity* of D. The area of D is given by

$$A = A(D) = \pi|a|^2\left[1 - \sum_{n=1}^{\infty} n|b_n|^2\right] \geq 0 \quad , \tag{4.2}$$

and therefore $\sum_{n=1}^{\infty} n|b_n|^2 \le 1$. This shows that $|b_n| \le n^{-\frac{1}{2}}$; n=1,2,... . We shall always assume that a is independent of t. In this case the evolution equation may be written as

$$\mathrm{Re}\left[\omega\phi'\dot{\bar{\phi}}\right] = \frac{\omega_0}{2}\,\mathrm{Im}\left\{\omega\phi'\left[\lambda\bar{\phi} + \frac{2}{m}\mu\frac{a^{m-1}}{\bar{a}}\phi^{m-1} - H(\phi)\right]\right\}, \quad \omega \in \Gamma, \tag{4.3}$$

where

$$\phi = \phi(\omega) \quad , \quad H(\phi) = H(\phi(\omega)) = \frac{1}{2\pi i}\int_{\Gamma} \frac{\overline{\phi(\tau)}}{\phi(\tau) - \phi(\omega)}\,\phi'(\tau)\,d\tau \ . \tag{4.4}$$

Here, in view of (4.1),

$$\dot{\bar{\phi}}(\omega) = \sum_{n-1}^{\infty} \dot{\bar{b}}_n \omega^n \ , \quad \omega \in \Gamma \quad ; \quad \dot{b}_n = b_n'(t) \ , \quad n=1,2,\ldots \ .$$

We seek solutions for which the area of the vortex D remains invariant with time. This is accomplished, in view of (4.2), by requiring that $b_n(t) = d_n e^{i\nu_n(t)}$, where the d_n are time independent and $\nu_n(t)$ are real valued functions of t. In this case

$$\dot{\phi}(\omega) = i\sum_{n=1}^{\infty} \dot{\nu}_n(t) b_n \omega^{-n} \ .$$

Especially, when $\nu_n(t) = \nu(t)$, n=1,2,..., equation (4.3) admits the form

$$\dot{\nu}(t)\,\mathrm{Im}\{\omega\phi'\left[\bar{\phi} - \omega^{-1}\right]\} = \frac{\omega_0}{2}\mathrm{Im}\left\{\omega\phi'\left[\lambda\bar{\phi} + \frac{2}{m}\mu\frac{a^{m-1}}{\bar{a}}\phi^{m-1} - H(\phi)\right]\right\}, \omega \in \Gamma \quad , \tag{4.5}$$

which can be also written as

$$\dot{\nu}(t)\left[\frac{1}{2}\frac{d}{|d\omega|}|\phi|^2 - \mathrm{Im}(\phi')\right] = \frac{\omega_0}{2}\frac{d}{|d\omega|}\left[\frac{\lambda}{2}|\phi|^2 + \frac{2}{m^2}\mathrm{Re}\left(\frac{a^{m-1}}{\bar{a}}\mu\phi^m\right) + \mathrm{Re}S(\phi)\right], \ \omega \in \Gamma \ , \tag{4.6}$$

with

$$S(\phi) = S(\phi(\omega)) = \frac{1}{2\pi i}\int_{\Gamma} \overline{\phi(\tau)}\,\log\left(1 - \frac{\phi(\omega)}{\phi(\tau)}\right)\phi'(\tau)\,d\tau \ . \tag{4.7}$$

The steady state equation is obtained from (4.6) by requiring that $\dot{\nu}(t)=0$ for all t.

For the integer $m \ge 2$ we consider

$$\phi_m(\omega) = \omega + b_{m-1}\omega^{-(m-1)} \quad ; \quad |b_{m-1}| \le (m-1)^{-1} \ . \tag{4.8}$$

The function $a\phi_m(\omega)$ maps E_Δ conformally onto the exterior E_{D_m} of the m-fold symmetric *hypotrochoid* D_m. When $|b_m| = (m-1)^{-1}$, D_m becomes an *hypocycloid* with m cusps. In the special case that m=2, D_2 is an ellipse. The area of D_m is given, in view of (4.2), by

$$A_m = A(D_m) = \pi a^2\left[1 - (m-1)|b_{m-1}|^2\right] ,$$

where in the hypocycloid case

$$A_m = \pi a^2 \frac{m-2}{m-1}, \qquad |b_{m-1}| = (m-1)^{-1}.$$

The trivial case $b_{m-1}=0$ corresponds to a disk centered at the origin and whose radius is $|a|$. In this case μ in (4.6) must be zero while λ and $\nu(t)$ are arbitrary. We therefore, assume that $b_{m-1}\neq 0$. A simple calculation shows that $S(\phi_m)$ of (4.7) admits the following expressions:

$$S(\phi_m) = -m^{-1}\,\bar{b}_{m-1}(\phi_m)^m, \qquad \phi_m = \phi_m(\omega) \quad . \tag{4.9}$$

Also

$$\mathrm{Im}(\phi_m') = -\frac{m-1}{m}\frac{d}{|d\omega|}\,\mathrm{Re}(\bar{b}_{m-1}\omega^m) \quad . \tag{4.10}$$

Upon substituting (4.8)-(4.10) in (4.6) we obtain

$$-\dot{\nu}(t)\frac{1}{m}\bar{b}_{m-1} = \frac{\omega_0}{2}\left[\lambda\bar{b}_{m-1} + \frac{1}{m}(K_m + \frac{m(m-1)}{2}\bar{K}_m\bar{b}^2_{m-1})\right], \quad m \geq 2 \quad , \tag{4.11}$$

with $K_m=0$ for $m\geq 3$. Here

$$K_m = \frac{2}{m}\frac{a^m}{|a|^2}\mu - \bar{b}_{m-1} \quad . \tag{4.12}$$

We first treat the case $m\geq 3$. In this case $K_m=0$, or

$$\mu = \frac{m}{2}\frac{|a|^2}{a^m}\bar{b}_{m-1} = \frac{m}{2}\frac{|a|^2}{a^m}\bar{d}_{m-1}\, e^{i\nu(t)} \quad ,$$

and furthemore

$$\dot{\nu}(t) = -\frac{m}{2}\lambda\omega_0 \quad .$$

Writing, $a=|a|e^{i\theta_a}$, $d_{m-1}=|d_{m-1}|e^{i\theta_b}=|b_{m-1}|e^{i\theta_b}$ we obtain the following result:

<u>*Theorem 2.*</u> - *Let* $m\geq 3$. *Then the m-fold symmetric hypotrochoid* D_m *whose boundary is described by* $z=a\phi_m(\omega)$ *with* $\phi_m(\omega)$ *as in* (4.8) *is an exact solution of Euler equation* (4.6). *Furthemore,*

$$\nu(t) = -\frac{m}{2}\omega_0\lambda t + C,$$

where C *is an arbitrary constant and* λ *is given by* (3.11) , *i.e.,* $\lambda=1-2\Omega/\omega_0$, *where* Ω *is an arbitrary constant. Also, using* (3.11),

$$e = \frac{m}{4}\omega_0\frac{|b_{m-1}|}{|a|^{m-2}}\cos(m\theta_a+\theta_b+\nu(t))$$

and

$$\alpha = \frac{m}{4}\omega_0 \frac{|b_{m-1}|}{|a|^{m-2}} \sin(m\theta_a + \theta_b + \nu(t)) \quad ,$$

and thus the strain and the shear are functions of t *with*

$$e^2 + \alpha^2 = \frac{m^2}{16}\omega_0^2 \frac{|b_{m-1}|^2}{|a|^{2m-4}} \leq \frac{1}{16}\left(\frac{m}{m-1}\right)^2 \omega_0 |a|^{-2m+4} \quad .$$

The above solution is also a steady state solution provided $\lambda=0$, *or* $\Omega=\omega_0/2$. *Moreover,*

$$\mu = \frac{m}{2}\frac{|a|^2}{a^m}\bar{b}_{m-1} \; ; \qquad \mu = \frac{2}{\omega_0}(e-i\alpha) \; ,$$

where b_{m-1} *is independent of* t.

When m=2 we have by virtue of (4.11) and (4.12),

$$-\dot{\nu}(t)\frac{1}{2}\bar{b} = \frac{\omega_0}{2}\left[\lambda\bar{b} + \tfrac{1}{2}(K_2 + \bar{b}^2\bar{K}_2)\right] \; ,$$

where

$$b = b_1 \; , \qquad K_2 = \frac{a^2}{|a|^2}\mu - \bar{b} \quad .$$

Therefore,

$$-\omega_0^{-1}\dot{\nu}(t) = \lambda + \tfrac{1}{2}(K_2\bar{b}^{-1} + \bar{K}_2\bar{b})$$

or, using (3.11),

$$-2\omega_0^{-1}\dot{\nu}(t) = 1 - |b|^2 - 4\frac{\Omega}{\omega_0} + \frac{1}{|a|^2}(a^2\mu\bar{b}^{-1} + \bar{a}^2\bar{\mu}\bar{b}) . \qquad (4.13)$$

This shows that $a^2\mu\bar{b}^{-1}+\bar{a}^2\bar{\mu}\bar{b}$ is a real number which implies that

$$a^2\mu b(1-|b|^2) = \bar{a}^2\bar{\mu}\bar{b}(1-|b|^2) \quad . \qquad (4.14)$$

Equation (4.14) gives two cases: (i) $|b|=1$, i.e., a vortex sheet and (ii) $|b|<1$, i.e., an elliptic vortex. In the first case, (4.14) is trivially satisfied and equation (4.13) reduces to

$$-2\omega_0^{-1}\dot{\nu}(t) = -4\frac{\Omega}{\omega_0} + \frac{1}{|a|^2} 2\mathrm{Re}(a^2\mu b) \quad .$$

This, using (3.11), simplifies to

$$\dot{\nu}(t) = 2\Omega - \frac{1}{|a|^2} 2\mathrm{Re}\left[(e-i\alpha)a^2 b\right] \; , \qquad |b| = 1 \quad . \qquad (4.15)$$

We are led, therefore, to the following theorem:

<u>*Theorem 3.*</u> - *The slit* $z=a(\omega+b\omega^{-1})$, $|b|=|\omega|=1$, *solves Euler equation* (4.6) *for* m=2 *and* $\phi(\omega)=\omega+b\omega^{-1}$, *provided* $\nu(t)$ *satisfies* (4.15) *with* Ω *being a real constant, and,* e *and* α *being functions of* t. *A steady state solution exists provided*

$$\Omega = \mathrm{Re}\left[(e-i\alpha)a^2|a|^{-2}b\right] , \qquad |b| = 1 \; ,$$

and thus $\Omega^2 \leq e^2+\alpha^2$.

We now treat the case (ii) of an elliptic vortex. Here, equation (4.14) gives that $a^2\mu b$ is real. Equation (4.13) is therefore reduced to

$$-2\omega_0^{-1}\dot{\nu}(t) = 1-|b|^2-4\frac{\Omega}{\omega_0}+\frac{1}{|a|^2|b|^2}(1+|b|^2)a^2b\mu, \quad |b|<1, \qquad (4.16)$$

with $\mathrm{Im}(a^2b\mu)=0$. Here only ν, b and μ can be functions of t. Let $b=|b|\cdot e^{i\theta_b+i\nu(t)}$ with θ_b being a constant. Also, let $a=|a|e^{i\theta_a}$, θ_a being a constant and let $\mu=\frac{2}{\omega_0}(e-i\alpha) = \frac{2}{\omega_0}\sqrt{e^2+\alpha^2}\,e^{i\rho(t)}$ with $\rho(t)$ being a function of t. The condition $\mathrm{Im}(a^2b\mu)=0$ shows that

$$2\theta_a + \theta_b + \nu(t) + \rho(t) \equiv n\pi (\mathrm{mod}\ 2\pi) \quad , \qquad (4.17)$$

where n is 0 or 1. This together with (4.16) gives

$$\dot{\nu}(t) = -\dot{\rho}(t) = 2\Omega - \frac{\omega_0}{2}(1-|b|^2) + (-1)^{n+1}\frac{1+|b|^2}{|b|}\sqrt{e^2+\alpha^2}, \qquad (4.18)$$

where it is assumed that $(e^2+\alpha^2)$ is a constant. We thus have:

<u>Theorem 4.</u> - *The ellipse* $z=a\phi(\omega)$, $|\omega|=1$, *with* $\phi(\omega)=\omega+b\omega^{-1}$, $|b|<1$, *solves Euler equation* (4.6) *for* $m=2$, *provided* (4.17) *and* (4.18) *are satisfied. Here,* $a=|a|e^{i\theta_a}$ *and* $b=|b|\cdot e^{i(\theta_b+\nu(t))}$ *with* θ_a *and* θ_b *being constants. Also* $e-i\alpha=(e^2+\alpha^2)^{\frac{1}{2}}e^{i\rho(t)}$ *with* $(e^2+\alpha^2)$ *being a constant and* $\rho(t)$ *a function of* t *satisfying* (4.17), *where* $n=0,1$. *The steady state solution is obtained when* $\dot{\nu}(t)=\dot{\rho}(t)=0$. *In this case we must have*

$$\omega_0^{-1}(-1)^{n+1}\sqrt{e^2+\alpha^2} = -2\frac{\Omega}{\omega_0}\frac{|b|}{1+|b|^2}+\tfrac{1}{2}\frac{|b|(1-|b|^2)}{1+|b|^2}\ , \quad |b| < 1\ , \qquad (4.19)$$

where, for $a=|a|e^{i\theta_a}$, $b=|b|e^{i\theta_b}$ *and* $e-i\alpha=\sqrt{e^2+\alpha^2}\,e^{i\rho}$,

$$2\theta_a + \theta_b + \rho \equiv n\pi(\mathrm{mod}\ 2\pi)\ ; \qquad \theta_a, \theta_b, \rho \in [0,2\pi)\ , \qquad (4.20)$$

for $n=0,1$.

This theorem yields the following corollaries, some of which are very classical.

<u>Corollary 1.</u> *(Kirchoff-1876) - Under the assumptions of Theorem 4, the ellipse is uniformly rotating with an angular velocity given by*

$$\Omega = \frac{\omega_0}{4}(1 - |b|^2)\ . \qquad (4.21)$$

<u>Proof.</u> This is obtained from (4.19) by letting $e=\alpha=0$. See also Lamb [5, p.232]. We also note that in this case we must have $0<4\Omega/\omega_0<1$.

<u>Corollary 2.</u> - *Let* $\alpha=e$ *and* $\Omega=\pm(-1)^{n+1}\sqrt{2}\alpha$ *where* $n=0,1$ *as in* (4.20). *Then the ellipses* $z=a(\omega+b\omega^{-1})$, $|b|<1$, *subject to* (4.20), *are always steady state solutions when* $\Omega=-(-1)^{n+1}\sqrt{2}\alpha$ *and* $(-1)^{n+1}\alpha/\omega_0>0$. *When* $\Omega=(-1)^{n+1}\sqrt{2}\alpha$ *the above ellipses are steady*

state solutions if and only if

$$0 < (-1)^{n+1} \frac{\alpha}{\omega_0} \leq \frac{\sqrt{2}}{4} (\sqrt{2} - 1)^2 \quad . \tag{4.22}$$

Proof. Equation (4.19) assumes, in this case, the form

$$(-1)^{n+1}\sqrt{2}\frac{\alpha}{\omega_0}(1\pm \frac{2|b|}{1+|b|^2}) = \tfrac{1}{2}|b|\frac{1-|b|^2}{1+|b|^2}$$

and therefore

$$(-1)^{n+1}\frac{\alpha}{\omega_0} = \frac{1}{2\sqrt{2}}|b|\frac{1+|b|}{1-|b|} \; ; \qquad \Omega = -(-1)^{n+1}\sqrt{2}\alpha \quad , \tag{4.23}$$

$$(-1)^{n+1}\frac{\alpha}{\omega_0} = \frac{1}{2\sqrt{2}}|b|\frac{1-|b|}{1+|b|} \; ; \qquad \Omega = (-1)^{n+1}\sqrt{2}\alpha \quad . \tag{4.24}$$

The function

$$f(s) = \frac{1}{2\sqrt{2}} s \frac{1+s}{1-s} \quad , \qquad s \in (0,1) \quad ,$$

is positive and strictly increasing in (0,1). This shows that in the case of (4.23) the only constraint on the steady state solution is that $(-1)^{n+1}\alpha/\omega_0 > 0$ thereby proving the first part of the corollary. On the other hand, the function

$$g(s) = \frac{1}{2\sqrt{2}} s \frac{1-s}{1+s} \quad , \qquad s \in (0,1) \quad ,$$

is positive in (0,1) and has a maximum at $s_0 = \sqrt{2}-1$. Therefore, in case of (4.24), a steady state solution exists if and only if

$$0 < (-1)^{n+1}\frac{\alpha}{\omega_0} \leq g(s_0) = \frac{\sqrt{2}}{4}(\sqrt{2}-1)^2 .$$

This completes the proof.

The proof of the following corollary is completely similar to the proof of Corollary 2:

Corollary 3. - Let $z=a(\omega+b\omega^{-1})$, $|b|<1$, *be the ellipse subject to condition* (4.20). *It is a steady state solution in the following cases:*

(i) $\alpha=0$, $\Omega= (-1)^{n+1}e$ *if and only if* $0 < (-1)^{n+1}\frac{e}{\omega_0} \leq \frac{1}{2}(\sqrt{2}-1)^2$,

(ii) $\alpha=0$, $\Omega=-(-1)^{n+1}e$ *if and only if* $(-1)^{n+1}\frac{e}{\omega_0} > 0$,

(iii) $e=0$, $\Omega= (-1)^{n+1}\alpha$ *if and only if* $0 < (-1)^{n+1}\frac{\alpha}{\omega_0} \leq \frac{1}{2}(\sqrt{2}-1)^2$,

(iv) $e=0$, $\Omega=-(-1)^{n+1}\alpha$ *if and only if* $(-1)^{n+1}\frac{\alpha}{\omega_0} > 0$.

Corollary 4. - *Let* $\Omega=0$ *and consider the ellipse* $z=a(\omega+b\omega^{-1})$, $|b|<1$, *subject to condition* (4.20). *It is a steady state solution if and only if*

$$0 < (-1)^{n+1} \frac{\sqrt{e^2+\alpha^2}}{\omega_0} \leq \frac{1}{2} \frac{(\sqrt{5}-2)^{\frac{1}{2}}(3-\sqrt{5})}{\sqrt{5}-1} .$$

Proof. In this case, equation (4.19) reduces to

$$(-1)^{n+1} \frac{\sqrt{e^2+\alpha^2}}{\omega_0} = \frac{1}{2} \frac{|b|(1-|b|^2)}{1+|b|^2}, \qquad |b| < 1 . \tag{4.25}$$

The function

$$h(s) = \frac{1}{2} \frac{s(1-s^2)}{1+s^2}, \qquad s \in (0,1) ,$$

is positive in (0,1) and has a maximum at $s_0=(\sqrt{5}-2)^{\frac{1}{2}}$. Therefore,

$$h(|b|) \leq h(s_0) = \frac{1}{2} \frac{(\sqrt{5}-2)^{\frac{1}{2}}(3-\sqrt{5})}{\sqrt{5}-1} .$$

The proof now follows from (4.25).

Remark. The last corollary agrees with the result of Moore and Saffman [7, p.345] which they obtained by matching on the boundary of the elliptic vortex the exterior and the interior stream functions. This was accomplished by appealing to classical methods of Kirchoff (see Lamb [5, p.232]) where Cartesian coordinates are used for the interior of the ellipse and elliptic coordinates describe its exterior. In the case of [7] α is always zero and this because the major axes of their ellipses are always parallel to the major axes of the Cartesian coordinate system, thereby allowing only two types of ellipses. In fact, in our case when we insist that a and b are real, i.e., in (4.20) $\theta_a,\theta_b \in \{0,\pi\}$, we will find that $\rho \equiv n\pi \pmod{2\pi}$ with $n \in \{0,1\}$. Since $\rho=\arg(e-i\alpha)$ we therefore conclude that $\alpha=0$. However, in our setting we, of course, have an infinity of ellipses which are steady state solutions with $\alpha \neq 0$. In [7], Moore and Saffman used the axis ratio θ of the ellipse, which in our notation is

$$\theta = \frac{1+|b|}{1-|b|}, \quad \theta \in (1,\infty); \quad |b| = \frac{\theta-1}{\theta+1}, \quad |b| \in (0,1) , \tag{4.26}$$

as a parameter instead of $|b|$. Their formula [7, p.344] reads, in case of $\alpha=0$ and $n=1$,

$$\frac{e}{\omega_0} = \frac{\theta(\theta-1)}{(\theta^2+1)(\theta+1)} ,$$

which, in view of (4.26), is equivalent to (4.25). The function on the right has a single maximum at θ_0 which is the unique root in $(1,\infty)$ of

$$\theta^4 - 2\theta^3 - 2\theta^2 - 2\theta + 1 = 0 .$$

They showed that θ_0 is approximately 2.9 and the maximum value is about 0.15. In our computation these values are exactly evaluated for,

$$\theta_0 = \frac{1+s_0}{1-s_0} = \frac{1+(\sqrt{5}-2)^{\frac{1}{2}}}{1-(\sqrt{5}-2)^{\frac{1}{2}}} \simeq 2.9$$

and $h(s_0) \approx 0.15$. This result may be also obtained by observing that

$$\theta^4-2\theta^3-2\theta^2-2\theta+1 = \tfrac{1}{4}(\theta+1)^4\left[\left(\frac{\theta-1}{\theta+1}\right)^4+4\left(\frac{\theta-1}{\theta+1}\right)^2-1\right] .$$

Other results similar to those of Corollaries 1-4 may be obtained by further specification of the bifurcation parameters α, e and Ω.

Another corollary is the following very classical result which is due to Love [6] who proved it by using different methods:

Corollary 5. - *An elliptic vortex, in a fluid which is at rest at infinity, with constant vorticity cannot move so as to remain elliptic unless it retains its shape and rotates with angular velocity given by Kirchoff's condition (4.23).*

Proof. In this case the stream function ψ in (3.5) is precisely ψ_0. Consequently, the evolution equation in (3.16) or (4.6) would read as

$$\mathrm{Re}\left[i\omega\phi' i\bar{\dot\phi}\right] = \frac{\omega_0}{2}\frac{d}{|d\omega|}\left[\tfrac{1}{2}|\phi|^2 + \mathrm{Re}\,S(\phi)\right] , \qquad \omega\in\Gamma , \tag{4.27}$$

where here $\phi=\omega+b\omega^{-1}$ and $b=b(t)$ is a function of t. According to (4.9) $S(\phi)=-2^{-1}\bar{b}\phi^2$. Substituting these in (4.27) and comparing powers of ω we obtain that

$$-i\dot{b} = \frac{\omega_0}{2}\, b(1-|b|^2) .$$

Since $\dot{b}/b$ is pure imaginary we have that $|b|^2$ is a constant and hence the unique solution of the above differential equation is $b=de^{i\nu t}$, where d is a constant with $|d|=|b|$ and

$$\nu = \frac{\omega_0}{2}(1-|b|^2) .$$

Thus, the ellipse rotates with a constant angular velocity $\Omega=\nu/2$ which is in agreement with (4.23). This completes the proof.

5. LINEARIZATION

We now consider the evolution equation (4.3) with m=2, thus

$$\mathrm{Re}\left[i\omega\phi' i\bar{\phi}\right] = \frac{\omega_0}{2}\frac{d}{|d\omega|}\left[\frac{\lambda}{2}|\phi|^2+\tfrac{1}{2}\mathrm{Re}\left(\frac{a^2}{|a|^2}\mu\phi^2\right)+\mathrm{Re}S(\phi)\right], \qquad \omega\in\Gamma, \tag{5.1}$$

where λ and μ are given in (3.11) and $S(\phi)$ as in (4.7). Here,

$$z=a\phi(\omega) \quad \text{with} \quad \phi(\omega)=\omega+\sum_{n=1}^{\infty} b_n\omega^{-n} \text{ and } \Gamma=\{\omega:|\omega|=1\} .$$

The above evolution equation may be written in terms of the so called "Faber polynomials" of the mapping ϕ. This and other related results will appear elsewhere. At this moment, however, we shall state the following theorem whose proof will appear in our forthcoming paper [2].

Theorem 5. - *The only solution* ϕ *of* (5.1) *among all mappings of the form* $\phi(\omega) = \omega + \sum_{n=1}^{N} b_n\omega^{-n}$, $N<\infty$, *is the elliptic solution* $\phi=\omega+b\omega^{-1}$, $|b|\leq 1$.

This theorem shows that a solution ϕ of (5.1) which is not of the form $\phi(\omega)=\omega+b\omega^{-1}$ must have an infinite expansion $\phi(\omega)=\omega+\sum_{n=1}^{\infty} b_n\omega^{-n}$. For sake of simplicity we may assume that the vortex is under no shear or strain, that is $\mu=0$ and thus the only possible vortex motions are those of pure rotation. There are two completely equivalent ways of looking at (5.1) with $\mu=0$. One way is to consider the steady state equation

$$\lambda|\phi|^2 + 2\mathrm{Re}S(\phi) = \text{const.}, \qquad \phi = \phi(\omega), \qquad \omega\in\Gamma , \tag{5.2}$$

where λ is the bifurcation parameter

$$\lambda = 1 - \frac{2\Omega}{\omega_0}$$

with Ω being interpreted as the angular velocity of the boundary of the vortex. The second way is to consider the evolution equation

$$\mathrm{Re}\left[i\omega\phi' i\bar{\phi}\right] = \frac{\omega_0}{2}\frac{d}{|d\omega|}\left[\tfrac{1}{2}|\phi|^2+\mathrm{Re}S(\phi)\right], \qquad \omega\in\Gamma , \tag{5.3}$$

where it is assumed that the vortex is initially in a fluid which is at rest at infinity. Clearly, in the case of (5.2) the vortex is not at rest at infinity, in fact $i(u-iv)=2\partial_z\psi(z)=\Omega\bar{z}$ near $z=\infty$. Of course, $\phi_0=\omega$ is the trivial solution for (5.2) and (5.3) while, by virtue of Corollaries 1 and 5, $\phi(\omega)=\omega+b\omega^{-1}$ solves (5.2) and (5.3) with the Kirchoff condition

(4.21).

The following theorem is a generalization of a result of Lamb [5, p.231].

Theorem 6. - Let $m\geq 2$ *be an integer and consider the hypotrochoids* $\phi_m(\omega)=\omega+b_{m-1}\omega^{-(m-1)}$ *with* $|b_{m-1}|^4<<1$ *for* $m\geq 3$. *Then* $\phi=\phi_m$, $m\geq 3$, *is an asymptotic solution of (5.2) and (5.3) with the angular velocity*

$$\Omega_m = \frac{\nu_m}{m} = \frac{\omega_0}{2}\frac{m-1}{m}(1-\frac{m}{2}|b_{m-1}|^2) \ ; \qquad |b_{m-1}|^4 << 1 \ , \quad m\geq 3 \ ,$$

where for $m=2$ *the above relation is exact provided* $|b_1|<1$.

Proof. In this case $S(\phi_m)$ is evaluated via (4.9). Therefore,

$$\begin{aligned} S(\phi_m) &= -\frac{1}{m}\bar{b}_{m-1}[\phi_m(\omega)]^m = -\frac{1}{m}\bar{b}_{m-1}\omega^m(1+b_{m-1}\omega^{-m})^m \\ &= -\frac{1}{m}\bar{b}_{m-1}\omega^m\sum_{k=0}^{m}\binom{m}{k}b_{m-1}^k\omega^{-mk} \\ &= -\frac{1}{m}\bar{b}_{m-1}\omega^m(1+mb_{m-1}\omega^{-m}+\binom{m}{2}b_{m-1}^2\omega^{-2m}) + O(|b_{m-1}|^4) \ , \end{aligned}$$

where the last equality is exact for m=2. Hence

$$S(\phi_m) = -\frac{1}{m}\bar{b}_{m-1}(\omega^m+\binom{m}{2}b_{m-1}^2\omega^{-m}+mb_{m-1}) + O(|b_{m-1}|^4) \ . \qquad (5.4)$$

We write $b=b_{m-1}$ and $\phi(\omega)=\phi_m(\omega)=\omega+b\omega^{-(m-1)}$. We consider equation (5.3) and assume that $b=b(t)$ is a function of t. Putting (5.4) in (5.3) and comparing powers of ω we obtain

$$-i\dot{b} = \frac{\omega_0}{2}(m-1)b[1-\frac{m}{2}|b|^2] \ .$$

Again, since $\dot{b}/b$ is pure immaginary we find that the solution of the above equation is $b=de^{i\nu t}$ where d is a constant with $|d|=|b|$ and

$$\nu = \nu_m = \frac{\omega_0}{2}(m-1)[1-\frac{m}{2}|b_{m-1}|^2] \ ; \ |b_{m-1}|^4<<1, \ m\geq 3 \ .$$

The angular velocity is, of course, $\Omega_m=\nu_m/m$. Similarly, for the steady state equation (5.2) we obtain, with the assumption that $|b_{m-1}|^4<<1$ for $m\geq 3$, that

$$\lambda = \lambda_m = 1 - \frac{2\Omega_m}{\omega_0} = \frac{1}{m}[1+\binom{m}{2}|b_{m-1}|^2]$$

and the theorem is proved.

The steady state equation (5.2) can be put in a structure which is familiar in the theory of nonlinear bifurcation problems. In fact, let

$$f(\omega) = \overline{\phi(\omega)} - \bar{\omega} = \sum_{n=0}^{\infty} \bar{b}_n \omega^n , \qquad \omega \in \Gamma ,$$

where, for convenience of the exposition we allow $b_0 \neq 0$ as an additional parameter in the mapping ϕ. We also define $Q(f)=S(\phi)$. A direct computation shows that

$$\left[Q(f)\right](\tau) = \frac{1}{2\pi i} \int_\Gamma \log\left[1 - \frac{\overline{f(\tau)+\tau}}{f(\omega)+\omega}\right] f(\omega)\,(1-\overline{\omega^2 f'(\omega)}\,)\,d\omega. \tag{5.5}$$

Also, equation (5.2) may be now written as

$$\lambda \mathrm{Re}(\omega f) = -\left[\mathrm{Re}Q(f) + \lambda \tfrac{1}{2}|f|^2\right] + \text{const.}, \qquad f = f(\omega), \quad \omega \in \Gamma. \tag{5.6}$$

Writing $e^{i\theta}=\omega$, $\theta \in [0,2\pi)$, and

$$f(\omega) = f(e^{i\theta}) = \sum_{n=0}^{\infty} a_n e^{in\theta}$$

we may regard Q as an operator from the $L_2[0,2\pi]$-closure of $\{e^{in\theta}\}_{n=0}^{\infty}$ into $L_2[0,2\pi]$. Equivalently, we let $H_2(\Gamma)$ be the usual *Hardy space* with the inner product

$$(f,g) = \frac{1}{2\pi} \int_\Gamma f(\omega)\overline{g(\omega)}\,|d\omega| = \sum_{n=0}^{\infty} a_n \bar{b}_n , \tag{5.7}$$

where

$$g(\omega) = \sum_{n=0}^{\infty} b_n \omega^n$$

and

$$\sum_{n=0}^{\infty} |a_n|^2 < \infty , \qquad \sum_{n=0}^{\infty} |b_n|^2 < \infty .$$

Clearly, $H_2(\Gamma)$ may be regarded as a closed subspace of $L_2(\Gamma)$ with respect to the arc-lenght measure $|d\omega|=d\theta$ on Γ. The space $H_2(\Gamma)$ is also identified with holomorphic functions inside the unit disk Δ whose boundary values are in $L_2(\Gamma)$. This identification shows that point evaluations in Δ are bounded linear functionals on $H_2(\Gamma)$. Also, the integration in (5.7) is carried over the boundary values (this refers to arbitrary non-tangential approach) of the holomorphic functions f and g in Δ.

From (5.5) we see that the operator Q maps $H_2(\Gamma)$ into $L_2[0,2\pi]$ and that $Q(0)=0$. Further, the Fréchet derivative of Q at $f=0$ is given by

$$[Q'(h)](\tau) = \frac{1}{2\pi i}\int_{\Gamma} \log\left(1-\frac{\tau}{\omega}\right) h(\omega)\,d\omega \ ; \quad \tau\in\Delta, \quad h\in H_2(\Gamma). \quad (5.8)$$

It is easy to see that Q_0' is a compact operator from $H_2(\Gamma)$ into itself with $Q_0'(1)=-\tau$ and also $\|Q_0'\|=1$.

Equation (5.6) can be viewed now as

$$\lambda \mathrm{Re}(e^{i\theta}f) = -\left[\mathrm{Re}Q(f) + \tfrac{1}{2}\lambda\,|f|^2\right]; \quad f = f(e^{i\theta}),\ f\in H_2(\Gamma), \quad (5.9)$$

where the left hand side of (5.9) defines an operator from $H_2(\Gamma)$ onto the Hilbert space of harmonic functions in Δ whose boundary values are in $L_2(\Gamma)$. This operator is, of course, real linear. The linearization of (5.9) about $f=0$ yields the following eigenvalue problem:

$$\mathrm{Re}(e^{i\theta}h) = -\mathrm{Re}Q_0'(h); \quad h\in H_2(\Gamma), \quad h = h(e^{i\theta}), \qquad (5.10)$$

where Q_0' is given in (5.8) and $\lambda=1-2\Omega/\omega_0$. Since both sides of (5.10) are harmonic inside Δ, (5.10) is completely equivalent to

$$\lambda(\omega h) = -\,Q_0'(h); \quad \omega\in\Delta, \quad h\in H_2(\Gamma). \qquad (5.10')$$

We now prove:

Theorem 7. - The complete set of eigenvalues and eigenfunctions in $H_2(\Gamma)$ *of (5.10') is* $\{\lambda_m, h_m\}$ *where*

$$\lambda_m = m^{-1}, \qquad h_m(\omega) = \omega^{m-1}; \qquad m = 1,2,\ldots\ .$$

Proof. Let $h\in H_2(\Gamma)$ with

$$h(\omega) = \sum_{n=1}^{\infty} c_n \omega^{n-1}, \qquad \omega\in\Delta \quad .$$

A use of (5.8) shows that

$$Q_0'(h) = -\sum_{n=1}^{\infty} \frac{1}{n} c_n \omega^n \quad .$$

This together with (5.10') shows that

$$\sum_{n=1}^{\infty} \left(\lambda - \frac{1}{n}\right)\omega^n = 0$$

and the result follows.

We should also remark that the above result can be also obtained by observing that (5.10) is equivalent to

$$(1-\lambda)h(\omega) = \omega h'(\omega) \ .$$

Also, by the preceding discussion and observing that the bifurcation problem (5.9) satisfies the usual smoothness properties as, for example, those listed in [4], we obtain:

<u>Corollary 6.</u> - *For any integer* $m \geq 1$ *there exist positive constants* ε_m *and* δ_m *such that for each* $\varepsilon \in (-\varepsilon_m, \varepsilon_m) - \{0\}$ *the problem* (5.9) *has a unique nontrivial solution* $\{\lambda_m^{(\varepsilon)}, h_m^{(\varepsilon)}\}$ *with* $h_m^{(\varepsilon)}$ *in the convex set*

$$A(\delta_m, \varepsilon) = \{h \in H_2(\Gamma) : h = \varepsilon(h_m + \varepsilon g), \ \|g\| \leq \delta_m, \ g \in B_m\}$$

where

$$B_m = \mathrm{Span}\left[h_{km}\right]_{k=2}^{\infty} = \{g \in H_2(\Gamma) : g(\omega) = \sum_{k=2}^{\infty} c_k \omega^{km-1}, \quad \omega \in \Delta\}.$$

Moreover, for some positive constants M_m *and* K_m *the solution has the form*

$$\lambda_m^{(\varepsilon)} = \lambda_m + \varepsilon\mu_\varepsilon, \quad |\mu_\varepsilon| \leq M_m, \quad (\lambda_m = m^{-1}),$$

$$h_m^{(\varepsilon)} = \varepsilon(h_m + \varepsilon g_\varepsilon), \quad \|g_\varepsilon\| \leq K_m \qquad g_\varepsilon \in B_m, \quad (h_m(\omega) = \omega^{m-1}).$$

Since $\lambda = 1 - 2\Omega/\omega_0$ we find that $\Omega_1 = \omega_0(1-\lambda_1)/2 = 0$ and therefore, we exclude this trivial case of $m=1$. If we now return to our original equation (5.2) we conclude the following important result:

<u>Corollary 7.</u> - *For any integer* $m \geq 2$ *there exist positive constants* ε_m *and* δ_m *such that for each* $\varepsilon \in (-\varepsilon_m, \varepsilon_m) - \{0\}$, *equation (5.2) has a unique m-fold symmetric uniformly rotating solution* $\phi_m^{(\varepsilon)}$ *with an angular velocity* $\Omega_m^{(\varepsilon)}$. *This solution has the form*

$$\phi_m^{(\varepsilon)}(\omega) = \omega + \varepsilon\omega^{-(m-1)} + \varepsilon^2 \sum_{k=2}^{\infty} c_k(\varepsilon)\omega^{-(km-1)},$$

with

$$\sum_{k=2}^{\infty} |c_k(\varepsilon)|^2 \leq \delta_m^2,$$

and

$$\Omega_m^{(\varepsilon)} = \frac{\omega_0}{2}\,\frac{m-1}{m}\left(1 - \frac{m}{2}\varepsilon\mu_\varepsilon\right),$$

where $|\mu| \leq M_m$ *for some positive constant* M_m.

This result coupled with Theorem 6 gives a somewhat more precise information about the nature of the steady state solution. We could also bifurcate the Kirchoff elliptic solution or we could continue with the procedure of Theorem 6 by adding more terms to ϕ_m. These and related results will be elaborated elsewhere.

BIBLIOGRAPHY

[1] Ahlfors, L., and Beurling, A., Conformal Invariants and function-theoretic null sets, Acta Math. 83 (1950), 101-129.

[2] Burbea, J., On stability of certain vortex motions, Proceedings on Nonlinear PDE in Engineering and Applied Science, Rhode Island, 1979, Marcel Dekker, to appear.

[3] Deem, G.S., and Zabusky, N.J., Vortex waves, stationary "V-states", interactions, recurrence and breaking, Phys.Rev.Letters 40 (1978), 854-862.

[4] Keller, H.B., and Langford, W.F., Iterations, perturbations and multiplicities for nonlinear bifurcation problems, Arch. Rational Mech.Anal. 48 (1972), 83-108.

[5] Lamb, H., *Hydrodynamics*, Dover Publications, New York, 1945.

[6] Love, A.E.H., On the stability of certain vortex motions, Proc.London Math.Soc. (1)25 (1893), 18-42.

[7] Moore, D.W., and Saffman, P.G., Structure of a line vortex in an imposed strain, "Aircraft Wake Turbulence and its Detection", Plenum Press (1971), 339-354.

[8] Zabusky, N.J., Coherent structures in fluid dynamics, "The Significance of Nonlinearity in the Natural Sciences", Plenum Press (1977), 145-205.

ANALYTICAL SOLUTIONS OF THE SINE-GORDON EQUATION AND THEIR APPLICATION TO JOSEPHSON TUNNEL JUNCTIONS

Bonaventura Savo

Istituto di Fisica, Università di Salerno, 84100 Salerno, Italy

INTRODUCTION

The Josephson junction is a physical system described (approximately) by the sine-Gordon equation. In its simplest form it is made of two superconductive films separated by a thin dielectric tunneling layer, as indicated in Fig. 1. The thickness of the films is typically several thousands of angstroms, that of the barrier layer of the order of 20 Å.

BASIC EQUATIONS

The electrical behaviour of a Josephson junction is governed by the equations [1]:

$$J = J_0 \sin\phi \quad ; \qquad \frac{d\phi}{dt} = \frac{2\pi}{\phi_0} V \quad , \tag{1a,b}$$

where J is the tunneling supercurrent density crossing the dielectric barrier, J_0 is a characteristic coefficient which depends on the materials and temperature, ϕ is the phase difference -or simply the phase- between the macroscopic quantum wave functions which characterize the two superconductors, V is the voltage difference across the junction and $\phi_0 = h/2e$ (h is the Planck's constant and e the electronic charge) is the magnetic flux quantum.

The subject of this lecture is confined to long junctions -see Fig. 1- in which one transverse dimension is very large compared with the

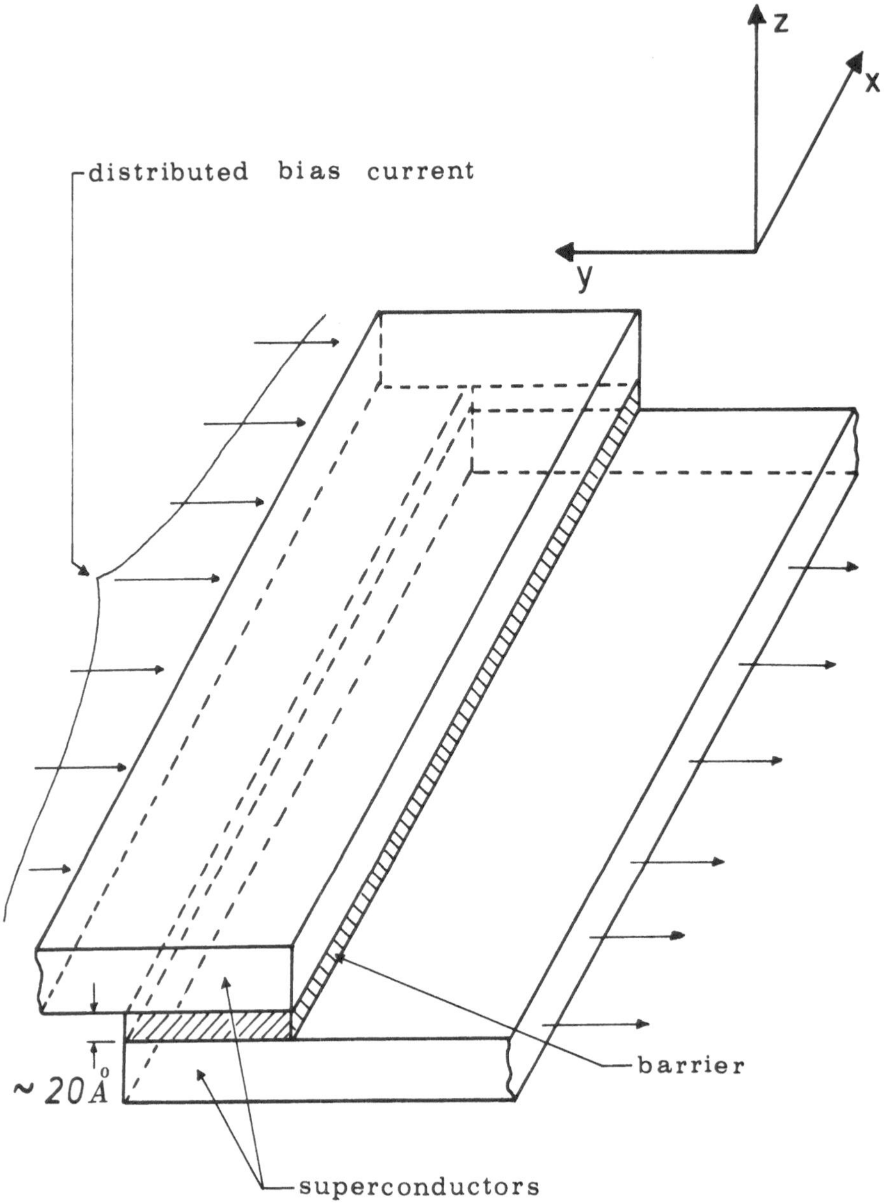

Fig. 1 - Schematic structure of a Josephson tunnel junction.

Josephson penetration lenght, whose typical order might be 100 μm, while the other is very small.

Such a junction may be modelled by the one-dimensional transmission line structure indicated in Fig. 2. In this model, besides the Josephson current J, which now has the dimensions of current per unit lenght, there is taken into account a uniform bias current J_B and a conductance G due to the quasiparticle tunneling current which, in approximate way, may be described by:

$$G = g_0 \,|V| \tag{2a}$$

where g_0 is a fairly complicated function of the temperature.

There exists also an effect due to the real part of the surface impedance of the superconductors [2]. It is not considered in this model.

The series inductance L and the shunt capacitance C (both expressed per unit lenght) of the strip-line are given by [3]:

$$L = \frac{\mu_0}{W}\,(d + \lambda_1 + \lambda_2) \ ; \qquad C = \frac{\kappa \varepsilon_0 W}{d} \tag{2b,c}$$

where μ_0 and ε_0 are, respectively, the permeability and the permettivity of free space, k and d, respectively, the relative dielectric constant and the thickness of the barrier layer, $\lambda_{1,2}$ the London penetration dephts of the two superconductors and W is the width of the junction in the y direction.

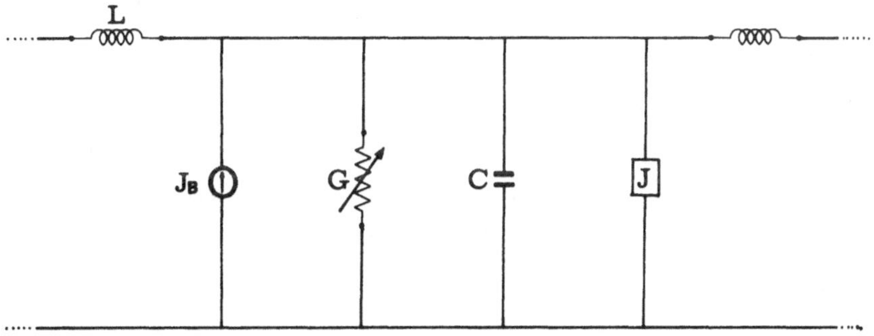

Fig. 2 - Transmission line model of a Josephson junction strip-line; the elements L, J_B, G, C, J are described by eqs. (1a,b), (2a,b,c).

Applying Kirchoff's law to the model of Fig.2 gives:

$$\frac{\partial V}{\partial x} = - L \frac{\partial i}{\partial t}$$

$$\frac{\partial i}{\partial x} = - C \frac{\partial V}{\partial t} - GV - J_0 \sin\phi + J_B \tag{3a,b}$$

and normalizing distance with respect to the Josephson penetration length, $\lambda_J \equiv (\phi_0 / 2\pi L J_0)^{\frac{1}{2}}$, and time with respect to the inverse of the Josephson plasma frequency, $\omega_J \equiv (2\pi J_0 / C\phi_0)^{\frac{1}{2}}$, one obtains, by means of eq.(1b), the basic equation for the Josephson junction:

$$\frac{\partial^2 \phi}{\partial x^2} - \frac{\partial^2 \phi}{\partial t^2} - \Gamma \left|\frac{\partial \phi}{\partial t}\right| \frac{\partial \phi}{\partial t} = \sin\phi - \gamma \tag{4}$$

in which

$$\Gamma \equiv g_0 \phi_0 / 2\pi C \qquad \text{and} \qquad \gamma \equiv J_B / J_0 \quad .$$

Whereas some particular analytical solutions of eq.(4) have been found for an infinite line [4,5], solutions for a finite line are not available up to now. A perturbation analysis seems reasonable since, in practice, Γ is a very small quantity and γ may also be a small quantity.

So, as first step, a more essential model of the Josephson junction may be considered neglecting the dissipative term $-\Gamma|\phi_t|\phi_t$ and the external (normalized) bias current γ; at this level eq.(4) is reduced to the sine-Gordon equation:

$$\frac{\partial^2 \phi}{\partial x^2} - \frac{\partial^2 \phi}{\partial t^2} = \sin \phi \quad . \tag{5}$$

DERIVATION OF THE SOLUTIONS OF EQ.(5)

Open circuit boundary condition seem to be a physically reasonable requirement for solutions of eq.(5) since the (linear) characteristic impedance of the line, $(LC)^{\frac{1}{2}}$, is usually very much less than the wave impedance of free space $(\mu_0/\varepsilon_0)^{\frac{1}{2}}$.

From eqs.(3) and (1b) such conditions correspond to setting:

$$\left.\frac{\partial \phi}{\partial x}\right|_{x=0} = 0 = \left.\frac{\partial \phi}{\partial x}\right|_{x=\ell} \tag{6a,b}$$

where ℓ is the normalized length of the line.

An exact analytical description of the fundamental oscillation modes, i.e. the plasma oscillation, the fluxon oscillation and the breather oscillation, is obtained by solutions of eq.(5) in terms of Lamb's ansatz [6]:

$$\phi = 4 \tan^{-1} \{f(x)\, g(t)\} \tag{7}$$

where

$$\left\{\frac{df}{dx}\right\}^2 = af^4 + (1+b)f^2 - c$$

$$\left\{\frac{dg}{dt}\right\}^2 = cg^4 + bg^2 - a \tag{8a,b}$$

and a, b and c are arbitrary constants.

In this derivation an interesting role was played by the comparison between some suitable solutions of eq.(5) and the results of essentially qualitative observations of Fulton [7] on a mechanical analog of the Josephson junction subject to the boundary conditions of eqs.(6).

This device is shown in its essential form in Fulton's sketches in Figs.3-4-5 and -6.

A generalization of eq.(7) has been obtained by the scaling:

$$f = pF; \quad g = qG; \quad \hat{x} = \beta(x-x_0); \quad \hat{t} = \Omega(t-t_0) \tag{9a,b,c,d}$$

where $p, q, \beta, \Omega, x_0, t_0$, are arbitrary constants. Under this scaling eqs.(8) become :

$$\left(\frac{dF}{d\hat{x}}\right)^2 = \frac{ap^2}{\beta^2} F^4 + \frac{1+b}{\beta^2} F^2 - \frac{c}{p^2\beta^2}$$

$$\left(\frac{dG}{d\hat{t}}\right)^2 = \frac{cq^2}{\Omega^2} G^4 + \frac{b}{\Omega^2} G^2 - \frac{a}{q^2\Omega^2} \quad . \tag{10a,b}$$

The analytical forms of the solutions of eq.(5) are determined by comparing eqs.(10) with the standard forms of the differential equations of the Jacobian elliptic functions [8].

In connection there are some general remarks:

i) the constants p and q appear always as the product pq, so they can be defined as a single constant $A \equiv pq$;

ii) the constants β, Ω, and A are connected by a relation which can be interpreted as a non linear dispersion relation; it comes out by imposing consistency between the parameters in eqs.(10);

iii) the constants x_0 and β are determined by the boundary conditions

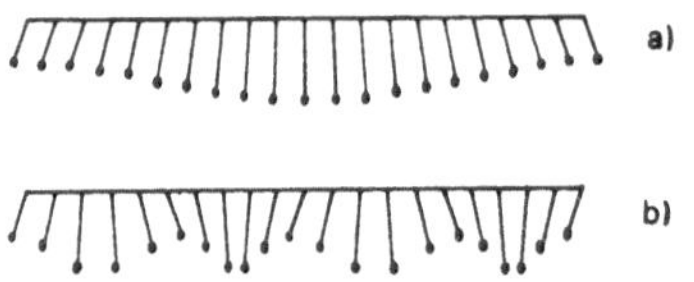

Fig.3 - Plasma oscillations on a mechanical line. This device consists of a series of mechanical pendula coupled by a torsion element. The angular displacement of pendula from equilibrium position is the analog of the phase in the Josephson junction. a): half wavelength; b): two wavelengths. *From Fig.20 of Fulton [7] (with permission).*

of eqs.(6).

1): Plasma oscillation

This solution is given by:

$$\phi = 4\tan^{-1}\left\{A\,\mathrm{cn}\left[\beta(x-x_0);k_f\right]\mathrm{cn}\left[\Omega(t-t_0);k_g\right]\right\} \tag{11}$$

where

$$k_f^2 = \frac{A^2\left[\beta^2(1+A^2)+1\right]}{\beta^2(1+A^2)^2} \quad ; \quad k_g^2 = \frac{A^2\left[\Omega^2(1+A^2)-1\right]}{\Omega^2(1+A^2)^2} \tag{12a,b}$$

It describes -see Fig.3- phase oscillations with the presence of spatial nodes.

The dispersion relation is:

$$\Omega^2 - \beta^2 = \frac{1-A^2}{1+A^2} \quad . \tag{13}$$

The boundary conditions set $x_0=0$ and:

$$\beta_n = \frac{2n}{\ell}K(k_f) \tag{14}$$

where n=1,2,... is the number of spatial nodes on the line and K(k) is the complete elliptic integral of first kind.

For n=0 eq.(11) is reduced to:

$$\phi = 4\tan^{-1}\left\{A\,\mathrm{sn}\left[\frac{t-t_0}{1+A^2}\ ;\ A^2\right]\right\} \quad . \tag{15}$$

It describes unison oscillations on the entire line.

2): Fluxon oscillation

This solution is given by:

$$\phi = 4\tan^{-1}\left\{A\,\mathrm{dn}\left[\beta(x-x_0);k_f\right]\,\mathrm{tn}\left[\Omega(t-t_0);k_g\right]\right\} \tag{16}$$

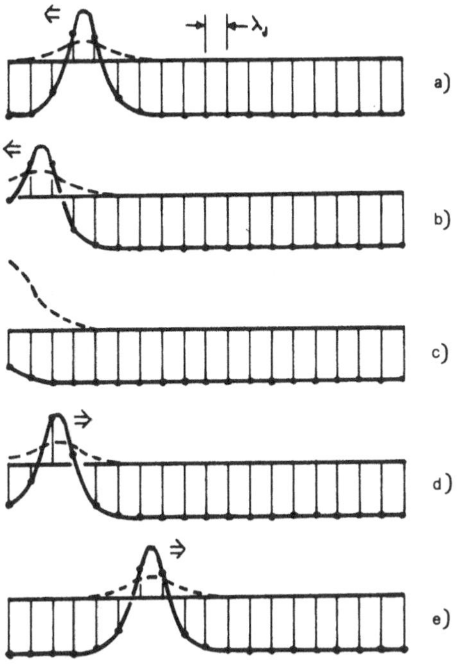

Fig.4 - Fluxon propagation on a mechanical line in five successive configurations: n=1 case in eq.(19c). Solid line: $-\cos\phi(x)$; dashed line $\propto \partial\phi(x)/\partial t$. *From Fig.25 of Fulton [7] (with permission).*

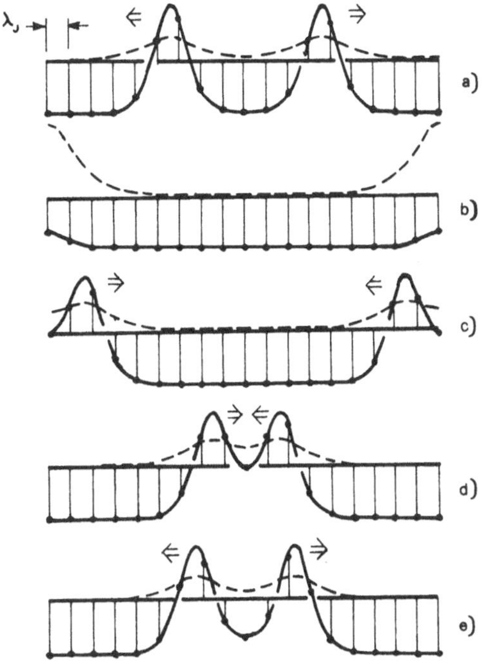

Fig.5 - Fluxon propagation on a mechanical line in five successive configurations: n=2 case in eq.(19c). Solid line: $-\cos\phi(x)$; dashed line $\propto \partial\phi(x)/\partial t$.

(From Fig.26 of Fulton [7] (with permission).

where

$$k_f = 1 - \frac{\frac{\beta^2}{A^2}(A^2-1)-1}{\beta^2(A^2-1)} \quad ; \quad k_g^2 = 1 - \frac{A^2\{\Omega^2(A^2-1)-1\}}{\Omega^2(A^2-1)} \quad . \tag{17a,b}$$

It describes -see Figs.4 and 5- monotonically increasing spatial and temporal phase changes that may be physically interpreted as a resonant motion of one or more quanta of magnetic flux, the so called fluxons. The dispersion relation is:

$$\beta = A\Omega \quad . \tag{18}$$

The boundary conditions set:

$$\text{a)}: \; x_0 = 0 \quad ; \qquad \text{b)}: \; \beta x_0 = K(k_f)$$

and

$$\text{c)}: \beta_n = \frac{n}{\ell} K(k_f) \tag{19a,b,c}$$

where n=1,2,... is the number of fluxons on the line.

There is no physical difference between cases a) and b): the one may be reduced to the other by appropriate change of the constant t_0.

For n=0 eq.(16) is reduced to:

$$\phi = 2 \sin^{-1} \left\{ \mathrm{sn}\left[\frac{t-t_0}{k} ; k\right] \right\} ; \qquad 0<k<1 \tag{20}$$

which describes only temporal phase changes in the absence of any magnetic flux.

A Josephson junction, in the approximation of the model described by the sine-Gordon equation with boundary conditions given by eqs.(6), is a conservative system. The relative hamiltonian density is given by:

$$H = \frac{\phi_t^2}{2} + \frac{\phi_x^2}{2} + 1 - \cos\phi \quad . \tag{21}$$

The (normalized) fluxon solution energy computed on this basis is

$$E_f = 8\,\beta\, n\, K(k_f) \quad ; \qquad n \neq 0 \tag{22}$$

where k_f is given by eq.(17a).

3): Breather oscillation

This solution is given by:

$$\phi = 4 \tan^{-1} \{A\, \mathrm{dn}[\beta(x-x_0);k_f]\, \mathrm{sn}[\Omega(t-t_0);k_g]\} \tag{23}$$

where

$$k_f^2 = 1 - \frac{1-\beta^2(1+A^2)/A^2}{\beta^2(1+A^2)}; \quad k_g = \frac{A^2[1-\Omega^2(1+A^2)]}{\Omega^2(1+A^2)} \quad . \tag{24a,b}$$

It describes -see Fig.6- phase oscillations smaller than 2π, but without spatial nodes, which may also be viewed as a fluxon-antifluxon resonant bound state.

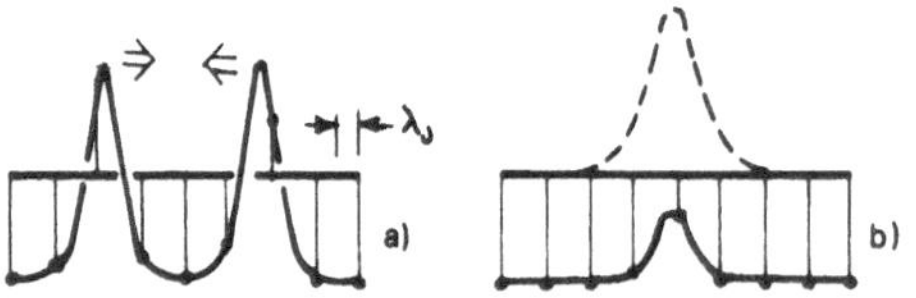

Fig.6 - Breather oscillations on an mechanical line for n=2 and boundary conditions (19), case a).
a): maximum separation;
b): one-quarter period later.
Solid line: $-\cos\phi(x)$;
dashed line $\propto \partial\phi(x)/\partial t$.
From Fig.28 of Fulton [7] (with permission).

The dispersion relation is again given by eq.(18) and the boundary conditions by eqs.(19). There exists, however, physical difference between cases a) and b) only when n is even: in case a) the breather is located near the center of the line, in case b) it is located near the ends of the line as a fluxon-virtual antifluxon bound state. When n is odd the two cases are identical under a reflection of the ends of the line.

The (normalized) breather solution energy computed on the basis of eq.(21) is:

$$E_b = 8\,\beta\, n\, K(k_f) \qquad ; \qquad n\neq 0 \tag{25}$$

where k_f is given by eq.(24a).

FLUXON-BREATHER-PLASMA OSCILLATION DECAY

A careful analysis [9] of solutions (11), (16) and (23) proves the following facts:

1) There exists a minimum value allowed for the breather amplitude A, in eq.(23), given by:

$$A_{min,b} = \frac{\ell}{n\pi} - \left[\frac{\ell^2}{n^2\pi^2} - 1\right]^{\frac{1}{2}} \tag{26}$$

provided that $(\ell/n)\geq\pi$. Setting $A\equiv A_{min,b}$, solution (23) is reduced to solution (15).

2) There exists a maximum value allowed for amplitude in eq.(23); it may be computed by solving the implicit equation:

$$\frac{A^2_{max,b}}{A^2_{max,b}+1} = \frac{n}{\ell}\, K(\{1-(1/A^4_{max,b})\}^{\frac{1}{2}}) \ . \tag{27}$$

3) Setting $\beta^2=A^2/(A^2-1)$, solution (16) is reduced to:

$$\phi = 4\tan^{-1}\left\{ A\,\mathrm{sech}\left[\frac{A(x-x_0)}{(A^2-1)^{1/2}}\right]\sinh\left[\frac{t-t_0}{(A^2-1)^{1/2}}\right]\right\} \tag{28}$$

which, with the identification $A\equiv\frac{1}{u}$ -u is the fluxon velocity- describes the fluxon-antifluxon collision on an infinite line. This leads to a physical interpretation for the fluxon parameter A which, moreover, is qualitatively consistent with the results of a more detailed analysis of eq.(16).

4) There exists a maximum value of A in eq.(16); it may be computed by solving the implicit equation:

$$\frac{A^2_{max,f}}{A^2_{max,f} - 1} = \frac{n}{\ell} K(\{1-(1/A^4_{max,f})\}^{\frac{1}{2}}) \ . \qquad (29)$$

This involves a minimum fluxon velocity allowed on a finite line.

The point 1) has a straightforward physical implication: when a breather approaches because of some dissipative mechanisme its minimum amplitude decays into the n=0 plasma oscillation mode described by solution (15). The points 2), 3) and 4) seem to suggest the existence of a decay mechanism for fluxon. However it can be proved that the highest breather energy is always smaller than the lowest fluxon energy, for fixed ℓ/n. This point makes not very acceptable the hypotesis of a direct decay between fluxons and breathers. It rather suggests a decay scheme less simple as, for example, between fluxons and breathers plus plasma oscillations.

FLUXON PROPAGATION

These investigations have implications of direct experimental significance. The object of interest is a phenomenon in the current-voltage characteristic of long junctions: it consists of a series of current spikes equally spaced in voltage -see Fig.7- usually called zero field steps, since they may be observed in zero

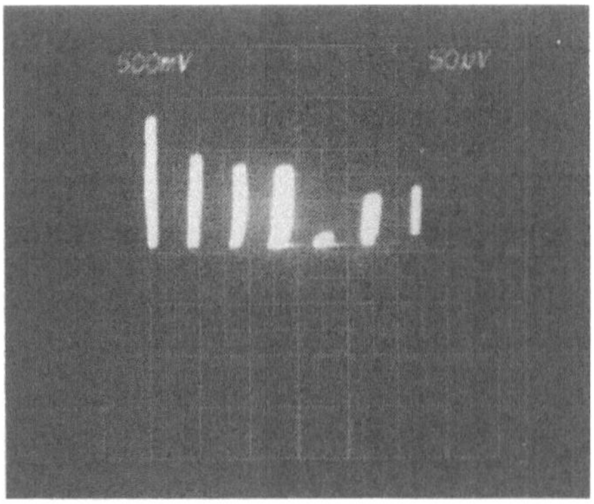

Fig. 7 - Zero-field-steps in a long Nb-Pb junction. Vertical: 500μA/ major div.; horizontal: 50μV/ major div.

applied magnetic field [10]. Fulton and Dynes [11] put forward the hypotesis that they are due to resonant fluxon motion on the junction. Accordingly one may use the solution (16) to describe them. However, since this phenomenon represents a state of non-zero power dissipation, its explanation requires a loss mechanism which is not described by eq.(5); therefore the theoretical current-voltage curve has been calculated by means of the power balance technique.

This consists of imposing

$$P_{in} = P_{loss} \tag{30}$$

where P_{in} is the average power input to the system furnished by the bias sources and P_{loss} is the average power lost through the dissipative mechanism.

They are given by:

$$P_{in} = \frac{\gamma}{T\ell} \int_{x=0}^{x=\ell} dx \int_{t=-T/2}^{t=+T/2} V\, dt$$

$$P_{loss} = 4\pi^2 \frac{\Gamma}{T\ell} \int_{x=0}^{x=\ell} dx \int_{t=-T/2}^{t=+T/2} V^3\, dt \tag{31a,b}$$

in which ℓ is the normalized length of the line, T is the temporal period of oscillation and $V \equiv \phi_t/2\pi$ is the normalized voltage. The result is obtained in parametric form applying eq.(30):

$$\gamma = 2A^2\Omega^2\Gamma\left[\frac{6E(k_f)}{K(k_f)} + \frac{2-k_g^2}{A^2}\right] \; ; \qquad \langle V\rangle = \Omega/K(k_g) \tag{32a,b}$$

in which E(k) is the complete elliptic integral of the second kind and k_f and k_g are given by eqs.(17).

Fig.8 shows the graph of γ versus $\langle V\rangle$. The power balance calculation explains sufficiently the voltage separation and the essential shape of the zero field steps. However this approximation is too rough to explain the finite and generally nonuniform height of the zero field steps and it does not take into account some experimental pecularities such as a voltage cut-off and a fine structure of the spikes [12].

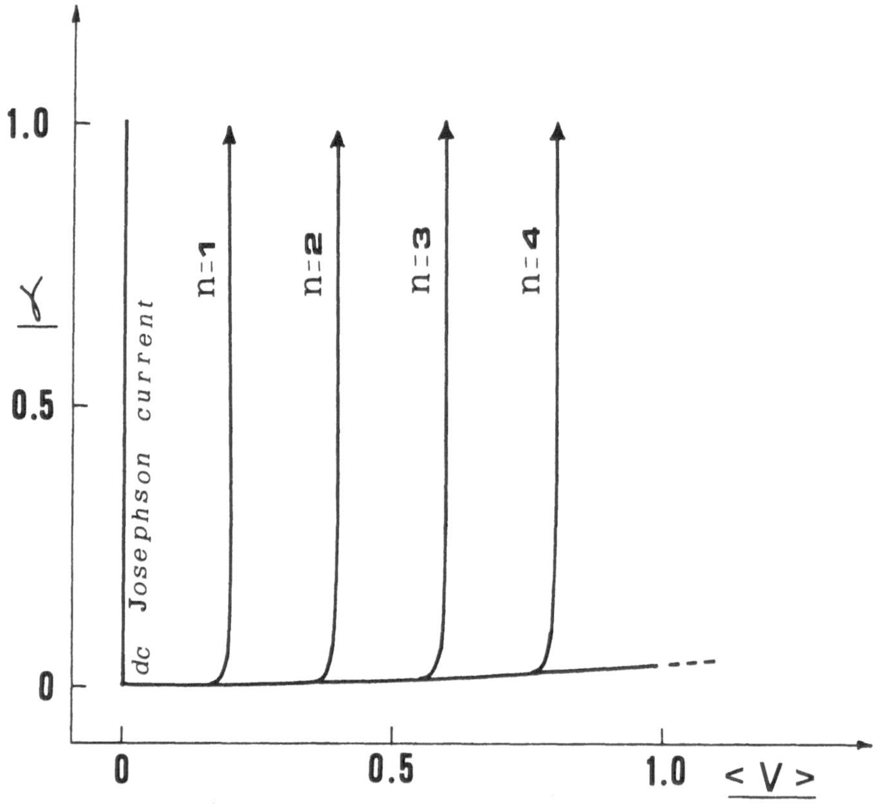

Fig. 8 - DC current singularities computed on the basis of eqs.(32a,b) in a junction having $\ell=5$ and $\Gamma=10^{-3}$.

ANALYSIS OF THE RADIATION

The frequency of the radiation from a Josephson junction biased on a zero field step is comprised in the microwave range.

This fact makes a long junction a device of interest in applications as a radiation generator.

An approximate analysis of the radiation from such a junction has been carried out [13] assuming that the radiation may be described in terms of the voltage V at the end points of the junction.

From eq.(16) this voltage is given by:

$$V(0,t) = \left.\frac{\phi_t}{2\pi}\right|_{x=0} = \frac{2A\Omega\,\mathrm{dn}(\Omega t;k_g)}{\pi\{1+(A^2-1)\,\mathrm{sn}^2(\Omega t;k_g)\}} = (2A\Omega/\pi)\,\mathrm{dn}(\Omega t;k_g)\ . \qquad (33)$$

The last approximation is valid in the part of the zero field steps of major practical interest.

An expansion of the Jacobian dn function in eq.(33) gives, respectively, for the amplitude and the frequency of the general Fourier component of order m:

$$V_m = \frac{2A\Omega}{K(k_g)} \operatorname{sech}\left[\frac{m\pi K(1-k_g^2)^{\frac{1}{2}}}{K(k_g)}\right]; \quad \omega_m = \frac{m\pi\Omega}{K(k_g)} \quad . \tag{34a,b}$$

The basic frequency calculated on this basis is 242 MHz/μV; it is consistent with the experimental results [14]. In Fig.9 is traced the resulting power spectrum as a function of the bias current γ for a particular set of junction parameters; in this graph γ is calculated from eq.(32a) and power is proportional to the squares of the V_m of eq.(34a).

COMMENTS

Further investigations of solutions of sine-Gordon equation could provide an analytical description of more complex oscillation modes which in part, have been already observed on a mechanical analog [15].

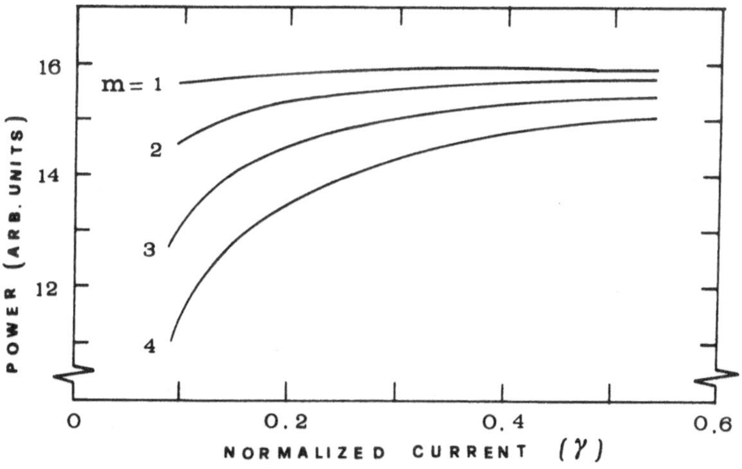

Fig. 9 - Power spectrum (lowest four harmonics) of the radiation of a junction having $\ell=5$, n=1, $\Gamma=10^{-3}$.

This research might facilitate the understanding of the decay phenomena.

The details of fluxon dynamics in the presence of loss and bias might be tachled in the spirit of a non linear perturbation theory [16]. In this connection the investigation of the behaviour of the zero field steps in external magnetic field suggests seeking solutions of eq.(4) with boundary conditions more appropriate than those expressed in eqs. (6). This last problem has its main objective in the application of a long junction as a microwave oscillator.

ACKNOWLEDGMENTS

The author wishes to express his appreciation to g.Perna for technical assistance in the preparation of the manuscript. Financial support was provided in part by the Progetto Finalizzato "Superconduttività" of the Consiglio Nazionale delle Ricerche.

REFERENCES

[1] B.D.Josephson, Advan.Phys. 14, 419 (1965).

[2] R.D.Parmentier, "Fluxon in long Josephson junctions", in *Solitons in Action*, K. Lonngren and A.Scott, eds. (Academic Press, New York, 1978), pp.173-199.

[3] J.C.Swihart, "Field solution for a thin-film superconducting strip transmission line", J.Appl.Phys. 32, 461-469 (1961).

[4] G.Costabile and R.D.Parmentier, "Analytic solution for fluxon propagation in Josephson junctions with bias and loss", in *Low Temperature Physics-LT 14*, vol.4, M.Krusius and M.Vuorio, eds. (North Holland, Amsterdam, 1975), pp.112-115.

[5] R.D.Parmentier and G.Costabile, "Fluxon propagation and dc current singularities in long Josephson junctions", Rocky Mountain J.Math.8, 117-124 (1978).

[6] G.Costabile, R.D.Parmentier, B.Savo, D.W.McLaughlin, and A.C.Scott, "Exact solutions of the sine-Gordon equation describing oscillations in a long (but finite) Josephson junction", Appl.Phys.Lett. 32, 587-589 (1978).

[7] T.A.Fulton, "Equivalent circuits and analogs of the Josephson effect", in *Superconductor Applications: SQUIDs and Machines*, B.B.Schwartz and S.Foner, eds.(Plenum Press, New York, 1977), pp.125-187.

[8] P.F.Byrd and M.D.Friedman, *Handbook of Elliptic Integrals for Engineers and Physicists.* (Springer-Verlag, Berlin, 1954).

[9] G.Costabile, R.D.Parmentier, and B.Savo, "Fluxon-breather-plasma oscillation decay in long Josephson junctions", J.Physique 39, Colloque C6, 567-568 (1978).

[10] J.T.Chen, T.F.Finnegan and D.N.Langenberg, "Anomalous DC current singularities in Josephson tunnel junctions", Physica 55, 413, (1971).

[11] T.A.Fulton and R.C.Dynes, "Single vortex propagation in Josephson tunnel junctions", Solid State Communications 12, 57 (1973).

[12] J.T.Chen and D.N.Langenberg, "Fine structure in the anomalous dc current singularities of a Josephson tunnel junction", in *Low temperature Physics-LT 13*, 3, K.D. Timmerhaus, W.J.O'Sullivan, and E.F.Hammel, eds., Plenum Press, New York, 1974, p.289.

[13] G.Costabile, R.D.Parmentier and B.Savo, "Josephson tunnel junctions generators of microwave radiation", in *Proceedings of the XV International Congress of Refrigeration,* Venezia, september 1979, paper A1/2-5.

[14] T.A.Fulton and L.N.Dunkleberger, "Vortex propagation and radiation emission in Josephson tunnel junctions", Revue de Physique Appliquée, 9, 299 (1974).

[15] M.Cirillo, "Mechanical models of long Josephson junctions", Graduation Thesis, Degree Program in Physics, University of Salerno, October 1979 (in italian).

[16] D.W.McLaughlin and A.C.Scott, "A multisoliton perturbation theory", in *Solitons in Action,* K.Lonngren and A.C.Scott, eds., (Academic Press, New York, 1978), pp. 201-256.

GEOMETRY OF THE AKNS - ZS INVERSE SCATTERING SCHEME

R. Sasaki

Niels Bohr Institute, Blegdamsvej 17,
DK-2100 Copenhagen Ø, Denmark

R.K. Bullough*

Nordita, Blegdamsvej 17,
DK-2100 Copenhagen Ø, Denmark

ABSTRACT

We review the geometrical theory of nonlinear evolution equations (NEEs) solvable by the AKNS[1]-generalised Zakharov-Shabat[2] scattering problem introduced by one of us previously[3,4]. We show how the theory contains within it the canonical structure known to be associated with integrable NEEs. We exploit the "gauge" transformations of the geometric theory to derive an infinite set of non-local Hamiltonian densities for the sine-Gordon equation. We show that it is from these that the hierarchy of Lax-type sine-Gordon equations can be derived. We summarise the relation between the geometric theory and the theory of prolungation structures due to Wahlquist and Estabrook.

* On leave from Department of Mathematics, U.M.I.S.T., P.O. Box 88, Manchester M60 1QD, England.

1. INTRODUCTION

This paper is primarily concerned to study the canonical structure of non-linear evolution equations (NEEs) solvable through the 2×2 AKNS[1]-generalized Zakharov-Shabat[2] scattering problem within the terms of the geometric approach to NEEs investigated by one of us[3,4] previously. In reference 4 abstract expressions for one-parameter families of conservation laws were obtained through the geometrical formalism but the connection with the canonical structure (cf. e.g. reference 5) was not established. In this paper we choose the well known sine-Gordon equation (s-G) as an instructive example and derive *four* infinite sets of conserved densities for it. We are primarily concerned with the geometrical structure, however, and make a number of points in this connection. The general theory extends to larger $n \times n$ scattering problems and results for this case will be given elsewhere[6].

Within the 2×2 scheme there have already been a number of different approaches to the geometry of soliton equations[7,8,9,10,11]. Our work is perhaps nearest to that cited in reference 7; but, as we shall indicate (and see references 6,12), it is perhaps more systematic and potentially more comprehensive and complete than these previous treatments. The work connects in a very natural way with the theory of prolongation structures due to Wahlquist and Estabrook[13] and we shall indicate this connection very briefly at the end of the paper, whilst further details on this aspect will be found elsewhere[6,12].

By the method used in the present paper one of us has already shown[3,4] that every solution of an NEE solvable by the 2×2 AKNS-ZS scheme represents a pseudo-spherical surface - that is a surface of constant negative Gaussian curvature - if the metric is appropriately chosen. From this point of view the groups SL(2,R) or SL(2,C) known[7] to be associated with the AKNS-ZS scheme arise as isometries of these surfaces, whilst it seems to be possible[4] to interpret Bäcklund transformations as "gauge" transformations from one pseudospherical surface to another. The gauge field interpretation of the theory is an important one and is stressed in the present paper. Some gauges offer greater facility for calculation than others and the usual AKNS gauge is in this respect not necessarily the best possible one. Indeed we shall show that the best gauge for calculating the conserved densities of the s-G is not that first used by AKNS[14] to solve this equation.

The sine-Gordon equation is well known and well known to be completely integrable. It is therefore surprising that, as far as we are aware, one half of its infinite set of conserved densities in involution has not been exhibited before expressed explicitly in terms of the sine-Gordon field. This half infinity is easy enough to express in terms of the scattering data (cf. e.g. reference 5). But as we shall show below in terms of the field the conserved densities become non-local, an almost trivial remark which however we have not noticed anywhere else before. It is these non-local densities which must be used as Hamiltonian densities in order to derive the infinite hierarchy of generalized s-G equations analogous to the infinite hierarchy of Korteweg-de Vries (KdV) equations first found by Lax[15].

In this paper we give the first of the generalised s-Gs we have found this way and indicate how the result can be extended to an arbitrary member of the s-G hierarchy. It seems to us that if one wishes to derive both all the involutive conserved densities and the equations of motion of the s-G hierarchy then the geometrical approach of this paper with its freedom to choose the best possible gauge is the best way to proceed. Of course we realise AKNS[16] gave a fairly general expression for NEEs solvable by their scheme some time ago - namely the NEEs

$$\sigma_3 \hat{u}_t + 2\Omega(L^+)\,\hat{u} = 0, \qquad (1.1)$$

in which

$$\hat{u} = \begin{pmatrix} r \\ q \end{pmatrix}, \quad L^+ = \frac{1}{2i}\begin{pmatrix} \partial/\partial x - 2r\int_{-\infty}^{x} dy\, q & , \; 2r\int_{-\infty}^{x} dy\, r \\ -2q\int_{-\infty}^{x} dy\, q & , \; -\partial/\partial x + 2q\int_{-\infty}^{x} dy\, r \end{pmatrix}, \qquad (1.2)$$

where q,r are the two AKNS potentials, $\sigma_3 = \begin{pmatrix} 1 & 0 \\ 0 & -1 \end{pmatrix}$ and $\Omega(\zeta)$ is a ratio of entire functions related to the linearised dispersion relations, $\omega_q(k)$, $\omega_r(k)$ by

$$\Omega(\zeta) = -\frac{i}{2}\omega_q(-2\zeta) = \frac{i}{2}\omega_r(2\zeta) \quad . \qquad (1.3)$$

We have used these expressions to check our generalized s-G equations[17] but apart from the case of the s-G itself the expression of (1.1) in terms of the sines and cosines of the single field

$$u = -2\int_{-\infty}^{x} q\,dx' = 2\int_{-\infty}^{x} r\,dx'$$

is not quite simple to establish. The main part of our paper, however, is concerned to include the canonical structure of the integrable NEEs inside the geometric picture in the way to be described in more detail elsewhere[17].

The paper is organised as follows: in §2 we show that there is a gauge field Ω*) and a curvature field θ associated with the AKNS inverse scattering scheme. The NEEs to be solved are represented by the condition of vanishing curvature, $\theta=0$, the integrability conditions. We show again that solutions of such equations are surfaces of constant negative Gaussian curvature for a chosen metric. In §3 we rederive the abstract expressions for the conservation laws and introduce the admissible gauge transformations; it is here that the gauge field and curvature field interpretations of Ω and θ become apparent. In §4 we derive both local and non-local conservation laws for the s-G: one half of the infinite involutive set consists of polynomial densities and one half is non-local; but in addition we find two other sets of conserved densities, specifically conserved densities of the s-G itself. From the involutive non-local set we derive equations of motion for the next member of the s-G hierarchy. The §5 briefly summarises the connection of the work with the theory of prolongation structures. The §6 is a short summary and discussion. We use the language of differential forms[18] throughout the paper, for this allows a simple and transparent mode of presentation. As noted the paper summarises work published in references 3,4,6,12 and 17, and especially that in the last: the reader is referred to these articles for further details.

---

*) The gauge field Ω is a traceless 2×2 matrix of three oneforms ω_1, ω_2, and ω_3 and has nothing to do with the (scalar) number $\Omega(\zeta)$ defined in (1.3). The notations have been used already in previous work. No confusion seems likely to arise in practice.

2. GEOMETRY OF SOLITON EQUATIONS

In this section we will give again[3,4] the simple geometric picture of all the soliton equations which can be solved in terms of the AKNS-ZS 2×2 scheme. In present terms this scheme is specified by associating a completely integrable Pfaffian system with the non-linear equation to be solved. This Pfaffian takes the form

$$dv = \Omega v \ , \qquad v = \begin{pmatrix} v_1 \\ v_2 \end{pmatrix} , \qquad \Omega = \begin{pmatrix} \omega_1 & \omega_2 \\ \omega_3 & -\omega_1 \end{pmatrix} , \tag{2.1}$$

where d denotes exterior differentiation[18]. The 2×2 matrix Ω is traceless and consists of a one-parameter family of one forms $(\omega_1,\omega_2,\omega_3)$ in the independent variables (x,t), the dependent variables q,r, and their derivatives; the parameter η plays the role of the eigenvalue of the scattering problem[1,2]. The integrability of eq.(2.1) requires that the following twoform θ vanishes:

$$\theta = d\Omega - \Omega\wedge\Omega \ , \tag{2.2}$$

where $\wedge$ denotes exterior product[18]. This, *by construction*, is the original non-linear equation to be solved. It is of course equivalent to the matrix Lax form of the equation

$$\hat{L}_t = [\hat{A},\hat{L}] \tag{2.3}$$

given for example, in reference 5; in this formalism

$$\hat{L}v = \zeta v \tag{2.4a}$$

is the scattering problem and

$$v_t = \hat{A}v \tag{2.4b}$$

is the time evolution of v: the eigenvalue ζ is related to η by $\zeta=-i\eta$.

The geometrical interpretation that all the AKNS-ZS systems describe pseudo-spherical surfaces rests on the observation that eq.(2.2) has the same form as the fundamental equations of such a surface[18]. The structure equations of a two dimensional Riemannian manifold are[18]

$$\left.\begin{aligned} dP &= \sigma^1\mathbf{e}_1 + \sigma^2\mathbf{e}_2 \ , \\ d\mathbf{e}_1 &= \omega\mathbf{e}_2 \ , \quad d\mathbf{e}_2 = -\omega\mathbf{e}_1 \ , \end{aligned}\right\} \tag{2.5}$$

in which the vectors $\mathbf{e}_1$ and $\mathbf{e}_2$ are chosen orthonormal and span the tangent plane of the manifold at each point; σ^1 and σ^2 are one forms dual to the $\mathbf{e}_i$; ω is the "connection" one form[18]. The integrability condition $d^2P = 0$ means

$$d\sigma^1 = \omega\wedge\sigma^2 \quad , \quad d\sigma^2 = -\omega\wedge\sigma^1 \quad . \tag{2.6a-b}$$

The Gaussian curvature K is defined by[18]

$$d\omega = -K\,\sigma^1\wedge\,\sigma^2 \quad . \tag{2.7}$$

For a pseudo-spherical surface, K=-1, and eqs.(2.6) and (2.7) can be written precisely in the form of eq.(2.2) by the particular choice

$$\Omega = \begin{pmatrix} -\frac{1}{2}\sigma^2 & \frac{1}{2}(\omega+\sigma^1) \\ \frac{1}{2}(-\omega+\sigma^1) & \frac{1}{2}\sigma^2 \end{pmatrix} \quad ; \tag{2.8}$$

notice that because (2.1) contains two potentials q and r in general the "surface" is not a simple one. Typically q and r are related by a symmetry, e.g. $r=-q=-q^*$ (for the s-G) or $r=-q^*$ (for the non-linear Schrödinger equation) and the surface is a simple real or complex one. Choices of Ω other than (2.8) are also possible[4,17] -see the different choice made in §4. The metric of the surface is given by

$$ds^2 = (\sigma^1)^2 + (\sigma^2)^2 \quad . \tag{2.9}$$

In terms of the eigenvalue, that is η, dependence of the AKNS-ZS form Ω we have a one-parameter family of pseudo-spherical surfaces for each solution of eq.(2.2). For the s-G, the KdV and the modified KdV equations the one-parameter families of solutions are generated by scale transformations although the R.M.B. equations[19], for example, with a more complicated structure, do not seem to have this property.

This geometric picture of soliton equations is a simple generalization of a well known result; by 1883 it was known (cf. e.g. references 20,22) that the s-G itself describes a pseudo-spherical surface. Further the SL(2,R) structure of soliton equations[7] acquires a simple geometric interpretation; SL(2,R) is the isometry group of real pseudo-spherical surfaces. The Bäcklund transformation for soliton equations apparently fits in well[4], for it was originally introduced[22] as a transformation from one pseudo-spherical surface to another. We make a brief further reference to Bäcklund transformations within the geometrical theory in §5.

3. CONSERVATION LAWS FOR THE AKNS-ZS SYSTEM

The existence of infinite numbers of conservation laws is one of the characteristic features of soliton equations. In this section we will show how these arise naturally within the geometrical structure[4] of the AKNS-ZS system. We derive expressions for the individual conserved densities in a particular case in §4.

We introduce pseudopotentials[13] Γ_1 and Γ_2 defined by

$$\Gamma_1 = \frac{v_2}{v_1} \quad , \qquad \Gamma_2 = \frac{v_1}{v_2} \tag{3.1}$$

The Pfaffians (2.1) are then rewritten as

$$0 = \varepsilon_1 \equiv d\Gamma_1 - \omega_3 + 2\Gamma_1\omega_1 + \Gamma_1^2\omega_2 \ , \tag{3.2a}$$

$$0 = \varepsilon_2 \equiv d\Gamma_2 - \omega_2 - 2\Gamma_2\omega_1 + \Gamma_2^2\omega_3 \ . \tag{3.2b}$$

From the integrability condition (2.2) the exterior derivatives of one-forms ε_1 and ε_2 can now be calculated to be

$$\begin{aligned} d\varepsilon_1 &= 2\,\varepsilon_1 \wedge (\omega_1 + \Gamma_1\omega_2) \quad , \\ d\varepsilon_2 &= 2\,\varepsilon_2 \wedge (-\omega_1 + \Gamma_2\omega_3) \quad , \end{aligned} \tag{3.3}$$

and these are necessary and sufficient conditions for the ε_i to be completely integrable. From closure $d^2\varepsilon_i = 0$ of each of equations (3.3) it follows that the one forms

$$\begin{aligned} \delta_1 &= \omega_1 + \Gamma_1\omega_2 \quad , \\ \delta_2 &= -\omega_1 + \Gamma_2\omega_3 \quad , \end{aligned} \tag{3.4}$$

are closed, that is

$$\begin{aligned} d\delta_1 &= 0 \quad , \\ d\delta_2 &= 0 \quad . \end{aligned} \tag{3.5}$$

These are actually the conservation laws. They are one-parameter families of conservation laws because of the η-dependence and are equivalent to an infinite family of parameter independent conservation laws. Expansion of the δ_i in inverse powers of η gives the well known polynomial conservation laws for the KdV, mKdV, s-G or other NEEs solvable by the AKNS-ZS scheme[23]. We show this for the s-G in §4. In all these cases the procedure provides a natural understanding of the rank of each conservation law. These were defined in reference 24 for the KdV and mKdV and in 25 for the s-G. In each case the ranks are the scale weights.

Note that eq.(3.3) shows that the functions Γ_i (i=1,2) are pseudopotentials in the sense of Wahlquist and Estabrook[13]. From eqs.(3.4) and (3.5) one can see that two other pseudopotentials can be introduced and these correspond to the y_3 in reference 13. This is done more completely in §5, where the connection of the present theory with the theory of prolongation structures is summarised. For more details see the papers of one of us[6,12].

It should be noted that the Pfaffian (2.1) together with its integrability condition (2.2) are form invariant under the following gauge transformation

$$\begin{aligned} v \to v' = Bv \quad , \quad \Omega \to \Omega' = B\Omega B^{-1} + dBB^{-1} \quad , \\ \theta \to \theta' = B\theta B^{-1} \quad , \end{aligned} \tag{3.6}$$

where B is an arbitrary (space-time dependent) 2×2 matrix of determinant unity,

$$\det B = B_{11}B_{22} - B_{12}B_{21} = 1 \quad . \tag{3.7}$$

The one-form Ω thus has an interpretation as a connection (gauge field), the two-form θ as a curvature (gauge field strenght) and the closure property for the NEE

$$d\theta = \Omega \wedge \theta - \theta \wedge \Omega \quad , \tag{3.8}$$

as Bianchi's identity. The transformation (3.6) is the gauge transformation of the 2×2 scattering theory described in §1. The gauge transformation for the pseudopotentials Γ_1 and Γ_2 turns out to be[17]

$$\Gamma_1' = \frac{B_{21} + B_{22}\,\Gamma_1}{B_{11} + B_{12}\,\Gamma_1} \quad , \quad \Gamma_2' = \frac{B_{11}\,\Gamma_2 + B_{12}}{B_{21}\,\Gamma_2 + B_{22}} \tag{3.9}$$

The closed one-forms δ_1 and δ_2 are consequantly gauge invariant up to a complete derivative[17], that is,

$$\left.\begin{aligned} \delta_1' &= \delta_1 + (B_{11} + B_{12}\,\Gamma_1)^{-1}\, d(B_{11} + B_{12}\,\Gamma_1) \quad , \\ \delta_2' &= \delta_2 + (B_{21}\,\Gamma_2 + B_{22})^{-1}\, d(B_{21}\,\Gamma_2 + B_{22}) \quad . \end{aligned}\right\} \tag{3.10}$$

Further the conservation laws obtained from the δ_i are closely related to those obtained from the scattering data. These are derived (cf. e.g. reference 5) from the trasmission coefficient $a(\zeta)$ (or $a(\eta)$) of the scattering problem (2.4a) through $\log a(\eta)$. The coefficient $a(\eta)$ can be expressed as the Wronskian of the two Jost function solutions (ϕ,ψ) of the scattering problem (2.4a)

$$a(\eta) \equiv W(\phi,\psi) = \phi_1\psi_2 - \phi_2\psi_1 \quad , \tag{3.11}$$

in which the indices refer to the components of ϕ and ψ. Because of the condition (3.7) $a(\eta)$ is gauge invariant. It is easy to demonstrate[17] that the following relationship exists between the gauge invariant quantities $a(\eta)$, δ_1 and δ_2,

$$\log a(\eta) = \int_{x=-\infty}^{\infty} (\delta_1 - d\beta_1) = \int_{x=-\infty}^{\infty} (\delta_2 - d\beta_2) \ , \tag{3.12}$$

where β_1 and β_2 are functions depending on the chosen gauge. Therefore the conservation laws obtained by expansion of $\log a(\eta)$ about an ordinary point $\eta=\mu$ $(a(\mu)\neq 0)$ or the point at ∞ $(\eta=\infty)$ are reproduced modulo complete derivatives by expansion of δ_i about the same point.

This result (3.12) establishes the connection between the canonical structure of the NEE solvable by the AKNS-ZS scheme and the geometrical theory.

4. LOCAL AND NON-LOCAL CONSERVATION LAWS FOR THE SINE-GORDON EQUATION

In this section we apply the present formalism to find infinite numbers of conservation laws for the s-G. This equation has been much studied[14,26,27,28,29] and it is well understood as an example of an infinite dimensional completely integrable Hamiltonian system[5,27,28,29]. Such a system has an infinite set of conserved quantities in involution; and from this or an appropriate infinite subset, an infinite set of Hamiltonians can in principle be found which determines an infinite hierarchy of generalized sine-Gordon equations analogous to the infinite hierarchy of KdV equations first found by Lax[15]. As noted in §1, as far as we are aware, this hierarchy of s-G equations has never been correctly exhibited beyond its first member. The only specific statement about even the second member of this s-G hierarchy is due to one of us

[29,30]. It was mentioned briefly in passing[29,30] and it is now clear that this brief remark was scarcely correct. The situation is a curious one and we analyse it in some detail in this section.

Let us start with the derivation of infinite numbers of conservation laws using the formalism presented in the previous section. A one-form matrix Ω for the s-G in lightcone coordinates

$$u_{xt} = \sin u \quad , \tag{4.1}$$

is given by[3,14,16]

$$\Omega = \begin{pmatrix} \eta dx + \frac{1}{4\eta}\cos u\, dt & -\frac{1}{2}u_x dx + \frac{1}{4\eta}\sin u\, dt \\ \frac{1}{2}u_x dx + \frac{1}{4\eta}\sin u\, dt & -\eta dx - \frac{1}{4\eta}\cos u\, dt \end{pmatrix} . \tag{4.2}$$

The η dependence can be understood as a consequence of the invariance of the original equation (4.1) under the scale transformation

$$x \to x' = \eta x \quad , \quad t \to t' = \eta^{-1} t \quad , \quad u \to u' = u \tag{4.3}$$

This becomes the Lorentz transformation in "laboratory coordinates"[25] and the system is Lorentz invariant in these coordinates. The s-G (4.1) is also symmetric under interchange of x and t, but the one-form Ω (4.2) is not. In order to simplify the calculation of the conservation laws we make full use of this symmetry by making the gauge transformation (3.6) with

$$B = \frac{1}{2}\begin{pmatrix} e^{-iu/4} & 0 \\ 0 & e^{iu/4} \end{pmatrix} \begin{pmatrix} 1+i & -(1-i) \\ 1+i & 1-i \end{pmatrix} . \tag{4.4}$$

This leads to

$$\Omega = \begin{pmatrix} \frac{i}{4}(u_x dx - u_t dt) & \eta e^{-iu/2} dx + \frac{1}{4\eta} e^{iu/2} dt \\ \eta e^{iu/2} dx + \frac{1}{4\eta} e^{-iu/2} dt & -\frac{i}{4}(u_x dx - u_t dt) \end{pmatrix} . \tag{4.5}$$

In terms of $\eta' = 2\eta$ we now get the x,t symmetric one-form

$$\Omega = \begin{pmatrix} \frac{i}{4}(u_x dx - u_t dt) & \frac{1}{2}\eta' e^{-iu/2} dx + \frac{1}{2\eta'} e^{iu/2} dt \\ \frac{1}{2}\eta' e^{iu/2} dx + \frac{1}{2\eta'} e^{-iu/2} dt & -\frac{i}{4}(u_x dx - u_t dt) \end{pmatrix} , \tag{4.6}$$

or

$$\omega_1 = \frac{i}{4}(u_x dx - u_t dt) \quad , \tag{4.7a}$$

$$\omega_2 = \frac{1}{2}\eta' e^{-iu/2}dx + \frac{1}{2\eta'}e^{iu/2}dt \quad , \tag{4.7b}$$

$$\omega_3 = \frac{1}{2}\eta' e^{iu/2}dx + \frac{1}{2\eta'}e^{-iu/2}dt \quad . \tag{4.7c}$$

In (4.6) and (4.7) and in what follows we omit the primes on η.

Note that the well known metric[20] of the pseudospherical surface associated with the s-G (4.1) is given by

$$ds^2 = (dx)^2 + 2\cos u\, dxdt + (dt)^2 \quad .$$

This corresponds to the choice (see (2.9))

$$\sigma^1 = \cos\frac{u}{2}(dx-dt), \quad \sigma^2 = \sin\frac{u}{2}(dx+dt), \quad \omega = \frac{1}{2}(u_x dx - u_t dt) \quad .$$

A possible identification with Ω (compare (2.8))

$$\Omega = \begin{pmatrix} \frac{i}{2}\omega & \frac{1}{2}(\sigma^1 - i\sigma^2) \\ \frac{1}{2}(\sigma^1 + i\sigma^2) & -\frac{i}{2}\omega \end{pmatrix}$$

together with the scale transformation (4.3) reproduces the one-form (4.6).

Next we solve the Riccati equations (3.2a,b) with the one-forms (4.7). Note the symmetry which the two equations (3.2a) and (3.2b) now have. For real values of the variables x,t and u and the parameter η we have that

$$\omega_1^* = -\omega_1 \quad , \qquad \omega_2^* = \omega_3 \quad , \tag{4.8}$$

and therefore the relationship

$$\Gamma_1^* = \Gamma_2 \quad , \tag{4.9}$$

($*$ denotes complex conjugate). In the same way the two closed one forms δ_1 and δ_2 are related by

$$\delta_1 = \delta_2^* \quad . \tag{4.10}$$

In the following we treat only Γ_1 and δ_1 and omit the suffices.

The Riccati equation (3.2a) is equivalent to

$$\Gamma_x - \frac{1}{2}\eta e^{iu/2} + \frac{i}{2}u_x\Gamma + \frac{1}{2}\eta e^{-iu/2}\Gamma^2 = 0 \quad , \tag{4.11a}$$

and

$$\Gamma_t - \frac{1}{2\eta}e^{-iu/2} - \frac{i}{2}u_t\Gamma + \frac{1}{2\eta}e^{iu/2}\Gamma^2 = 0 \quad . \tag{4.11b}$$

These two equations are symmetric under the operation

$$u \leftrightarrow -u, \quad x\text{-derivative} \leftrightarrow t\text{-derivative}, \quad \eta \leftrightarrow \eta^{-1} . \qquad (4.12)$$

We solve equations (4.11a,b) in two different ways by substituting the alternative asymptotic expansions

$$\Gamma = \sum_{k=0}^{\infty} \eta^{-k} \gamma_k , \qquad (4.13a)$$

and

$$\Gamma = \sum_{k=0}^{\infty} \eta^{k} \tilde{\gamma}_k . \qquad (4.13b)$$

The coefficients $\gamma_k(\tilde{\gamma}_k)$ are determined from eq.(4.11a) ((4.11b)) by equating like powers of $\eta^{-1}(\eta)$. Consistency of this procedure is guaranteed by the complete integrability of eq.(3.2a). Because of the symmetry (4.12) we have the relationship

$$\tilde{\gamma}_k = \gamma_k(u \to -u,\ u_x \to -u_t,\ u_{xx} \to -u_{tt}, \ldots) . \qquad (4.14)$$

Therefore we have to determine only the γ_k's. The first coefficient γ_o is calculated to be

$$\gamma_o = \pm e^{iu/2} .$$

Here we choose without loss of generality

$$\gamma_o = - e^{iu/2} . \qquad (4.15)$$

Next we get

$$\gamma_1 = -i\, e^{iu/2} u_x, \quad \gamma_2 = e^{iu/2} (\tfrac{1}{2} u_x^2 - i\, u_{xx}) , \qquad (4.16)$$

and in general the following recursion formula for the γ_k,

$$\gamma_k = (\gamma_{k-1})_x + \frac{i}{2} u_x \gamma_{k-1} + \frac{1}{2} e^{-iu/2} \sum_{\ell=1}^{k-1} \gamma_{k-\ell}\, \gamma_\ell . \qquad (4.17)$$

Introducing new coefficients $\xi_k(\tilde{\xi}_k)$ by

$$\gamma_k = e^{iu/2}\, \xi_k , \quad (\tilde{\gamma}_k = e^{-iu/2}\, \tilde{\xi}_k) , \qquad (4.18a,b)$$

we obtain a simple recursion formula for ξ_k

$$\xi_k = (\xi_{k-1})_x + \frac{1}{2}\sum_{\ell=2}^{k-2} \xi_{k-\ell}\,\xi_\ell\ , \quad k\geq 3\ , \tag{4.19}$$

and

$$\xi_0 = -1\ , \qquad \xi_1 = -iu_x\ , \qquad \xi_2 = \frac{1}{2}u_x^2 - iu_{xx}\ . \tag{4.20}$$

From the recursion formula we see that $\xi_k (k\geq 3)$ is a polynomial in ξ_2 and its x-derivatives. If we denote

$$\xi_2 = M = \frac{1}{2}u_x^2 - iu_{xx}\ , \tag{4.21}$$

the first several terms are written as

$$\xi_3 = M_x\ , \qquad \xi_4 = M_{xx} + \frac{1}{2}M^2\ , \qquad \xi_5 = M_{xxx} + (M^2)_x\ ,$$

$$\xi_6 = M_{xxxx} + \frac{3}{2}(M^2)_{xx} + \frac{1}{2}M^3 - \frac{1}{2}(M_x)^2\ , \quad \ldots \tag{4.22}$$

From (4.13a) and (4.18a) into (3.4), we have after collecting like powers of η^{-1} that

$$\delta = -\frac{1}{2}\eta dx - \frac{i}{4}du + \frac{1}{2}\sum_{k=1}^{\infty}\eta^{-k}\,p_k\ , \tag{4.23a}$$

with

$$p_k = \xi_{k+1}\,dx + e^{iu}\,\xi_{k-1}\,dt\ , \quad k\geq 1\ , \tag{4.23b}$$

and from (3.5), $d\delta=0$,

$$dp_k = 0\ . \tag{4.23c}$$

Equation (4.23c) represents the infinite set of polynomial conservation laws

$$(\xi_{k+1})_t - (e^{iu}\,\xi_{k-1})_x = 0\ . \tag{4.24}$$

Explicitly

$$\mathrm{Re}\,\xi_2 = \frac{1}{2}u_x^2\ , \qquad \mathrm{Re}\,\xi_4 = \frac{1}{8}u_x^4 - \frac{1}{2}u_{xx}^2\ , \quad \text{and}$$

$$\mathrm{Re}\,\xi_6 = \frac{1}{16}u_x^4 - \frac{5}{4}u_{xx}^2\,u_x^2 + \frac{1}{2}u_{xxx}^2\ , \tag{4.25}$$

after removal of a perfect derivative. Thus the $\mathrm{Re}\,\xi_{2n}$ are equivalent the T^{2n} given in reference 25 in a completely "reduced" form. The $\mathrm{Im}(\xi_{2n})$ prove to be perfect derivatives and are trivial: they must be trivial because they change sign under the symmetric operation of the s-G equation

$$u \to -u \ .$$

Nevertheless they play an important role in the recurrence relation. The ξ_{2n+1} also emerge as perfect derivatives: it is proved in reference 25 that the whole infinite set of these consists of trivial conservation laws.

An alternative set of non-polynomial conservation laws follows from the Riccati equation (4.11b) expanded in a power series in η by (4.13b). We find in this case[17] that the first several members of a set of nontrivial conserved densities in the form $\mathrm{Re}\,\{e^{-iu}\tilde{\xi}_{2n}\}$ are

$$-\cos u \ , \quad \frac{1}{2}u_t^2 \cos u + u_{tt}\sin u \ , \quad \ldots \qquad . \tag{4.26}$$

The conserved densities (4.26) are certainly conserved densities for the s-G itself, but are not the involutive set of densities we wish to find for the hierarchy of s-G equations. This point can be seen as follows. It is known from previous work[5,29] that the Hamiltonian $\bar{H}_{2m+1}$ obtained by expanding $\log a$ in powers of η

$$\log a(\eta) = \sum_{m=0}^{\infty} \bar{H}_{2m+1}\,\eta^{2m+1} \ , \tag{4.27}$$

expressed *in terms of scattering data* generates an equation of motion of the member of the s-G hierarchy which has a linearized dispersion relation

$$\omega = (-1)^m\, k^{-(2m+1)} \ . \tag{4.28}$$

The structure of the generalized Marchenko equation also tells us[5] that the equation has the single soliton (kink) solution

$$u(x,t) = 4\tan^{-1}\exp(\eta x + \eta^{-(2m+1)}t + \delta) \ , \tag{4.29}$$

where δ is a constant.

If the first member of the conserved densities (4.26), $-\cos u$, or rather $1-\cos u$, is adopted as a Hamiltonian it gives the s-G (4.1) as the equation of motion[5]. However, the second member of (4.26) gives, through Hamilton's principle, an equation of motion

$$u_{xt} = -\frac{1}{2}\{u_t^2 \sin u - 2u_{tt}\cos u\} \ , \tag{4.30}$$

which is the one suggested by one of us previously[29,30]. This equation linearises to

$$u_{xt} = u_{tt} \quad , \tag{4.31}$$

and has a linearised dispersion relation

$$\omega = -k \quad , \tag{4.32}$$

which is certainly different from (4.28) for m=1. It is also easy to check that equation (4.30) does not have the kink solution (4.29) for m=1. Therefore we have to conclude that the set (4.26) is not the involutive set which generates the s-G hierarchy.

The sets of conserved densities we require are found by alternative expansions in powers of η and η^{-1} of the Riccati equation derived from the scattering problem (2.4a) $\hat{L}v = \zeta v$. The calculations are carried out in reference 17 and involve the solution of a system of differential equations instead of (4.17). This system of equations can be solved for the boundary conditions for u, namely,

$$u \to 0 \pmod{2\pi} \;; \qquad u_x,\, u_{xx},\, \text{etc.} \to 0 \text{ as } |x| \to \infty \quad . \tag{4.33}$$

The constants of integration are chosen so as to be compatible with equation (4.11b). The first several terms analogous to (4.22) are

$$\tilde{\xi}_0 = -1, \quad \tilde{\xi}_1 = i\int_{-\infty}^{x} \sin u\, dx, \quad \tilde{\xi}_2 = i\int_{-\infty}^{x} e^{-iu} dx_1 \int_{-\infty}^{x_1} \sin u\, dx_2,$$

$$\tilde{\xi}_3 = i\int_{-\infty}^{x} e^{-iu}\, dx_1 \int_{-\infty}^{x_1} e^{-iu}\, dx_2 \int_{-\infty}^{x_2} \sin u\, dx_3$$

$$+ \int_{-\infty}^{x} e^{-iu}\, dx_1 \int_{-\infty}^{x_1} \sin u\, dx_2 \int_{-\infty}^{x_2} \sin u\, dx_3, \quad \tilde{\xi}_4 = \ldots \tag{4.34}$$

We find instead of (4.23a) that

$$\delta = -\frac{1}{2\eta} dt + \frac{i}{4} du + \frac{1}{2}\sum_{k=1}^{\infty} \eta^k q_k \quad , \tag{4.35a}$$

with

$$q_k = e^{-iu}\tilde{\xi}_{k-1}\, dx + \tilde{\xi}_{k+1}\, dt\, , \qquad k \geq 1 \quad , \tag{4.35b}$$

whilst instead of (4.24)

$$(e^{-iu}\tilde{\xi}_{k-1})_t - (\tilde{\xi}_{k+1})_x = 0 \tag{4.36}$$

We show next that the quantities $\mathrm{Re}(e^{-iu}\tilde{\xi}_{2k})$ defined by (4.34) have the properties of the members of the infinite set of conserved densities we have been looking for. The quantities $\int_{x=-\infty}^{\infty} \mathrm{Re}(q_k)$ are

$$\overline{H}_1 = \int_{-\infty}^{\infty} (1-\cos u)\, dx, \quad \overline{H}_2 = \int_{-\infty}^{\infty} \sin u\, dx \int_{-\infty}^{x} \sin u\, dx_1,$$

$$\overline{H}_3 = \int_{-\infty}^{\infty} \sin u\, dx \int_{-\infty}^{x} \cos u\, dx_1 \int_{-\infty}^{x_1} \sin u\, dx_2 + \int_{-\infty}^{\infty} \cos u\, dx \int_{-\infty}^{x} \sin u\, dx_1 \int_{-\infty}^{x_1} \sin u\, dx_2,$$

$$\begin{aligned}\overline{H}_4 = & \int_{-\infty}^{\infty} \sin u\, dx \int_{-\infty}^{x} \cos u\, dx_1 \int_{-\infty}^{x_1} \cos u\, dx_2 \int_{-\infty}^{x_2} \sin u\, dx_3 \\ & + \int_{-\infty}^{\infty} \cos u\, dx \int_{-\infty}^{x} \sin u\, dx_1 \int_{-\infty}^{x_1} \cos u\, dx_2 \int_{-\infty}^{x_2} \sin u\, dx_3 \\ & -2\int_{-\infty}^{\infty} \sin u\, dx \int_{-\infty}^{x} \sin u\, dx_1 \int_{-\infty}^{x_1} \sin u\, dx_2 \int_{-\infty}^{x_2} \sin u\, dx_3 \\ & +2\int_{-\infty}^{\infty} \cos u\, dx \int_{-\infty}^{x} \cos u\, dx_1 \int_{-\infty}^{x_1} \sin u\, dx_2 \int_{-\infty}^{x_2} \sin u\, dx_3,\end{aligned}$$

$$\overline{H}_5 = \ldots, \quad \ldots \quad . \tag{4.37}$$

We shall use $\overline{H}_1$, $\overline{H}_2$, $\overline{H}_3$, etc as Hamiltonians and check that they have the right properties for the hierarchy of s-G's.

Note first of all that since the sequence (4.37) is obtained by expanding $\log a$ in powers of ζ the quantities

$$\overline{H}_2 = \int_{-\infty}^{\infty} \sin u\, dx \int_{-\infty}^{x} \sin u\, dx_1 = \frac{1}{2}\left(\int_{-\infty}^{\infty} \sin u\, dx\right)^2 ,$$

$$\overline{H}_4 = \ldots , \qquad \overline{H}_6 = \ldots , \qquad \ldots , \tag{4.38}$$

all actually vanish. This follows from the result demonstrated in reference 5 (see Table 3) that by expanding $\log a$ expressed in terms of scattering data in powers of ζ (the case of $\mu=0$ of Table 3) all the $\overline{H}_{2m}$ vanish. Thus in particular

$$\int_{-\infty}^{\infty} \sin u\, dx = 0 \tag{4.39}$$

for every member of the s-G hierarchy.

It is easy to check that for the s-G itself with boundary conditions (4.33) relations (4.39), $\overline{H}_2=0$, and $\overline{H}_4=0$ are all true. But $\overline{H}_{2m}$ are infinite in number and apparently distinct, and so large a sequence of constraints on every number of the s-G hierarchy is rather unexpected. Note, however, that this set of constraints is not actually imposed but arises naturally through the argument. In this sense the constraints are not contraints at all. Notice that at first sight these constraints are nevertheless unreasonable in that even for the boundary conditions (4.33) arbitrary initial data will not satisfy $\overline{H}_{2m}=0$. The point here however is that initial data for the s-G hierarchy cannot be chosen in an arbitrary manner: arbitrary initial data will in general evolve under the equation of motion and break the boundary conditions. The set of initial data satisfying $\overline{H}_{2m}=0$ (together with the further constraints we describe next) is just the selected set of initial data which is compatible with the boundary conditions (4.33) and the equation of motion.

The further constraints are these. Because of the symmetry argument that under the operation $u\to -u$, the s-G is invariant, the quantities $\int_{x=-\infty}^{\infty} \mathrm{Im}\, q_k$ are also trivial Hamiltonians. These quantities form the sequence

$$\int_{-\infty}^{\infty}\sin u\, dx,\quad \int_{-\infty}^{\infty}\cos u\, dx\int_{-\infty}^{x}\sin u\, dx_1,$$

$$\int_{-\infty}^{\infty}\cos u\, dx\int_{-\infty}^{x}\cos u\, dx_1\int_{-\infty}^{x_1}\sin u\, dx_2-\int_{-\infty}^{\infty}\sin u\, dx\int_{-\infty}^{x}\sin u\, dx_1\int_{-\infty}^{x_1}\sin u\, dx_2,$$

$$\ldots\ , \tag{4.40}$$

and these must also vanish. Evidently the sets (4.38) and (4.40) are not independent.

From this analysis it follows that the first few nontrivial Hamiltonians are

$$\overline{H}_1 = \int_{-\infty}^{\infty}(1-\cos u)\, dx\ ,$$

$$\overline{H}_3 = \int_{-\infty}^{\infty}\sin u\, dx\int_{-\infty}^{x}\cos u\, dx_1\int_{-\infty}^{x_1}\sin u\, dx_2 + \int_{-\infty}^{\infty}\cos u\, dx\int_{-\infty}^{x}\sin u\, dx_1\int_{-\infty}^{x_1}\sin u\, dx_2,$$

$$\overline{H}_5 = \ldots,\quad \ldots\ . \tag{4.41}$$

The equations of motion which derive from these are (*)

$$u_{xt} = \sin u \ ,$$

$$u_{xt} = \sin u \int_{-\infty}^{x} \sin u \, dx_1 \int_{-\infty}^{x_1} \sin u \, dx_2 - \cos u \int_{x}^{\infty} \cos u \, dx_1 \int_{-\infty}^{x_1} \sin u \, dx_2 ,$$

$$u_{xt} = \ldots , \quad \ldots \quad . \tag{4.42}$$

Notice that the s-G itself is exceptional in being invariant under the scale transformation (4.3). It follows that there is no Lorentz invariant form of any equation of the hierarchy except this first member. The set is therefore presumably without interest to field theorists.

To see that the Hamiltonians (4.41) have the properties required of them note first that the linearised dispersion relation of the second member of (4.42) is precisely $\omega = -k^3$. For from the vanishing of the second integral in the sequence (4.40) the second equation in (4.42) is equivalent to

$$u_{xt} = \left(\cos u \int_{-\infty}^{x} \cos u \, dx_1 + \sin u \int_{-\infty}^{x} \sin u \, dx_1 \right) \int_{-\infty}^{x_1} \sin u \, dx_2 , \tag{4.43}$$

which linearises to

$$u_{xt} = \int_{-\infty}^{x} dx_1 \int_{-\infty}^{x_1} u \, dx_2 , \tag{4.44a}$$

or

$$u_{xxxt} = u , \tag{4.44b}$$

with the dispersion relation $\omega = -k^{-3}$ as indicated. One also finds that the equation (4.43) has the kink solution (4.29) for m=1.

We have not been able to construct a general proof that the $\bar{H}_{2m+1}$ in the sequence (4.41) yield linearised dispersion relations $\omega = (-1)^m \cdot k^{-(2m+1)}$ and non-linear equations of motion (**) which have the kink solution (4.29). Nevertheless we have found two sequences of conserved densities by expanding $\log a$ about $\zeta = 0$ and $\zeta = \infty$ ($\eta = 0$ and $\eta = \infty$) via the

---

(*) Problems of convergence of integrals arising in executing the functional derivatives appear to go away because of the constraints provided by the trivial Hamiltonians (4.38) and (4.40). This again illustrates their significance.

(**) In this connection however see the Appendix of reference 17.

Riccati equation of the (gauge transformed) scattering problem (2.4a). We must conclude from this that these densities are the involutive set of densities.

It remains to ask therefore what the set $\mathrm{Re}(e^{-iu}\tilde{\xi}_k)$ which leads[17] to (4.26) consists of. To answer this observe that the sequence (4.41) of proposed Hamiltonians for the s-G hierarchy can be reduced to

$$\bar{H}_1' = H_1 = \int_{-\infty}^{\infty} (1-\cos u)\,dx\,;\quad \bar{H}_3' = \int_{-\infty}^{\infty} (\tfrac{1}{2}u_t^2 \cos u + u_{tt} \sin u)\,dx\,,$$

$$\bar{H}_5' = \ldots\,, \tag{4.45}$$

by appeal to the equation of motion of the s-G

$$u_{xt} = \sin u \quad .$$

This is exactly the sequence given in (4.26) which conversely can be put in the form (4.41) by appeal to the equation of motion of the s-G. We conclude from this that it is an accidental consequence of the form of the s-G itself that the alternative set of conservation laws (4.26) arises: notice that there is also a sequence of conserved quantities for the s-G derivable from (4.23) which are non-local in time. This set arises by setting

$$u_x = \int_{-\infty}^{t} \sin u \, dt' \tag{4.46}$$

into the sequence (4.25). The same sequence is obtained from (4.11b) by expanding it in inverse powers of η by (4.13a). It seems that given an involutive set of conserved densities and consequently a hierarchy of equations of motion it could always be possible to find further sets of conserved densities for a particular equation of motion by utilising that equation as we have done at (4.45).

5. CONNECTION WITH THE THEORY OF PROLONGATION STRUCTURES

In this section we establish the connection between the geometric theory and the theory of prolongation structures due to Wahlquist and Estabrook[13] by indicating the relation between their pseudopotentials and the pseudopotentials which arise naturally within the geometric theory. For more details the reader is referred to reference 6 and 12.

By analogy with the work of Wahlquist and Estabrook[13] pseudopotentials $\{Z_i\}$ $(i=1,2,3,...)$ can be defined by Pfaffians

$$\alpha_i = 0, \qquad \alpha_i = dZ_i + F_i dx + G_i dt \,, \tag{5.1}$$

provided that the derivatives $d\alpha_i$ are contained in the ring spanned by the components of θ

$$\left.\begin{aligned} \theta_1 &= d\omega_1 - \omega_2 \wedge \omega_3 \,, \\ \theta_2 &= d\omega_2 - 2\omega_1 \wedge \omega_2 \,, \\ \theta_3 &= d\omega_3 + 2\omega_1 \wedge \omega_3 \,, \end{aligned}\right\} \tag{5.2}$$

and the $\{\alpha_i\}$:

$$d\alpha_i = \sum_{k=1}^{3} a_{ik}\theta_k + \sum_j b_{ij} \wedge \alpha_j \,. \tag{5.3}$$

In equations (5.1) and (5.3) the F_i, G_i and the a_{ik} are functions of the Z_i and the variables appearing in Ω, and the b_{ij} are one-forms. The functions $Z_1 \equiv v_1$ and $Z_2 \equiv v_2$ appearing in the linear problems (2.1) (that is (2.4a) and (2.4b)) are by definition pseudopotentials for (2.1) is

$$\alpha_i = 0 \,, \qquad \alpha_i = dZ_i - \Omega_{ij} Z_j \,, \qquad i,j = 1,2,$$

and

$$d\alpha_i = -\theta_{ij} Z_j + \Omega_{ij} \wedge \alpha_j \,, \qquad i,j=1,2 \,. \tag{5.4}$$

We call them linear pseudopotentials. The next pseudopotentials are $Z_3 \equiv \Gamma_1 = v_2/v_1$, $Z_4 \equiv \Gamma_2 = v_1/v_2$ defined in (3.1). For by applying d to equation (3.2) we get, after writing $Z_3 \equiv \Gamma_1$, $Z_4 \equiv \Gamma_2$, $\alpha_3 \equiv \varepsilon_1$, $\alpha_4 \equiv \varepsilon_2$, that

$$\left.\begin{aligned} d\varepsilon_1 &\equiv d\alpha_3 = 2\alpha_3 \wedge (\omega_1 + Z_3\omega_2) - \theta_3 + 2Z_3\theta_1 + Z_3^2\theta_2 \,, \\ d\varepsilon_2 &\equiv d\alpha_4 = 2\alpha_4 \wedge (-\omega_1 + Z_4\omega_3) - \theta_2 - 2Z_4\theta_1 + Z_4^2\theta_3 \,. \end{aligned}\right\} \tag{5.5}$$

These expressions reduce to (3.3) on the solution manifold for which $\theta=0$. These quantities Z_3 and Z_4 are the quadratic pseudopotentials discussed by many authors[7,13,31,32] and the SL(2,R) structure[7,32] and the connection interpretation[7] are based on equation (3.2). It has been shown (cf.Crampin[7]) that the pseudopotentials y_8 of reference 13 is related to Z_3 and Z_4 for the original AKNS choice of $\{\omega_i\}$ for the KdV by a gauge transformation and redefinition of the parameter. By specifying A, B, C in (2.4b) as functions of q and r and using the quadratic pseudopotentials in an *appropriate gauge*, as shown in reference 13, Crampin[7], and 31, one obtains the familiar forms of Bäcklund transformation. An attempt to generalize the Bäcklund transformation in a gauge covariant way is made in reference 4.

The next pseudopotentials Z_5 and Z_6 are suggested by equation (5.5). We define

$$\left.\begin{aligned} \alpha_5 &= dZ_5 - (\omega_1 + Z_3\omega_2) \quad , \\ \alpha_6 &= dZ_6 - (-\omega_1 + Z_4\omega_3) \quad . \end{aligned}\right\} \qquad (5.6)$$

Applying d we get

$$\left.\begin{aligned} d\alpha_5 &= -\alpha_3\wedge\omega_2 - \theta_1 - Z_3\theta_2 \quad , \\ d\alpha_6 &= -\alpha_4\wedge\omega_3 + \theta_1 - Z_4\theta_3 \quad . \end{aligned}\right\} \qquad (5.7)$$

The relationship between (Z_3,Z_4) and (Z_5,Z_6) is the same as that of y_8 and y_3 in reference 13. Note that from (3.4)

$$\left.\begin{aligned} \alpha_5 &= dZ_5 - \delta_1 \quad , \\ \alpha_6 &= dZ_6 - \delta_2 \quad . \end{aligned}\right\} \qquad (5.8)$$

On the solution manifold $\alpha_5=\alpha_6=0$ and the closure $d\delta_1=d\delta_2=0$ of (3.5) follows immediately. Thus, contrary to the introductory remarks in reference 13, the pseudopotentials Z_5 and Z_6 (that is y_3 in reference 13) are closely related to the conventional conservation laws. Evidently the conservation laws arise naturally out of the prolongation structure.

The next pseudopotentials Z_7 and Z_8 are obtained by generalizing the relationship between y_3 and y_2 in reference 13. We define

$$\left.\begin{aligned} \alpha_7 &= dZ_7 - e^{-2Z_5}\,\omega_2 \quad , \\ \alpha_8 &= dZ_8 - e^{-2Z_6}\,\omega_3 \quad , \end{aligned}\right\} \qquad (5.9)$$

for which

$$\left.\begin{aligned} d\alpha_7 &= 2e^{-2Z_5}\alpha_5\wedge\omega_2 - e^{-2Z_5}\theta_2 \quad , \\ d\alpha_8 &= 2e^{-2Z_6}\alpha_6\wedge\omega_3 - e^{-2Z_6}\theta_3 \quad . \end{aligned}\right\} \qquad (5.10)$$

The significance of these pseudopotentials Z_7 and Z_8 is still to be explored.

The potential y_5 of reference 13 does not seem to be generated by our geometrical method. It corresponds to a kind of non-local conservation law. These various results otherwise completely establish the relationship of the geometrical method to the theory of prolongation structures as these were introduced in reference 13. The structure extends in a thoroughly natural way to a 3×3 scattering problem and thence by implication to the n×n scattering problem based on the present 2×2 analysis.

6. SUMMARY AND DISCUSSION

We have shown that the geometrical interpretation of the AKNS-ZS inverse scattering scheme introduced previously by one of us[3,4] provides a natural structure for the systematic investigation of the familiar properties of NEEs solvable by the scheme -namely their canonical structure and complete integrability, their prolongation structure, and (by implication) their Bäcklund transformation. We have formulated the canonical structure and prolongation structure in a gauge covariant way -hence the power of this geometrical approach- and incidentally exploited this to make a small additional contribution to the theory of the already much studied sine-Gordon equation. Notice that the NEEs expressed as the vanishing of the curvature two-form $\theta=0$ naturally associate the group SL(2,R) (or SL(2,C)) with the AKNS-ZS scheme, since by (5.2) the equations $\theta_i=0$ are formally just the Maurer-Cartan equations of an SL(2,R) or SL(2,C) Lie algebra (cf.Crampin[7]). Of course an SU(2) Lie algebra can equally be chosen[6]. Naturally our approach, which starts from the AKNS-ZS scattering scheme, does not solve the main problem in soliton theory -the discovery of the linear problem like (2.1) for an arbitrary NEE. However our geometrical approach does extend naturally to n×n scattering problems and may cover all equations with soliton solution in 1+1 dimensions. We show elsewhere[6] how these n×n problems can be reduced to 2×2 sub-problems. Moreover it is already clear that much of the analysis of the AKNS-ZS problem applies to other 2×2 problems not clearly of AKNS-ZS type: the recent work on the Ernst equations of general relativity[33,34] has yielded a linear scattering problem[33] and a Bäcklund transformation[34] which appear to be contained within our geometrical scheme.

REFERENCES

1. M.J.Ablowitz, D.J.Kaup, A.C.Newell and H.Segur, Phys.Rev.Lett. 31 (1973) 125.
2. V.E.Zakharov and A.B.Shabat, JETP (Sov.Phys.) 34 (1972) 62.
3. R.Sasaki, Phys.Lett. 71A (1979) 390.
4. R.Sasaki, Nucl.Phys. B154 (1979) 343.
5. R.K.Dodd and R.K.Bullough, Phys.Scripta, 1979, in press.
6. R.Sasaki, Niels Bohr Inst. prep.,NBI-HE-79-31,Sept.1979,to be publ.in Proc.Roy.Soc.
7. R.Hermann, Phys.Rev.Lett. 36 (1976) 835,
   M.Crampin, Phys.Lett. 66A (1978) 170.
8. F.Lund, Phys.Rev.Lett. 38 (1977) 1175.
9. G.L.Lamb, Jr., Phys.Rev.Lett. 37 (1976) 235.
10. M.Lakshmanan, Phys.Lett. 64A (1978) 354.
11. A.Sym and J.Corones, Phys.Rev.Lett. 42 (1979) 1099.
12. R.Sasaki, Phys.Lett. 73A (1979) 77.
13. H.D.Wahlquist and F.B.Estabrook, J.Math.Phys. 16 (1975) 1.
14. M.J.Ablowitz, D.J.Kaup, A.C.Newell and H.Segur, Phys.Rev.Lett. 30 (1973) 1262.
15. P.D.Lax, Comm.Pure and Appl.Math. 21 (1968) 467.
16. M.J.Ablowitz, D.J.Kaup, A.C.Newell and H.Segur, Stud.in Math.Phys. 53 (1974) 249.
17. R.Sasaki,R.K.Bullough,Niels Bohr Inst.prep.NBI-HE-79-32,to be publ.in Proc.Roy.Soc.
18. H.Flanders, "Differential Forms" (Academic Press, New York, 1963).
19. R.K.Bullough,P.W.Kitchenside,P.M.Jack and R.Saunders, Phys.Scripta,1979 in press.
20. L.P.Eisenhart, "A treatise on the differential geometry of curves and surfaces" (Dover Publ., New York, 1960).
21. "Solitons", Springer Topics in Modern Physics Series, R.K.Bullough and P.J.Caudrey eds. (Springer-Verlag, Heidelberg) In press. To appear early 1980.
22. A.V.Bäcklund, Lund Univ.Arsskrift, 10 (1875) and 19 (1883).
23. K.Konno, H.Sanuki and Y.H.Ichikawa, Prog.Theor.Phys. 52 (1974) 886.
    M.Wadati, H.Sanuki and K.Konno, Prog.Theor.Phys. 53 (1975) 419.
24. R.M.Miura, C.S.Gardner and M.D.Kruskal, J.Math.Phys. 9 (1968) 1204.
25. R.K.Dodd and R.K.Bullough, Proc.Roy.Soc. A352 (1977) 481.
26. P.J.Caudrey,J.D.Gibbon,J.C.Eilbeck,R.K.Bullough,Phys.Rev.Lett. 30 (1973) 237.
27. L.A.Takhtadzhyan and L.D.Faddev, Theor.and Math.Phys. 21 (1974) 1041.
28. H.Flaschka and A.C.Newell, in "Dynamical Systems Theory and Applications", J.Moser ed. (Springer, Heidelberg 1975).
29. R.K.Bullough and R.K.Dodd, in "Synergetics. A Workshop", H.Haken ed. (Springer, Heidelberg 1977).

30. R.K.Bullough and P.J.Caudrey, in "Nonlinear evolution equations solvable by the spectral transform", F.Calogero ed. (Pitman, London 1978).
R.K.Bullough, in "Nonlinear equations in physics and mathematics", A.O.Barut ed. (Reidel.Publ.Dordrecht, Holland 1978).

31. H. H. Chen, Phys.Rev.Lett. 33 (1974) 925.
R.Konno and M.Wadati, Prog.Theor.Phys. 53 (1975) 1652.

32. J.Corones, J.Math.Phys. 18 (1977) 163,
H.C.Morris, J.Math.Phys. 18 (1977) 533,
R.K.Dodd and J.D.Gibbon, Proc.Roy.Soc. A359 (1978) 411.

33. D.Maisson, Phys.Rev.Lett. 41 (1978) 521.

34. B.K.Harrison, Phys.Rev.Lett. 41 (1978) 1197.

FUNDAMENTAL EQUATIONS OF CLASSICAL AND OF QUANTUM PHYSICS AS SPECIAL CASES OF A NONLINEAR EQUATION*

S. N. Bagchi

Department of Physics,
Concordia University, Montreal, Canada

At the outset I would like to thank sincerely the organizers of this conference for giving me the opportunity to present here a rather unconventional approach to the problems of micro-physics. I have chosen this topic of my lecture in order to pay my homage to the memory of Albert Einstein whose birth centenary physicists are celebrating this year. The topic is not only important for me personally but, I think, also for the physicists in general. As you will see below, this work, guided mainly by Einstein's concept of Physics, has already vindicated *quantitatively* his opinion that it is possible to formulate a deterministic continuum approach to physics, including the extraordinarily successful quantum theory, at least so far as a single particle Schroedinger and Dirac equations are concerned. At the end of my lecture I had ventured to suggest a programme, unfortunately not yet worked out quantitatively, in conformity with Einstein's latest ideas, as how one can perhaps tackle more successfully the problems which are troubling quantum physicists today. Recalling the prophecy of W.K.Clifford[1] about the role of "space" in physics, later verified by Einstein's general theory of relativity, I think, I do not need an apology for including my speculations in printed words, although they may not lead to the anticipated progress.

---

* Lecture delivered at the International Meeting on "Nonlinear Evolution Equations and Dynamical Systems" held in Lecce, Italy, during June 20-23 1979.

1. NONLINEARITY IN PHYSICS

It is generally believed that physics is essentially governed by nonlinear mathematical equations and any measurement implies a nonlinear interaction. But modern physicists being preoccupied with the linear formalism of the conventional quantum mechanics do not as yet fully realize (in spite of the fact that Heisenberg devoted later years of his life to develop a nonlinear theory for "elementary" particles) that the days of linear theory are over. According to him[2] "it is rather a controversial question whether it (quantum theory) will finally be a linear or nonlinear theory". If the problems now at the frontiers of physics and theoretical biology are to be tackled in a mathematically rigorous way to yield useful new results, I believe, one must obtain explicitly all possible (including singular) solutions of nonlinear partial differential equations (PDE). As yet one usually deals with nonlinear PDE where the differential operators are still linear. But, I think, for "trans-quantum" physics as well as for theoretical biology PDE where the operators themselves are nonlinear would prove to be more valuable. Of course, the difficulties are enormous particularly because of the fact that the required mathematics is not yet developed. Only the mathematicians can help us physicists in such unknown and uncharted paths.

Before proceeding to derive the fundamental equation on two, (almost self-evident), postulates, permit me here to say a few words about the problems facing us today, as I see it. I think these remarks will persuade you to investigate the equations derived below more seriously.

Essentially there is only one fundamental problem facing theoretical physicists today:

How to develop a unified field theory which would encompass not only electromagnetism and gravitational but also "elementary particles"?

Many would perhaps outright discard such an ambitious programme. However, if we wish to retain this universally accepted goal of physicists, two problems must be settled first.

1) How to derive a nonlinear PDE for a *nonsymmetric* tensor field?
2) How to bring the present day linear quantum mechanical formalism in harmony with the notion of "space-time" and covariance with respect to any arbitrary coordinate transformation?

Re (2): As is well known even the problem of quantization in Riemannian space of the general theory of relativity had led to unsurmountable difficulties. Further, Wigner[3] had pointed out the conceptual difficulty in harmonising the concept of space-time in general relativity with that of present day quantum theory. It seems to me that such conceptual difficulties arise mainly because of the acceptance of Heisenberg's Uncertainty Relation as a basic *principle* of nature.

If we agree that any dynamical quantity must *in principle* be describable as a function x,y,z,t, our work[4,5*] had *proved* that Heisenberg's Uncertainty Relation is the price of our representation. For example, instead of treating momentum as a function of $\vec{x}$ and t, in quantum theory one represents it in terms of the Fourier components of the wave field. Consequently, Uncertainty Relation proves the general validity of the mathematical relations that exist between the coordinates of physical and its reciprocal space.

It need not be accepted as a *fundamental principle* of nature. Note also that Dirac[6] had come to the conclusion, albeit from an entirely different point of view, that Planck's constant is not likely to remain as a *fundamental constant* of nature.

Re (1): If we accept the point of view of Einstein that all fundamental laws of physics must be covariant with respect to any arbitrary coordinate transformation, it is possible to circumvent these difficulties by discovering a nonlinear PDE of the nonsymmetric tensor field. It is the wave field of the particle, which one might very well call "World aether" whose properties are involved in the process of measurement. In order to circumvent the conceptual difficulties discussed by Wigner and others, perhaps it would be more convenient if we retain the space-time framework as a scaffold just for the description of the properties of this world aether which alone really determines the outcome of operational procedures used to obtain the observable properties of

---

* The earlier work (1953-56) on scalar wave field by Hosemann and Bagchi as well as the later work (1964) of Bagchi on vector field had been summarized and discussed extensively in the paper[5]. I regret very much that this paper(5) for reasons beyond my control contains a few, (though almost obvious), printing mistakes.

the particles. In practice, apart from convenience, this distinction between the 4-dimensional space-time world with a geometry endowed with curvature and torsion and the same properties attributed to the wave field while keeping the usual space-time frame will reduce to a question of semantics without any physical content. Nevertheless, for the philosophy of Science[7] such a distinction might be useful.

Anyway, what I wish to emphasize here is that if one wants to develop post-quantum physics, one should try to obtain a PDE valid for a nonsymmetric tensor field endowed with curvature and torsion, (see below). In order to achieve this we have to abandon the notion that the wave field of a particle does not represent a *real physical quantity* but merely serves as a mathematical tool for calculating the probability of finding objects. But Renninger's thought experiment[8], as I have analyzed elsewhere[5], clearly demonstrates that the wave field of a photon (electron) represents a physical reality. Consequently, following the viewpoint of Einstein, we must accept that a potential or a field component (i.e. the world aether) is a real natural physical quantity which changes according to definite deterministic laws. Bohm-Aharonov effect appears to confirm this for electromagnetic field. Consequently, restricting ourselves to a second order differential equation, the desired equation is likely to be a hyperbolic one. But unfortunately, as emphasized by Dirac[9], "methods based on the equations of motion, so necessary for low-energy physics, have been largely abandoned in the field of high-energy physics". He strongly believes that in the interest of the unity of physics, we should use the equations of motion.

Finally, I would like to bring to your attention the fact that solutions of nonlinear equations not only can resolve many unsolved problems but also can open up a new vista. As concrete illustrations, let me cite here two examples from my own work.

(1) For almost a century one believed in Kirchhoff's Theorem, (which had become almost an axiom of Physics), that one cannot determine the density distribution in a substance from the scattered intensity *alone* without the knowledge of the "phases". The entire edifice of crystal structure analysis was therefore based on guessing correctly the phases of the intensity spectra. But our work[10] on the actual solution of the nonlinear integral equation without a kernel, namely

$$Q(\vec{x}) = \int \rho(\vec{y})\, \rho(\vec{x} + \vec{y})\, dv_y$$

disproved the general validity of this theorem and opened up a new field for the direct structure determination of crystals and, in particular, of disordered structures where the standard methods of crystal structure analysis cannot be applied[11].

(2) An explicit analytic solution[12] of the important nonlinear equation of the generalized Poisson-Boltzmann type

$$\nabla^2 \lambda = f(\lambda)$$

proved beyond doubt that the well known Debye-Hueckel theory for strong electrolyties cannot be accepted even as a limiting theory[13]. But the general validity[13] of Debye's ion-atmosphere concept had led me to formulate an alternative and promising statistical mechanical theory of liquids and of molten salts, (to be communicated shortly). It might be noted that as of today there is no satisfactory statistical mechanical theory of liquids. Both the approaches, from the standpoint of distribution function theory and of Mayer's cluster theory so vigorously pursued now-a-days, suffer from serious mathematical and physical difficulties.

2. HYPOTHESES AND POSTULATES

In this lecture I shall deal only with a single particle whose wave field can be represented either as a scalar function or as a vector function.

Hypothesis:

The fundamental hypothesis is based on the *empirical* fact that a particle possesses *simultaneously* both corpuscular and wave properties. Both are physical realities. They need not be interpreted as a manifestation of wave-corpuscle dualism in the sense of Copenhagen School. Consequently, if we assume that ultimately the wave field of the particle *alone* will enable us to determine all the characteristic properties of the particle, there must be a definite relation between Newton-Einstein corpuscular properties and Huygens-Maxwell wave properties of the particle.

Postulate I:

Our first postulate is that this relation, following Hamilton-de Broglie pilot principle, can be obtained by connecting the 4-momentum of the particle, (in the corpuscular sense), entirely in terms of its wave field. For brevity, we shall call the postulate I as "*Pilot Principle*".

In the sense of a scalar field this is given by the gradient of the phase of the wave-field which can easily be converted into an expression containing only the wave function. But in the case of the vector field a similar expression will have to be introduced as a postulate.

Further, in order to account for the diffraction effects of a material particle, it would be necessary to *redefine* the dynamical mass of the particle as a function of the amplitude of the wave field also.

Postulate II:

The desired equation for the wave field, (in both cases), is obtained from the (almost self-evident) postulate that over the entire domain the energy-momentum is conserved. This can also be looked upon as the continuity equation.

3. NOTATIONS

Since we will be dealing here quantitatively only with scalar and vector wave fields of the particle in a covariant form with respect to Lorentz transformation, all the equations will first be derived as world equations. Corresponding nonrelativistic expressions will be obtained as special cases. For convenience, we shall be using the 4-dimensional vector analysis in the form introduced by Sommerfeld[14].

Our world is the Minkowski space with the coordinates $x_1, x_2, x_3, x_0 = ict$ and signature $+++-$.

Examples:

$$\text{4-distance vector} \quad \underline{x} = \vec{x} + ict.\ \vec{s}_0 = \sum_{j=o}^{3} x_j \vec{s}_j \tag{1}$$

$\vec{s}_j$'s are the four mutually orthogonal unit vectors.

$$\text{4-velocity} \quad \underline{v} = \frac{d\underline{x}}{d\tau} = \kappa(\vec{v} + ic\vec{s}_0) \tag{2}$$

where the local time t is related to the proper time τ by

$$dt = k\,d\tau \quad , \tag{3}$$

$$k = (1 - v^2/c^2)^{-1/2} \quad , \tag{4}$$

$$\underline{v}\cdot\underline{v} = -c^2 \quad . \tag{5}$$

The generalized 4-momentum of the particle

$$\underline{p} = \underline{p}_N + \underline{p}_e \equiv \vec{p} + \frac{i}{c} H \vec{s}_0 \tag{6}$$

where

$$\underline{p}_e = e\underline{\Phi} \tag{7}$$

represents the field momentum due to an external field of electromagnetic type whose 4-potential can be written as

$$\underline{\Phi} = \vec{\Phi} + i\Phi_0 \vec{s}_0 \tag{8}$$

e represents the invariant charge of the particle.

$$\underline{p}_N = M_0(\underline{x})\,\underline{v} \quad ; \tag{9}$$

$$M_0 = \mu(\underline{x})\, m_0 \tag{10}$$

$\underline{p}_N$ represents the kinetic 4-momentum of the particle, m_0 its conventional rest mass and μ is the mass factor which depends on the amplitude of the wave field. Consequently, the dynamical mass M of the particle is given by

$$M = \mu\, m_0\, \kappa \tag{11}$$

H and U represent the total and potential energy of the particle respectively.

4-nabla operator: $\underline{\nabla} = \vec{\nabla} + \frac{\partial}{ic\partial t} \vec{s}_0$ (12)

D'Alembertian operator: $\Box = \nabla^2 - \frac{\partial^2}{c^2\partial t^2}$. (13)

4. THE DIFFERENTIAL EQUATION GOVERNING THE SCALAR WAVE FIELD OF A PARTICLE

The wave function associated with a particle is represented as:

$$\varepsilon(\underline{x}) = a(\underline{x}) \exp i(W/\hbar) \tag{14}$$

a and W are real and $\hbar=h/2\pi$, where h is Plank's constant.

The mass factor μ is defined as

$$\mu = \left[1 - \frac{\Box a}{a}\left(\frac{\hbar}{m_0 c}\right)^2\right]^{1/2} \qquad (15)^*$$

The usual expressions of point mechanics as well as those of geometrical optics are obtained from the condition

$$\Box a = 0 \; ; \qquad \mu = 1 \quad . \qquad (16)^{**}$$

According to our Postulate I, the generalized 4-momentum of the particle is given by

$$\underline{p} = \underline{\nabla} W = \frac{\hbar}{2i}\left(\frac{\underline{\nabla}\varepsilon}{\varepsilon} - \frac{\underline{\nabla}\varepsilon^*}{\varepsilon^*}\right) \qquad (17)$$

and the energy-momentum density of the wave field by

$$\varepsilon\varepsilon^*\underline{p} = \frac{\hbar}{2i}\left[\varepsilon^*\underline{\nabla}\varepsilon - \varepsilon\underline{\nabla}\varepsilon^*\right] \qquad (18)$$

where $\varepsilon^*(\underline{x})=a(\underline{x}) \exp -i\left[W(\underline{x})/\hbar\right]$.

We were gratified to know (private communication) that Einstein liked this formulation of the mass in terms of the wave function alone without increasing the order of the differential operator. He even anticipated, (what I found later to be correct), the power of equation (18). He joked about its mysterious nature by remarking that as if it came out of "witch's grandmother's kitchen!" Following his remarks we later found that the dynamical mass can be obtained from the wavefield alone if we can solve the corresponding differential equation completely. Thus,

$$m_0 = \frac{1}{c}\left[\frac{1}{c^2}\left(\frac{\partial W}{\partial t} + U\right)^2 - \left(\vec{\nabla} W - \vec{p}\right) + \frac{\Box a}{a}\hbar^2\right]^{1/2} \qquad (19)$$

$$M_0 = \frac{1}{c}\left[\frac{1}{c^2}\left(\frac{\partial W}{\partial t} + U\right)^2 - \left(\vec{\nabla} W - \vec{p}_e\right)^2\right]^{1/2} \qquad (20)$$

---

* This relation was obtained first by de Broglie[15] in 1927. In the language of the de Broglie school μ is called "quantum potential".

** This important special case was first pointed out to me by (late) Prof.S.N.Bose in 1954. I would like to take this opportunity to express again my deep gratitude to Profs. M.von Laue, S.N.Bose, L.de Broglie and A.Einstein for their interest and encouragement while the scalar theory was being developed by Hosemann and myself during 1953-56.

By the way, the eqs. (19) and (20) answers the criticisms of Brillouin[16]: "one completely ignores any possibility of mass connected with the external potential energy, (p.14)".

It must however be noted here that in order to get the mass spectra of unstable "elementary particles", which possibly arise from the interactions of wave fields associated with different particles within their vortical domains, thus generating turbulence (see ref. 5), we are faced with unsurmountable mathematical difficulties as well as completely unknown physical laws of the vortices when they interpenetrate into one another within this domain of extremely high energies. In order to circumvent these difficulties I had at the end of my lecture talked, (albeit vaguely for the moment), about the mechanism of the generation of unstable "elementary" particles from the standpoint of a nonlinear PDE involving the nonsymmetric tensor field.

Finally, in order to obtain the desired equation we use the Postulate II, namely

$$\underline{\nabla}\cdot(\varepsilon\varepsilon^{*}\underline{p}) = 0 \tag{21}$$

Now, substituting (18) in (21) we get

$$\varepsilon^{*}\Box\varepsilon - \varepsilon\Box\varepsilon^{*} = 0 \tag{22}$$

Note that the space-time curvature of the wave field is proportional to the wave function itself, just as in the meson equation. But unlike Klein-Gordon equation, eq.(22) itself is not only nonlinear but cannot be reduced in the general case to an equation involving only one wave-function. This is true for all the general equations (22-24) derived below.

One can by using the expression (17) arrive at other equivalent expressions, namely,

$$\Box\varepsilon + \left[\underline{p}^2/\hbar^2 - \Box a/a\right]\varepsilon = 0 \tag{23}$$

$$\Box\varepsilon - \left[\frac{1}{4}\left(\frac{\underline{\nabla}\varepsilon}{\varepsilon} - \frac{\underline{\nabla}\varepsilon^{*}}{\varepsilon^{*}}\right)^2 + \frac{\Box a}{a}\right]\varepsilon = 0 \tag{24}$$

Unfortunately, all the equations (22-24) are too general to be of any practical use, at least at the present stage of our knowledge, though they reveal some interesting mathematical properties.

First, eq.(22), (at least formally*), looks like the difference of two adjoint systems where the divergence of current density automatically vanishes (cf.17). But unlike the usual cases of linear DE here we have not to postulate that the linear operators separately vanish. In fact, $\Box\varepsilon=0$ or $\Box\varepsilon^*=0$ cannot be valid in the general case (see later). In our case the eq.(22) comes as a consequence of (18) and (21), though the differential equations governing the wave field (eqs. 23-24) involve nonlinear operators and both ε and ε^*. They cannot be separated into an equation involving only either ε or ε^* except under special restrictions (see below). Mathematically, this implies that the most general solution of such an equation is nonanalytic. Our work on the Kepler problem[18] strongly suggests that nonanalytic and singular solutions of some linear equations would be of great physical interest. One should also note carefully the remark of Sommerfeld[17] that even in the wave mechanical case, although the linear wave equation can be expressed in terms of one wave function alone, "the wave mechanical problem is not determined by one wave equation but by a pair of equations, the original and its adjoint, p. 228".

The equations (22-24) also show that every property of the particle can be derived from the knowledge of the wave function. The external field itself can also be expressed in terms of the wave function of the source. As will be shown later that to get the "spin" of the particle we will have to use a vector field**. The scalar wave field is valid for "spinless" particles. The "spin" is related to the "intrinsic" angular momentum around the singularity of a vector field endowed with vorticity, (see later). As yet we do not know how the invariant charge of the particle is related to the wave field. My conjecture is that the mathematical representation of the wave function must always be a complex quantity even for a neutral particle, (see the remark of Sommerfeld mentioned above). The sign and the magnitude of the charge would probably depend on the torsion of the field, in general a nonsymmetric tensor field. I

---

* Sommerfeld[17] expresses the opinion that adjoint systems have meaning only for linear operators.

** Note that a spinor or a bispinor can be expressed in terms of a 4-vector.

am as yet not quite sure as to the physical nature of the world aether. As I have discussed elsewhere (see refs.5,7) it possibly represents a continuum of energy density whose topological distorsions and fluctuations manifest as observable phenomena ascribed to the particle.

In order to obtain the fundamental equations of classical and of quantum physics from any of the above general equations, we have to incorporate in them explicitly the specific characteristic properties of a particle, (e.g. rest mass, charge, etc.). One can easily obtain such an equation by using equations (6-10, 15) in eq. (23) and noting the relation (cf.6)

$$(\underline{p} - \underline{p}_e)^2 = \underline{p}_N^2 = -(\mu m_0 c)^2 \quad . \tag{25}$$

Thus our desired equation is

$$\Box\varepsilon + \frac{1}{\hbar^2}\left[2(\underline{p}\cdot\underline{p}_e) - \underline{p}_e^2 - m_0^2 c^2\right]\varepsilon = 0 \quad . \tag{26}$$

Using the analogous equation for ε^*, we can rewrite (26) in the symmetric form:

$$\varepsilon^*\Box\varepsilon + \varepsilon\Box\varepsilon^* - \frac{2i}{\hbar^2}\left[(\varepsilon^*\underline{\nabla}\varepsilon - \varepsilon\underline{\nabla}\varepsilon^*)\cdot\underline{p}_e\right] - \frac{2}{\hbar^2}\left[m_0^2c^2 + \underline{p}_e^2\right]\varepsilon\varepsilon^* = 0 \tag{27}$$

We assert that the eq.(26) or (27) represents the most general differential equations governing the motion of a particle whose wave-field can be represented by a scalar function under a given external field of electromagnetic type. The equation (26) contains explicitly the given characteristic properties of the particle.

This assertion gets its full *a posteriori* justification from the fact that it reduces to the well known fundamental equations of classical and quantum physics under well defined restricted conditions. As can be checked readily that the eq.(26) reduces to a linear equation in the absence of an external field as well as under certain special cases noted below even when the external field is present.

Further, as remarked by Einstein (private communication) these waves are running waves. Later our work on the Kepler problem of the hydrogen atom[18] showed that Schroedinger's ψ function are obtained from the superposition of two pilot wave functions. Consequently, with ψ-function we cannot account for the trajectory of the particle. We have to fall back upon the statistical interpretation to interpret the observed results.

Therefore it should be particularly noted that the ε-function should not be identified with quantum mechanical ψ-function and the dynamical $\underline{p}$ should not be set as $\underline{p}_{op} = \frac{h}{i}\underline{\nabla}$ without further restrictions (see later). But $\underline{p}$ can always be represented by the expression (17).

Further, our analysis of the fundamental solution[19] of the Klein-Gordon equation suggests that these are essentially Huygens' elementary waves in "world aether".

Stationary state solutions arise out of the superposition of running waves.

5. GENERALIZED HAMILTON-JACOBI EQUATION

If we insert equations (15) and (17) in the eq.(25), we get

$$(\underline{\nabla}W - \underline{p}_e)^2 + m_0^2c^2 = \hbar^2 \frac{\Box a}{a} \quad . \tag{28}$$

Eq.(28) for the case of point mechanics, i.e. $\Box a=0$, reduces to the well known relativistic H-J equation (29)

$$(\vec{\nabla}W - \vec{p}_e)^2 - (\frac{H-U}{c})^2 + m_0^2c^2 = 0 \tag{29}$$

It is extremely important to note here that we get the equations of point mechanics also if we set h=0. In macrophysics h is no doubt negligible but, strictly speaking, it is never zero. Nevertheless, even for $h \neq 0$, classical mechanics (as well as geometrical optics), remains exact if $\Box a=0$. The only limitation of classical mechanics is that it does not take the amplitude of the wave function into account although its phase became a powerful tool in the hands of Jacobi. The magnificent analytical structure of classical mechanics retains its validity for quantum mechanics also provided we modify the definition of dynamical mass slightly and consider the amplitude of the "matter wave" as well. Our definition of dynamic mass is perfectly in harmony with the consequences of the general theory of relativity.

Further, if H is a system constant and

$$H - m_0c^2 = E = \text{constant} \tag{30}$$

we can set

$$W(\vec{x},t) = S(\vec{x}) - Ht \ ;$$

to get the nonrelativistic H-J equation (31):

$$\frac{(\vec{\nabla}S - \bar{p}_e)^2}{2m_0} + U = E; \quad (\Box a=0, \ E=\text{constant}, \ \left|\frac{m-m_0}{m_0}\right| << 1) \tag{31}$$

Thus we are perfectly justified to characterize point mechanics as well as geometrical optics by the condition (16).

It must however be noted that in order to solve the generalized H-J equation (28) ant to determine the trajectory of the particle in the general case, we need to know the amplitude as well as the phase of the wave-field as a function of space and time. So this theory in its generality differs fundamentally from classical mechanics. The initial value problems of classical mechanics has been changed to the boundary value problems.

It is obvious that the eqs. (28)(29) for the phase of the wave function is a differential equation of the first order. Nevertheless, Schroedinger changed this into a second order differential equation by using the operator formalism:

$$\frac{\hbar}{i} \underline{\nabla} = \underline{p}_{op} \tag{32}$$

As discussed before (see refs.4,5) this clearly demonstrates that $\underline{p}_{op}$ of conventional quantum mechanics is a function of the Fourier space and the linear operator (32) comes as a particular case of the eq.(17) when the Schroedinger condition

$$\underline{p}_e \cdot \underline{\nabla} a = 0 \tag{33}$$

is satisfied, (see below).

6. IMPORTANT SPECIAL CASES OF THE GENERALIZED WAVE EQUATION

(i) Klein-Gordon Equation:

The generalized wave eq.(26) reduces to K.G. equation

$$\Box\varepsilon - \left(\frac{m_0 c}{\hbar}\right)^2 \varepsilon = 0 \tag{34}$$

either in the absence of an external field, (i.e. $\underline{p}_e=0$), or for a neutral particle whose wave function can be described by a scalar function.

The difficulties which cropped up in the physical interpretation of the K-G equation were mainly due to the basic formalism of conventional quantum mechanics, which has been proved by our work (see refs. 4 and 5) to be valid only under restricted conditions. They do not arise in the theory presented here.

(ii) The Wave Equation of Optics:

For photons $m_0 = e = 0$, hence,

$$\nabla^2\varepsilon - \frac{1}{c^2}\frac{\partial^2\varepsilon}{\partial t^2} = 0 \quad ; \tag{35}$$

from the generally valid eq.(26) we can conclude that eq.(35) governs the motion of not only photons but also of any spinless neutral particle of mass zero.

Mathematically, eq.(26) also reduces to the equation (35) if

$$2(\underline{p}\cdot\underline{p}_e) - \underline{p}_e^2 - m_0^2c^2 = 0 \quad . \tag{35a}$$

But obviously this relation cannot be satisfied generally unless the rest mass of the particle is zero.

(iii) Schroedinger Equation:

In wave mechanics the differential equations are derived by using the operator formalism. But in this theory $\underline{p}$ as a function of $\vec{x}$ and t is related with the wave function, also a function of $\vec{x}$ and t by the eq.(17). The usual operator formalism of wave mechanics results from (17) when the wave function $\varepsilon(\vec{x},t)$ satisfies the Schroedinger condition:

$$\left(\underline{p}_e\cdot\frac{\underline{\nabla}\varepsilon}{\varepsilon}\right) = -\left(\frac{\underline{\nabla}\varepsilon^*}{\varepsilon^*}\cdot\underline{p}_e\right) \tag{33a}$$

or equivalently,

$$\underline{p}_e\cdot\underline{\nabla}a = 0 \quad . \tag{33}$$

Under this restriction the generalized wave equation (26) reduces to the relativistic Schroedinger-Gordon equation:

$$\Box\varepsilon + \frac{1}{\hbar^2}\left[\frac{2\hbar}{i}(\underline{\nabla}\underline{p}_e\cdot\underline{\nabla}\varepsilon) - \underline{p}_e^2\varepsilon - m_0^2c^2\varepsilon\right] = 0 \tag{36}$$

For the stationary state of the wave field, we have

$$\frac{\partial^2\varepsilon}{\partial t^2} = -(H/\hbar)^2\varepsilon \; ; \qquad \left[\frac{\partial a}{\partial t}=0;\; \frac{\partial H}{\partial t}=0\right] \quad . \tag{37}$$

Consequently eq.(37) becomes

$$\nabla^2\varepsilon + \frac{1}{\hbar^2}\left[\left(\frac{H-U}{c}\right)^2 - p_e^2 - m_0^2c^2 - 2(\vec{p}\cdot\vec{p}_e)\right] = 0 \quad . \tag{38}$$

For an electrostatic potential, (i.e. $\vec{p}_e=0$), eq.(38) is the relativistic Schroedinger equation used by Sommerfeld in investigating the fine structure of the hydrogen spectra.

Finally, using the relations (33a) and (37) and the nonrelativistic approximation $\left|\frac{M-m_0}{m_0}\right|<<1$ as well as assuming $|\vec{p}_e|<<|\vec{p}|$, we get from (38) the time-independent Schroedinger equation

$$\nabla^2\varepsilon + \frac{2m_0}{\hbar^2}(E-U)\varepsilon - \frac{2i}{\hbar}(\vec{p}_e\cdot\vec{\nabla}\varepsilon) = 0 \tag{39}$$

where $E=H-m_0c^2$.

It should again be emphasized that the wave function $\varepsilon(\vec{x},t)$ should not be identified even in these cases with Schroedinger's Ψ-function without further qualifications.

Hence we repeat:

Einstein's convinction that it is possible to explain quantum phenomena deterministically is proved beyond doubt *quantitatively* at least for a single particle whose wave field can be described by a scalar function.

7. EXTENSION OF POINT MECHANICS

As shown before (see refs.4 and 5) the entire formalism of classical mechanics, namely, Lagrangian and Hamiltonian mechanics, Hamilton-Jacobi equation, etc. can not only be retained in this theory but also permits us to follow the trajectory of a quantum particle. One might wonder how the classical formalism can explain the diffraction phenomena. The answer is that the mass factor μ produces a new type of force, the so called diffraction force, which deviates the trajectories of the particles even in the absence of an external field whenever it passes through a slit and always deterministically according to the pilot principle, eq.(17). For quantitative proofs and further physical interpretations, see refs. 4 and 5. The only defect of the pre-quantum classical mechanics was the fact that it did not consider the amplitude of the wave function which inevitably accompanies the particle. If this new *empirical* fact, namely, a particle has both the corpuscular and wave properties is taken into account, the initial value problems of classical prequantum mechanics is changed to boundary value problems, just as in wave mechanics. The relation between pre-quantum classical mechanics and quantum mechanics is exactly analogous to that between geometrical and physical optics. The pre-quantum mechanics is *strictly* valid as long as $\Box a = 0$. The new Lagrangian and Hamiltonian, of course, contains μ, but nevertheless satisfies the usual Lagrangian and Hamiltonian equations.

For example

$$L(\vec{x},\vec{v},t) = \frac{\underline{v}\cdot\underline{p}}{\kappa} = -\frac{\mu m_0 c^2}{\kappa} + (\vec{v}\cdot\vec{p}_e) - U \tag{40}$$

$$H(\vec{x},\vec{p},t) = U + c\left[(\mu m_0 c)^2 + (\vec{\nabla}W - \vec{p}_e)^2\right]^{\frac{1}{2}} \tag{41}$$

An interesting consequence of this theory, which perhaps might be verified by direct experiments, is the fact that in the Fraunhofer Zone the velocity of a photon is c, but in the Fresnel Zone it is less than c. This does not contradict the second postulate of the special theory of relativity.

One can also show that the intensity of the diffracted beam in the Fresnel Zone is not equal to the square of the amplitude of the wave function but depends on its phase also:

$$I = \varepsilon\,\varepsilon^* \,|\vec{\nabla} W/\hbar\,| \quad . \tag{42}$$

8. CURRENT DENSITY

We now show that for particles whose rest mass is not zero, the usual expression for the quantum mechanical current density satisfies the equation of continuity only under certain special conditions noted below.

Let us consider a collection of incoherent particles with ensemble normalization and set $\gamma\varepsilon=\Psi$;

$$\frac{\gamma^2\,\varepsilon\,\varepsilon^*\underline{p}_N}{m_0} = \underset{\sim}{J} \tag{43}$$

γ is the normalization constant (cf.refs 4 and 5). From equations (6),

(7) and (18) we get the usual quantum mechanical expression of the 4-current density

$$\underline{J} = \frac{\hbar}{4\pi i m_0}(\Psi^*\underline{\nabla}\Psi - \Psi\underline{\nabla}\Psi^*) - \frac{e}{m_0}\Psi\Psi^*\underline{\Phi} \quad . \tag{44}$$

Now, from the eqs.(6)(17)(21) and by utilising the Lorentz condition for the 4-potential we see that the 4-divergence of the 4-current is given by

$$\underline{\nabla}\cdot\underline{J} = -\frac{2|\Psi|e}{m_0}(\underline{\Phi}\cdot\underline{\nabla}|\Psi|) \quad . \tag{45}$$

This becomes zero only when any of the following conditions are fulfilled:

(i) There is no external field, i.e. $\underline{\Phi}=0$;

(ii) The particle in neutral;

(iii) There exists an inertial system for which $\frac{\partial a}{\partial t} = \vec{\Phi}\cdot\vec{\nabla}a = 0$ for all $\vec{x}$ and t;

(iv) There exists an inertial system for which $\Phi_0 = |\vec{\nabla}\Psi| = 0$ for all $\vec{x}$ and t.

Evidently, $\varepsilon\varepsilon^*\underline{p}_N$ is a measure of the current density of the energy-momentum of the particle registered by the measuring instrument under normal experimental conditions. Considering a world tube parallel to $\underline{p}_N$ and subjecting the number of particles within this tube to the ensemble normalization, we find that only under the above mentioned four conditions $\Psi\Psi^*$ is a measure of the density of the particles. It should be noted that the fact that the 4-current density is not in general divergence free does not imply that the number of particles (neglecting creation and annihilation of particles), is not conserved. It only states that under these circumstances there is an interaction and exchange of energy and momentum between the corpuscle and its associated wave. Eq.(21) assures us that everywhere and under all circumstances the energy-momentum is conserved and the corresponding 4-current density is always divergence free.

9. THE EQUATION FOR A VECTOR FIELD OF A PARTICLE

(i) Definitions and Postulates:

Let the wave function associated with a particle be represented by the 4-vector $\underline{\varepsilon}$ in Minkowski space:

$$\underline{\varepsilon}(\underline{x}) = \underline{A}(\underline{x}) \exp i\left[W(\underline{x})/\hbar\right] \tag{46}$$

Analogous to the eq.(18), we define the 4-momentum density $\underline{\varepsilon}\,\underline{\varepsilon}^*\,\underline{P}(\underline{x})$ of the vector field by

$$\underline{\varepsilon}\,\underline{\varepsilon}^*\underline{P} = \frac{\hbar}{2i}\left[(\underline{\varepsilon}^*\cdot\underline{\nabla})\underline{\varepsilon} - (\underline{\varepsilon}\cdot\underline{\nabla})\underline{\varepsilon}^*\right] \tag{47}$$

where

$$\underline{\varepsilon}^*(x) = \underline{A}(\underline{x})\ \exp -i\left[W(\underline{x})/\hbar\right] \tag{48}^*$$

Now, using 4-dimensional vector calculus we can rewrite (47) in the form:

$$\underline{\varepsilon}\,\underline{\varepsilon}^*\,\underline{P} = \frac{\hbar}{2i}\left[\underline{\varepsilon}^*(\underline{\nabla}\cdot\underline{\varepsilon}) - \underline{\varepsilon}(\underline{\nabla}\cdot\underline{\varepsilon}^*)\right] \tag{49}$$

or,

$$\underline{A}^2\cdot\underline{P} = \underline{A}^2\cdot\underline{p} + \underline{A}^2\cdot\underline{\omega} \tag{50}$$

i.e.,

$$\underline{P} = \underline{p} + \underline{\omega} \tag{51}$$

where $$\underline{p} = \underline{\nabla}W \tag{52}$$

and $$\underline{\omega} = \left[\underline{s}_A\left[\underline{s}_A\underline{p}\right]\right] \tag{53}$$

$\underline{s}_A$ is the unit vector in the direction of $\underline{A}$.

It can be easily verified that the vector $\underline{P}$ is parallel to $\underline{A}$ and from eqs.(50) and (53) we note that it contains an irrotational part $\underline{p}$ and a rotational part $\underline{\omega}$, in conformity with the general property of a vector.

As before, we identify $\underline{p}=\underline{\nabla}W$ (17) as the translatory 4-momentum of the particle, i.e.,

$$\underline{p} = \underline{p}_N + \underline{p}_e$$

where $\underline{p}_e=e\underline{\Phi}$ (7) and $\underline{p}_N=\mu m_0\underline{v}$ (9).

But now the scalar mass factor μ should have to be expressed as

$$\mu = \left[1 - \frac{(\underline{A}\cdot\Box\underline{A})}{\underline{A}^2}\left(\frac{\hbar}{m_0 c}\right)^2\right]^{\frac{1}{2}} \tag{54}$$

Evidently, the function $\underline{\omega}$ had then to be identified with the linear 4-momentum of the particle due to the vortical motion of its wave field, so that $[\underline{r}\ \underline{\omega}]$ represents the *intrinsic* 4-angular momentum of the particle. $\underline{r}$ is the 4-distance of the field point from the centre of the vortex.

---

* It should be noted that $\underline{\varepsilon}^*$ of eq.(48) is not the mathematical complex conjugate of $\underline{\varepsilon}$. Since the scalar product of two 4-vectors must be Lorentz invariant one cannot write $\underline{\varepsilon}^*=\underline{A}^*\exp -iW/\hbar$ where $\underline{A}^*$ is the mathematical complex conjugate of $\underline{A}$, i.e. if $\underline{A}=\vec{A}+iA_0\vec{s}_0$, then $\underline{A}^*=\vec{A}-iA_0\vec{s}_0$.

For a particle at rest the centre remains stationary, although the field is always circulating around this centre, (the singularity of the field), with the velocity of light.*

We now postulate again that the energy-momentum of the particle is conserved, i.e.,

$$\underline{\nabla}(\underline{\varepsilon}\cdot\underline{\varepsilon}^{*}\underline{P}) = 0 \ . \qquad (55)$$

(ii) Differential Equations for the Vector Wave Field:

Utilizing the relations (47-53) in the eq.(55) we get various equivalent forms of PDE:

$$\underline{\varepsilon}^{*}\Box\underline{\varepsilon} + \underline{\varepsilon}^{*}\left[\underline{\nabla}\left[\underline{\nabla}\,\underline{\varepsilon}\right]\right] - \underline{\varepsilon}\Box\underline{\varepsilon}^{*} - \underline{\varepsilon}\left[\underline{\nabla}\left[\underline{\nabla}\,\underline{\varepsilon}^{*}\right]\right] = 0 \qquad (56)$$

$$\underline{\varepsilon}^{*}\Box\underline{\varepsilon} = \underline{A}\Box\underline{A} - \frac{\underline{A}\,\underline{p}^{2}}{\hbar^{2}} \qquad (57)$$

Substituting in (57). The relations

$$\underline{p}_{N}^{2} = (\underline{p} - \underline{p}_{e})^{2}$$

or

$$\underline{p}^{2} - 2(\underline{p}\cdot\underline{p}_{e}) - \mu^{2}m_{0}^{2}c^{2}$$

and the value of μ given by (54), we obtain

$$\underline{\varepsilon}^{*}\Box\underline{\varepsilon} + \frac{1}{\hbar^{2}}\left[2(\underline{p}\cdot\underline{p}_{e}) - \underline{p}_{e}^{2} - m_{0}^{2}c^{2}\right]\underline{\varepsilon}\,\underline{\varepsilon}^{*} = 0 \ . \qquad (58)$$

Since $\underline{\varepsilon}^{*}$ and $\underline{\varepsilon}$ are not mutually (pseudo) orthogonal, we can write the wave equation (58) in the form (59) containing the specific properties (e.g. rest mass and charge) of a given particle:

$$\Box\underline{\varepsilon} + \frac{1}{\hbar^{2}}\left[2(\underline{p}\cdot\underline{p}_{e}) - \underline{p}_{e}^{2} - m_{0}^{2}c^{2}\right]\underline{\varepsilon} = 0 \qquad (59)^{**}$$

Finally, using the relation (51) we obtain the desired equation (60) for a single particle whose wave function can be represented by a 4-vector

$$\Box\underline{\varepsilon} + \frac{1}{\hbar^{2}}\left[2(\underline{P}\cdot\underline{p}_{e}) - \underline{p}_{e}^{2} - m_{0}^{2}c^{2}\right]\underline{\varepsilon} = \frac{2}{h}(\underline{\omega}\cdot\underline{p}_{e})\,\underline{\varepsilon} \qquad (60)^{**}$$

---

* For a moving particle, if necessary, we have to distinguish between the phase velocity and group velocity of the field.

** Although the equations (59) and (60) look superficially as containing only $\underline{\varepsilon}$, but because of the definition of $\underline{P}$, $\underline{p}$ and $\underline{\omega}$ they contain both $\underline{\varepsilon}$ and $\underline{\varepsilon}^{*}$ in the general case.

I believe that eq.(60) represents the most general equation for any single particle, (containing explicitly its rest mass and charge), in an external field of electromagnetic type whose wave field is a vector function of the Minkowski space. This is fully justified as the various known equations of physics can be obtained from this equation under well defined specific restrictions.

10. IMPORTANT SPECIAL CASE

(i) Equation for the Wave Field of a Photon:

For photons, $m_0=e=0$, eq.(60) becomes the well known equation of electromagnetic theory if we identify $\underline{\varepsilon}$ with the 4-potential of the electromagnetic field:

$$\Box \underline{\varepsilon} = 0 \qquad (61)^*$$

(ii) Proca Equation:

For $\underline{p}_e=0$ we get the equation of Proca type:

$$\Box \underline{\varepsilon} - \left(\frac{m_0^2 c^2}{\hbar^2} \right) \underline{\varepsilon} = 0 \qquad (62)$$

It should be noted that $\underline{\omega}$ does not explicitly enter into this eq. (62). Consequently, it is reasonable to assume that this equation (62) is valid for all charged particles, (including electrons)**, in the absence of an external field as well as for neutral particles of any "spin" whose rest masses are not zero. In order to prove this we shall convert the eq.(60) to the "iterated" Dirac equation of Sommerfeld[17].

(iii) Iterated Dirac Equation :

For simplicity, let us assume that for a particle at rest the vorticity of the field *practically* extends to a small radius $\underline{r}_0$, of the dimension of the Compton wavelenght (cf.ref.5), around the centroid of the particle (i.e. the singularity of the vortex field) lying at $\underline{x}_a$. That means, the observable energy-momentum of the corpuscle is practically

---

* Nevertheless, at the present moment it would be premature to identify $\underline{\varepsilon}$ with the self 4-potential of the particle in the general case.

** It should be noted that 4-vector can be expressed in terms of spinors and bispinors.

given by the average value of the field over the dimension of the vortex. We can then convert the scalar coefficient at the right hand side of eq.(60) by

$$\frac{2}{\hbar^2} \oint (\underline{p}_e \cdot \underline{\omega})\, d\underline{s} = \frac{1}{\hbar} \left[\underline{\nabla}\, \underline{p}_e \right] \cdot \left[s_{ij} \right] \tag{63}$$

provided the circulation Γ is quantised in units of h, i.e.,

$$\Gamma = \oint \underline{v}\, ds = \frac{nh}{m_0} ; \qquad (n = \text{integer}) \tag{64}$$

s_{ij} is the unit ij-th component of the antisymmetric tensor $\left[s_{ij}\right]$, the so called six vector produced by the 4-rotation of the wave field. It might be noted that it has a physical significance and formally these components have the same algebraic properties as Dirac matrices and the hypercomplex quantities.

The eq.(60) under these conditions, i.e., under the assumption that the field averaged over its finite vortical domain essentially represents the corpuscle, reduces to

$$\Box \underline{\varepsilon} + \frac{1}{\hbar^2} \left[(2\underline{P} \cdot \underline{p}_e) - \underline{p}_e^2 - m_0^2 c^2 \right] \underline{\varepsilon} = \frac{1}{\hbar} \left[\underline{\nabla}\, \underline{p}_e \right] \cdot \left[s_{ij} \right] \underline{\varepsilon} =$$

$$= \frac{e}{\hbar} \sum_{\alpha,\beta=1}^{3} (F_{\alpha\beta} \cdot s_{\alpha\beta}) \underline{\varepsilon} - \frac{e}{\hbar c}\, i \sum_{\alpha=1}^{3} (F_{\alpha o} \cdot s_{\alpha o}) \underline{\varepsilon} \quad ; \tag{65}$$

$F_{\alpha\beta}$'s are the components of the field tensor $\left[\underline{\nabla}\, \underline{\Phi} \right]$.

Eq.(65) is completely equivalent to the *iterated* Dirac equation of Sommerfeld, provided we replace $\underline{P}$ by the corresponding quantum mechanical operator* and the hypercomplex quantities γ_α, γ_β by $s_{\alpha\beta}$.

Sommerfeld[17] had shown that the iterated Dirac equation gives the same eigenvalues and the "spin" of the particle as the original linearized Dirac equations. Equation (65), however, being a second order linear PDE (for the case of quantum mechanical linear operator formalism), may in more general cases provide solutions which need not be contained in the linearized first order Dirac equations, (cf.20).

---

* Corresponding to the Schroedinger condition (33), we have here the Dirac condition for the validity of the operator formalism of quantum theory:

$$(\delta \underline{p}_e \cdot \underline{\omega}) \underline{\varepsilon} - \frac{\hbar}{i} \exp iW/\hbar \underline{p}_e \cdot \underline{\nabla}\, \underline{A} = 0 \tag{66}$$

It is interesting to note that in the early days of Dirac equation, many distinguished authors[21] tried to find the corresponding equivalent wave equation, i.e. a second order hyperbolic equation. Sommerfeld[17] seems to prefer the iterated second order equation to the original first order Dirac equations.

Further, in our case $\underline{P}$ in (65) must be replaced, not by $\underline{P}_{op}$ of quantum mechanics, but by eq.(47) or (49). Consequently, even the eq. (65) is in general nonlinear and its singular as well as nonanalytic solutions might be of physical interest. Nevertheless, it should contain all the solutions obtainable from the original linearized Dirac equations as special cases.

It should also be noted that in this theory the "intrinsic" angular momentum of a particle as a function of $\vec{x}$ and t is given by $n\hbar$ (n=integer) whereas the quantum mechanical "spin" obtainable from the eigenvalues of the iterated Dirac equation (65) as proved by Sommerfeld[17], is given by $\hbar/2$. Hence there exists no contradiction in this provided we carefully distinguish between the meanings of "intrinsic angular momentum" and the quantum mechanical "spin" of a particle, (for detailed physical discussion, see ref.5).

11. CONCLUDING REMARKS

In order to obtain the finer details of the properties of a stable elementary particle, we have to solve completely any of the equations (56) to (60). Further, for exploring the consequences of mutual interactions of particles, which may generate unstable elementary particles, one is confronted with as yet unsolved physical and mathematical problems when individual quantum vortices interact with and interpenetrate into one another, probably producing a turbulent behaviour of the composite wave field. Evidently, the possibility of any rigorous solution of the problem at the present state of our knowledge is very remote.

Nevertheless, assuming a reasonable physical model based on this new approach, namely, the physical properties of a particle are essentially given by the vortex of its field, (cf.Kelvin's vortex atoms[22]), and the known laws of quantum mechanics and hydrodynamics one can understand without any inner contradiction whatsoever and often quantitatively many of the, as yet unexplained and uncomprehended, mysterious quantum phenomena, namely

(i) Zitterbewegung model of the electron proposed by Schroedinger;

(ii) the physical meaning of the relation $E=h\nu$;

(iii) the strange property of the Dirac electron viz, the electron *at rest* has the velocity c;

(iv) the nature of nuclear forces.

For quantitative details of this *model* of the particle, see ref.5.

Finally, in order to circumvent the horrible difficulties mentioned above for exploring the properties of "elementary" particles when they are subjected to collisions at high energies, I would like to suggest here the following programme based on:

(i) The hypothesis that the properties of any particle are completely given by a second order hyperbolic PDE of its associated wave field. In the general case, this wave field is a nonsymmetric tensor field.

(ii) The *empirical* physical fact that there are only two *primarily given stable* particles namely, the electron and the proton. All other particles, including the stable photon, are created when these "primary" particles suffer collisions by highly energetic particles.

(iii) The criteria of stability are given by the restriction that the wave field of a stable particle has a constant curvature, constant torsion and the same sense of torsion and circulation throughout space and time.

(iv) When this wave field is subjected to an arbitrary coordinate transformation, (kinematic equivalent to the phenomenological force of impact), the resulting equation governing the wave field becomes unstable and returns to the equilibrium stable states in one or many intermediate steps.

The final stable states may represent the "primary" particles as well as any other stable particles, like photons, neutrinos, antiparticles.

All of these stable particles are also characterized by the fact that their respective wave fields have constant curvature and torsion and the same sense of torsion and circulation. They differ from one another in the degrees of curvatures, torsions and the sense of torsion and circulation.

Please excuse me for presenting here this qualitative programme. For many personal reasons, I am afraid that I will not be able to test this programme by formulating and solving it mathematically. Consequently in the interest of progress of Science, I thought it desirable to take this opportunity to share with you my vague feelings hoping that they will attract the notice of theoretical physicists who want to look at their problems in a way different from the current fashionable one.

REFERENCES

1. Clifford, W.K. (1876) Proc.Camb.Phil.Soc. 2, 157.
2. Heisenberg, W. (1966): Nonlinear Problems in Physics, Proc.of the Int.School of Nonlinear Mathematics and Physics, Munich. Springer Verlag. (1968). See also Physics Today (1967), 20, 27.
3. Wigner, E. (1955): Jubilee of Relativity Theory, Bern, 1955. Helv.Phys.Acta Supplement IV (1956). See also, Rev.Mod.Phys. (1957) volume 29.
4. Hosemann, R. & Bagchi, S.N. (1955) Z.Phys. 142, 334, 347, 363.
5. Bagchi, S.N. (1975) Proc.Ind.Association for the Cultivation of Science. Calcutta 58, 21.
6. Dirac, P.A.M. (1963) Scientific American 208, 45.
7. Bagchi, S.N. (1975) Philosophy of Science, Some Problems. Bull.Ramkrishna Mission Institute of Culture, Calcutta June & July.
8. Renninger, M. (1953) Z.Phys. 136, 251.
9. Dirac, P.A.M. (1970) Physics Today..., 29.
10. Hosemann, R. & Bagchi, S.N. (1952) Acta Cryst. 5, 749; (1953) Acta Cryst.6,315,404. See also Direct Analysis of Diffraction by Matter, North Holland (1962).
11. Bagchi, S.N. (1972) Proc.of the Int.Biophysics Congress, Moscow, p.755.
12. Bagchi, S.N. & Plischke (1968) J.Ind.Chem.Soc. 45, 925.
13. Bagchi, S.N. (1974) Proc.Int.Conf.on Statistical Physics, Calcutta.
14. Sommerfeld, A. (1910) Ann.der Physik 32, 749 and 33, 650.
15. de Broglie, L. (1953) La physique Quantique restera indeterministe?, Gauthier-Villars, Paris.
16. Brilloüin, L. (1970) Relativity Re-examined, Academic Press.
17. Sommerfeld, A. (1951) Atombau und Spektrallinien Bd. II, Friedr.Vieweg, Braunschweig.
18. Hosemann, R. & Bagchi, S.N. (1956) Z.Phys. 145, 65.
19. Hosemann, R. & Bagchi, S.N. (1954) Z.Phys. 139, 1.
20. Courant, R. & Hilbert, D. (1965) Methods of Math.Phys. vol.II, p.13, Interscience Publishers.
21. Darwin, C.G. (1928) Proc.Roy.Soc.London 118A, 654.

    Eddington, A.S. (1928) ibid 121A, 524.
    Sauter, F. (1930) Z.Phys. 63, 803; 64, 295.
22. Lord Kelvin (1910) Mathematical and Physical Papers vol.IV, Cambridge University Press.

NOETHER'S THEOREM AND INFINITIES OF POLYNOMIAL CONSERVED DENSITIES

Mark J. M[c]Guinnes

Department of Mathematical Physics

University College, Belfield

Dublin 4, Ireland

ABSTRACT

The infinities of polynomial conserved densities known to exist for a class of nonlinear evolution equations are investigated using Noether's theorem. The densities are identified as canonical energy or momentum densities on enveloping solution sets of higher-order integrodifferential nonlinear equations. Proofs appear elsewhere.

1. NOETHER'S THEOREM

Noether's theorem is a means of associating a conservation equation with an infinitesimal transformation[1]. The transformation will be one which leaves the action integral invariant,

$$\mathscr{I} = \int_R \mathscr{L} dx \quad , \tag{1.1}$$

where $\mathscr{L}$ is the Lagrangian density of the system, and the integral is over four-space.

The usual variational approach[2] and the application of Hamilton's principle, give the equation of motion (there may be more than one) of the system in terms of the Lagrangian density $\mathscr{L}$, the Euler-Lagrange equations

$$E(\mathscr{L}) \equiv \sum_{a=0}^{\infty} (-1)^a d_{\mu_1} d_{\mu_2} \ldots d_{\mu_a} \frac{\partial \mathscr{L}}{\partial \phi_{\mu_1 \ldots \mu_a}} \triangleq 0 \quad , \tag{1.2}$$

where

$$d_{\mu_1} \equiv \frac{d}{dx^{\mu_1}} \quad , \qquad \phi_{\mu_1} \equiv \frac{\partial\phi}{\partial x^{\mu_1}} \quad ,$$

and for example

$$\{x^{\mu_i}\} = \{x^0, x^1, x^2, x^3\},$$
$$= \{t, x, y, z\} \quad , \tag{1.3}$$

and where $\stackrel{\circ}{=}$ means "equals for solutions".

Noether applied a similar variational approach to obtain the Noether Relation, which holds if $\mathscr{I}$ is invariant under the transformation $(\delta\phi, \delta x)$,

$$-\bar{\delta}\phi \; E(\mathscr{L}) = \frac{dJ^\mu}{dx^\mu} \quad , \tag{1.4}$$

where

$$J^\mu = \Pi^\mu(\mathscr{L}) \, \bar{\delta}\phi - \mathscr{L}\delta x^\mu \quad , \tag{1.5}$$

$$\Pi^\mu(\mathscr{L}) \equiv \sum_{a=0}^{\infty} \sum_{b=0}^{a} (-1)^b \, d_{\mu_1} \ldots d_{\mu_b} \left(\frac{\partial}{\partial \phi_{\mu_1 \ldots \mu_a \mu}} \right) d_{\mu_{b+1}} \ldots d_{\mu_a} \quad , \tag{1.6}$$

and where to first order in $\delta x = x' - x$,

$$\phi'(x') - \phi(x) = \delta\phi = \bar{\delta}\phi + \delta x^\mu \phi_\mu \quad . \tag{1.7}$$

From the Noether Relation (1.4), it follows that

$$\frac{dJ^\mu}{dx^\mu} \stackrel{\circ}{=} 0 \tag{1.8}$$

which is the conservation equation associated via Noether's theorem with the inveriance of $\mathscr{I}$ under $\bar{\delta}\phi$.

A generalization of Noether's theorem due to Rosen[1] *starts* with Noether's Relation (1.4), and bypasses the variational procedure. If one starts with the equation of motion $F \stackrel{\circ}{=} 0$, and multiplies by some expression $\bar{\delta}\phi$, then *if* it is possible to rearrange the product into a 4-divergence,

$$-\bar{\delta}\phi \; F = d_\mu J^\mu \quad , \tag{1.9}$$

it follows that $\bar{\delta}\phi$ is the transformation associated with the conserved vector J^μ, and if a Lagrangian density exists for F, the action integral is invariant under $\bar{\delta}\phi$. The two approaches, generalised and conventional Noether's theorem, are equivalent.

Since Noether's theorem associates a conservation law with an infinitesimal transformation, it may be used to *identify* a conserved quantity by its associated transformation (note that this correspondance may not be unique[3]). For example, common associations are time translation invariance with energy conservation, space translation invariance with conservation of momentum, and rotational invariance about some axis with conservation of angular momentum in the direction of that axis.

If a Lagrangian density can be derived for an equation, those of the well-known conservation laws that hold may be written down by inspection of that Lagrangian density. For example, if the Lagrangian density has no explicit time-dependence, the action integral is invariant under an infinitesimal time translation and energy is conserved and is written down as

$$\int_{3\text{-space}} \left[\Pi^t(\mathscr{L})\, \phi_t - \mathscr{L} \right] dx\, dy\, dz \quad . \tag{1.10}$$

The generalised Noether's theorem due to Rosen does not have this advantage, but it may be used if a Lagrangian density does not exist for the system.

2. NONLINEAR SYSTEMS

A number of nonlinear evolution equations are known to possess an infinity of conservation laws, and are also solvable exactly by an inverse scattering method. For a large class of these equations, each of the infinity of conserved densities may be identified as a canonical *energy* or *momentum* density of a higher-order *enveloping* equation, as follows.

The class of equations may be written

$$\underline{F}_\circ \equiv \begin{pmatrix} r \\ -q \end{pmatrix}_t + \left(\underline{\underline{L}}^+\right)^m \begin{pmatrix} r \\ q \end{pmatrix} \triangleq 0 \quad , \tag{2.1}$$

where

$$\underline{\underline{L}}^{+} = \begin{pmatrix} d_x - 2r\int^x q \ , & 2r\int^x r \\ -2q\int^x q \ , & -d_x + 2q\int^x r \end{pmatrix} . \tag{2.2}$$

The integrals over x have a lower boundary on which the field variables r, q are assumed to vanish, together with their derivatives. Equations (2.1) are solvable by the inverse spectral method[5], and different choices of r, q and the integer m give many different equations, including the KdV, the modified KdV and the nonlinear Schrodinger equations. The method described briefly in the following pages extends to the equations

$$\begin{pmatrix} r \\ -q \end{pmatrix}_t + A_0\left(\underline{\underline{L}}^{+}\right)\begin{pmatrix} r \\ q \end{pmatrix} \triangleq 0 \tag{2.3}$$

when A_0 is entire in its argument, and under certain assumptions[4] extends to equations (2.3) when A_0 is a ratio of entire functions.

The higher-order enveloping equations are given by

$$\underline{F}_n \equiv \underline{\underline{\sigma}}\left(\underline{\underline{L}}^{+}\right)^n\left[\begin{pmatrix} r \\ -q \end{pmatrix}_t + \left(\underline{\underline{L}}^{+}\right)^m\begin{pmatrix} r \\ q \end{pmatrix}\right] \triangleq 0 , \quad n=1, 2, \ldots \tag{2.4}$$

where

$$\underline{\underline{\sigma}} = \begin{pmatrix} 0 \ , & 1 \\ 1 \ , & 0 \end{pmatrix} ,$$

which inverts the vector of equations to obtain them in the correct order for a variational approach. Note that the solution sets of these integrodifferential equations contain that of the original class $\underline{F}_0$. Hence conservation laws derived for solutions to equations $\underline{F}_n \triangleq 0$ must also hold for solutions to the original class, $\underline{F}_0 \triangleq 0$.

The derivation of a set of canonical energy densities for the equations $\underline{F}_n \triangleq 0$ will be discussed here. The infinitesimal transformation associated with energy conservation is the time translation,

$$\delta t = \varepsilon , \quad \delta x = 0 , \quad \delta \underline{r} = \underline{0} , \tag{2.5}$$

where ε is an infinitesimal parameter, that is,

$$\bar{\delta}\underline{r} = -\varepsilon\, \underline{r}_t , \tag{2.6}$$

where

$$\underline{r} \equiv \begin{pmatrix} r \\ q \end{pmatrix} \quad . \tag{2.7}$$

To obtain energy densities for the equations $\underline{F}_n \triangleq 0$, we look for T_n in the Noether Relation

$$\underline{r}_t \cdot \underline{F}_n = d_t T_n + d_x X_n \quad . \tag{2.8}$$

The main features of the proof of equation (2.8) will be discussed here. A detailed proof is to be published elsewhere[4].

A novel feature of the proof of relation (2.8) is the splittings of the problem into two parts, an integrodifferential part

$$\underline{r}_t \cdot \left[\underline{\underline{\sigma}} \left(\underline{\underline{L}}^+\right)^n \begin{pmatrix} r \\ -q \end{pmatrix}_t \right] = d_x P_n \quad , \tag{2.9}$$

and a partial differential part

$$\underline{r}_t \cdot \left[\underline{\underline{\sigma}} \left(\underline{\underline{L}}^+\right)^{n+m} (\underline{r}) \right] = d_t T_n + d_x (X_n - P_n) \quad . \tag{2.10}$$

The integrodifferential part, which contributes only to the energy flux, is proved by using the generalised Noether's theorem due to Rosen, since no Lagrangian density can be found. That is, equation (2.9) is proved by induction.

Conventional Noether's theorem may be used to prove equation (2.10), since the l.h.s. of equation (2.10) can be shown to be partial differential[4], and a Lagrangian density for the equation

$$\underline{r}_t \cdot \left[\underline{\underline{\sigma}} \left(\underline{\underline{L}}^+\right)^{n+m} \underline{r} \right] \triangleq 0 \tag{2.11}$$

may be derived as

$$\mathscr{L}_{n+m} = \left(\underline{\underline{\sigma}}\underline{r}\right) \cdot \int_0^1 \underline{L}_{n+m} \left(\lambda \underline{r}\right) d\lambda \quad , \tag{2.12}$$

where λ is a parameter, and

$$\underline{L}_{n+1} = \underline{\underline{L}}^+ (\underline{L}_n) \quad ,$$

$$\underline{L}_\circ = \underline{r} \quad . \tag{2.13}$$

Note that the integral over λ merely introduce numbers, so that $\mathscr{L}_k$ is a polynomial in $\underline{r}$ and its x-derivatives. Since the Lagrangian density (2.12) has no explicit time-dependence, energy is conserved for equation (2.11), and the energy conservation relation (Noether's Relation) is equation (2.10), with

$$T_n = \mathscr{L}_{n+m} \quad . \tag{2.14}$$

The sum of equations (2.9) and (2.10) yields relation (2.8). Since T_n is a conserved energy density for the higher-order enveloping equation $\underline{F}_n \triangleq 0$, it must be a conserved density for the original set $\underline{F}_o \triangleq 0$. Hence the set of densities $\{T_n \; ; \; n=0,1,\ldots\}$ constitutes an infinite set of polynomial conserved densities for the equations $\underline{F}_o \triangleq 0$.

The first four of the set $\{T_n\}$ have been compared (for n=0) to the first four of the set of densities obtained by Konno et al[6], and have been found to be equivalent. Hence the formula (2.12) is likely to be equivalent to the different formula obtained by Konno et al[6].

ACKNOWLEDGMENT

I am grateful to the Education Department in Ireland for the Research Fellowship and Travel Assistance that made it possible to attend this conference.

REFERENCES

1 J.Rosen, Int.J.Theor.Phys. 4, 287 (1971).
Ann.Phys.(N.Y.) 69, 349 (1972).
Ann.Phys.(N.Y.) 82, 54 & 70 (1974).
E.L.Hill, Rev.Mod.Phys. 23, 253 (1951).

2 E.J.Saletan & A.H.Cromer, Theoretical Mechanics (Wiley), N.Y.(1971).

3 H.Stendel, Ann.Phys. 32, 214 (1975).

4 M.J.McGuinnes, submitted to J.Math.Phys.

5 M.Ablowitz et al., Stud.Appl.Maths. 53, 249 (1974).

6 K.Konno et al., Prog.Theor.Phys. 52, 886 (1974).

Selected Issues from

Lecture Notes in Mathematics

Vol. 662: Akin, The Metric Theory of Banach Manifolds. XIX, 306 pages 1978.

Vol. 665: Journées d'Analyse Non Linéaire. Proceedings, 1977. Edité par P. Bénilan et J. Robert. VIII, 256 pages. 1978.

Vol. 667: J Gilewicz, Approximants de Padé. XIV, 511 pages. 1978.

Vol. 668: The Structure of Attractors in Dynamical Systems. Proceedings, 1977. Edited by J. C. Martin, N. G. Markley and W. Perrizo. VI, 264 pages. 1978.

Vol. 675: J. Galambos and S. Kotz, Characterizations of Probability Distributions. VIII, 169 pages. 1978.

Vol. 676: Differential Geometrical Methods in Mathematical Physics II, Proceedings, 1977. Edited by K. Bleuler, H. R. Petry and A. Reetz. VI, 626 pages. 1978.

Vol. 678: D. Dacunha-Castelle, H. Heyer et B. Roynette. Ecole d'Eté de Probabilités de Saint-Flour. VII-1977. Edité par P. L. Hennequin. IX, 379 pages. 1978.

Vol. 679: Numerical Treatment of Differential Equations in Applications, Proceedings, 1977. Edited by R. Ansorge and W. Törnig. IX, 163 pages. 1978.

Vol. 681: Séminaire de Théorie du Potentiel Paris, No. 3, Directeurs: M. Brelot, G. Choquet et J. Deny. Rédacteurs: F. Hirsch et G. Mokobodzki. VII, 294 pages. 1978.

Vol. 682: G. D. James, The Representation Theory of the Symmetric Groups. V, 156 pages. 1978.

Vol. 684: E. E. Rosinger, Distributions and Nonlinear Partial Differential Equations. XI, 146 pages. 1978.

Vol 690 W J J Rey. Robust Statistical Methods VI. 128 pages 1978

Vol. 691: G. Viennot, Algèbres de Lie Libres et Monoides Libres. III, 124 pages. 1978.

Vol. 693: Hilbert Space Operators, Proceedings, 1977. Edited by J. M. Bachar Jr. and D. W. Hadwin. VIII, 184 pages. 1978.

Vol. 696: P. J. Feinsilver, Special Functions, Probability Semigroups, and Hamiltonian Flows. VI, 112 pages. 1978.

Vol. 702: Yuri N. Bibikov, Local Theory of Nonlinear Analytic Ordinary Differential Equations. IX, 147 pages. 1979.

Vol. 704: Computing Methods in Applied Sciences and Engineering, 1977, I. Proceedings, 1977. Edited by R. Glowinski and J. L. Lions. VI, 391 pages. 1979.

Vol. 710: Séminaire Bourbaki vol. 1977/78, Exposés 507–524. IV, 328 pages. 1979.

Vol. 711: Asymptotic Analysis. Edited by F. Verhulst. V, 240 pages. 1979.

Vol. 712: Equations Différentielles et Systèmes de Pfaff dans le Champ Complexe. Edité par R. Gérard et J.-P. Ramis. V, 364 pages. 1979.

Vol. 716: M. A. Scheunert, The Theory of Lie Superalgebras. X, 271 pages. 1979.

Vol. 720: E. Dubinsky, The Structure of Nuclear Fréchet Spaces. V, 187 pages. 1979.

Vol. 724: D. Griffeath, Additive and Cancellative Interacting Particle Systems. V, 108 pages. 1979.

Vol. 725: Algèbres d'Opérateurs. Proceedings, 1978. Edité par P. de la Harpe. VII, 309 pages. 1979.

Vol. 726: Y.-C. Wong, Schwartz Spaces, Nuclear Spaces and Tensor Products. VI, 418 pages. 1979.

Vol. 727: Y. Saito, Spectral Representations for Schrödinger Operators With Long-Range Potentials. V, 149 pages. 1979.

Vol. 728: Non-Commutative Harmonic Analysis. Proceedings, 1978. Edited by J. Carmona and M. Vergne. V, 244 pages. 1979.

Vol. 729: Ergodic Theory. Proceedings 1978. Edited by M. Denker and K. Jacobs. XII, 209 pages. 1979.

Vol. 730: Functional Differential Equations and Approximation of Fixed Points. Proceedings, 1978. Edited by H.-O. Peitgen and H.-O. Walther. XV, 503 pages. 1979.

Vol. 731: Y. Nakagami and M. Takesaki, Duality for Crossed Products of von Neumann Algebras. IX, 139 pages. 1979.

Vol. 733: F. Bloom, Modern Differential Geometric Techniques in the Theory of Continuous Distributions of Dislocations. XII, 206 pages. 1979.

Vol. 735: B. Aupetit, Propriétés Spectrales des Algèbres de Banach. XII, 192 pages. 1979.

Vol. 738: P. E. Conner, Differentiable Periodic Maps. 2nd edition, IV, 181 pages. 1979.

Vol. 742: K. Clancey, Seminormal Operators. VII, 125 pages. 1979.

Vol. 755: Global Analysis. Proceedings, 1978. Edited by M. Grmela and J. E. Marsden. VII, 377 pages. 1979.

Vol. 756: H. O. Cordes, Elliptic Pseudo-Differential Operators – An Abstract Theory. IX, 331 pages. 1979.

Vol. 760: H.-O. Georgii, Canonical Gibbs Measures. VIII, 190 pages. 1979.

Vol. 762: D. H. Sattinger, Group Theoretic Methods in Bifurcation Theory. V, 241 pages. 1979.

Vol. 765: Padé Approximation and its Applications. Proceedings, 1979. Edited by L. Wuytack. VI, 392 pages. 1979.

Vol. 766: T. tom Dieck, Transformation Groups and Representation Theory. VIII, 309 pages. 1979.

Vol. 771: Approximation Methods for Navier-Stokes Problems. Proceedings, 1979. Edited by R. Rautmann. XVI, 581 pages. 1980.

Vol. 773: Numerical Analysis. Proceedings, 1979. Edited by G. A. Watson. X, 184 pages. 1980.

Vol. 775: Geometric Methods in Mathematical Physics. Proceedings, 1979. Edited by G. Kaiser and J. E. Marsden. VII, 257 pages. 1980.